Avoiding Attack

# Avoiding Attack

## The evolutionary ecology of crypsis, warning signals, and mimicry

Graeme D. Ruxton

Thomas N. Sherratt

Michael P. Speed



OXFORD
UNIVERSITY PRESS

OXFORD
UNIVERSITY PRESS

Great Clarendon Street, Oxford OX2 6DP

Oxford University Press is a department of the University of Oxford.
It furthers the University's objective of excellence in research, scholarship,
and education by publishing worldwide in

Oxford New York

Auckland Bangkok Buenos Aires Cape Town Chennai
Dar es Salaam Delhi Hong Kong Istanbul Karachi Kolkata
Kuala Lumpur Madrid Melbourne Mexico City Mumbai Nairobi
São Paulo Shanghai Taipei Tokyo Toronto

Oxford is a registered trade mark of Oxford University Press
in the UK and in certain other countries

Published in the United States
by Oxford University Press Inc., New York



First published 2004



A catalogue record for this title is available from the British Library

Library of Congress Cataloging in Publication Data
Data available

ISBN 0 19 852859 0 (hbk)
ISBN 0 19 852860 4 (pbk)

10 9 8 7 6 5 4 3 2 1

Typeset by Newgen Imaging Systems (P) Ltd., Chennai, India
Printed in Great Britain
on acid-free paper by Antony Rowe, Chippenham

To Hazel, Alison and Rachel

# Contents

# Introduction

## Scope

Biology, the pursuit of understanding of living organisms, is a huge field. Hence, few researchers would consider themselves biologists, pure and simple, without qualification. We tend to specialize, sacrificing breadth of knowledge for depth in one aspect or other. We all know that there is a cost to this approach. When struggling with a problem, there is the nagging fear that your struggle is caused by your failure to consider factors beyond your self-imposed field of interest. However, we scientists generally feel that we must pay this cost in order to have sufficient depth of knowledge in at least one field to make a contribution to the advancement of that field. We try to mitigate the adverse effects of specialism mainly by communication with others with different specialisms, and despite the risks we continue to specialize.

The perceptive reader will have identified that the purpose of our opening paragraph was to lay the logical foundations for our excuses for the things that are not in this book. All three of us, would be comfortable being described as 'evolutionary biologists', in that our primary interest is in how natural selection has acted, and continues to act, to produce organismal traits. For example, when confronted with the dark and light stripes of a zebra, we are likely to ask 'what selection pressures have lead to the evolution and maintenance of these stripes?' We are less likely to ask the equally valid biological questions: 'by what physical and biochemical means are these stripes formed?' or 'how do stripes develop during the processes involved in the growth of a single zebra?' or 'What parts of the genome of the zebra affect the production of these stripes.' One commonly cited evolutionary explanation for zebra's stripes is that they influence potential predators of the zebra either making it harder for the predator to detect zebras against the background, or making it harder to select one zebra from a fleeing pack, or to judge a final leap at a fleeing zebra. These are the sort of mechanisms that concern us in this book; we are interested in how prey can influence their predators' senses in a way that reduces their likelihood of being attacked and/or the likelihood that an attack will be successful.

The zebra illustrates the danger of our strategy of specializing in predator–prey issues. It could be that the stripes have evolved principally for reasons unconnected with reducing the vulnerability of zebra to lions. For example, another hypothesis for the zebra's stripes is that the different radiative properties of the dark and light regions act to induce strong convection currents, which aid in the cooling of the animal in the hot African sun. An alternative hypothesis is that they function as social signals to other zebra to encourage grooming. Hence the search for an understanding of the evolution and maintenance of the zebra's stripes entirely within the realm of predatory interactions may ultimately be fruitless. To further complicate things, it is unlikely that selection for stripes has been influenced by thermoregulation (say) and has been completely selectively neutral with respect to predation. In general, we would expect traits as substantial as zebra stripes to have been the result of a complex of several evolutionary pressures. Why then, have we limited ourselves to consideration of predator–prey associations? Our reason is purely pragmatic. We took 24 months to write this book. A book that considered all the evolutionary pressures on even just the outward visual appearance of organisms (ignoring other sensory pathways), would have either been a much more superficial summary than we were interested in, or would have been an undertaking that would have cost us careers, the understanding of our wives and what sanity we have left, to produce a book that no publisher would touch. We *had* to specialize.

Why did we choose to specialize in predator–prey considerations? The honest answer is that these were simply the things we were most interested in. We do not deny that social signalling has a very strong influence on how organisms look, sound, and smell. However, we feel strongly that signalling of quality to potential mates has come to dominate many scientists' thinking of the evolutionary pressures on what we might describe as the sensory ecology of individuals (i.e. how they are perceived by other organisms that can influence their fitness). The adult elephant has no concern about hiding from predators or prey, and we are happy to believe that the way an adult male elephant looks, sounds, and smells is likely to be relatively uninfluenced by the considerations presented in our book. However, there are very few species that have no regard to how they are perceived by predators and/or prey. Anti-predator adaptations occur in every biome of the world and in almost every major taxonomic group and, as R. A. Fisher argued, their very presence tells us that predation is a phenomenon of great ecological and evolutionary significance. Indeed even when an adaptation has been developed for display of quality to members of the same species, as seems to be the case for the peacock's tail, that adaptation has significant influence on that bearer's risk of predation. Indeed, some theories for the significance of the tail to peahens require that the tail does have a detrimental effect which might include the peacock's risk of predation. We feel that the recent neglect of predator–prey influences on sensory ecology is a significant impediment to our understanding of ecology. Our, rather immodest, aim in this book is to demonstrate that interesting and important questions in how prey are perceived by predators (and vice versa) remain to be answered. In doing this, we hope to spark a renewed interest in this too long neglected area of biology.

Before describing the structure of the book, we need to make a few further general points about our aims and scope. Even though we necessarily discuss various aspects of sensory ecology in isolation, it is very important to remember that the evolution of traits that influence the perception of one organism by its predators or prey is likely to be driven by more than one mechanism. Although our primary focus is on predator–prey interactions, we must sometimes introduce other factors. For example, an understanding of the effectiveness of transparency as a means of reducing ease of detection by predators or prey can only be understood in the context of some of the physical costs of transparency, such as vulnerability to damage from ultraviolet light. Even within the context of predator–prey considerations, more than one mechanism will often be in play. Although we stress that background matching and disruptive coloration are separate mechanisms that can both serve to reduce a viewer's ability to visually detect an organism, we would not generally expect the disruptive mechanism to operate completely in isolation from background matching. The natural world is intrinsically complex, and we should never forget that our models (both verbal and quantitative) represent considerable simplifications.

Our approach is based on mechanisms rather than species or ecosystems. That is, we have sections called 'warning signals' or 'countershading' rather than 'the leopard's spots' or 'camouflage in the desert'. By taking this approach we wish to encourage the search for general underlying principles to predatory–prey aspects of sensory ecology. Although we do, of course, draw on specific examples from a range of species and ecosystems, we have not tried to be encyclopaedic in scope. For example, you will hear no more about zebras and peafowl in this book. Rather we hope to introduce all the relevant general principles that will allow you to evaluate the likely selection pressures on the appearance of these (and any) taxa due to their sensory relationships with predators and prey.

Of course potential prey species have evolved a variety of ways to reduce their likelihood of attack by predators. First, and most simply, they may evolve means to avoid detection. We will deal with this collection of adaptations in Part I. Second, prey species may evolve traits that render themselves unprofitable to predators, for instance, by evolving spines, stings, and toxins. As we will see, generation of these forms of defence has also led to the evolution of signalling in prey to advertise their defensive qualities to predators. We consider these phenomena in Part II. Finally, in Part III. we see how these signalling systems and others have been exploited by other species.

Some authors have found it helpful to distinguish between a 'primary defence' and a 'secondary defence'. By definition, primary defences operate before a predator initiates any prey-catching behaviour (Robinson, 1969; Edmunds, 1974) and their function is to prevent pursuit. As such, primary defences include not only crypsis (Part I), but also warning signals (Chapter 6) and mimicry (Chapters 9 and 10). Conversely, secondary defences are said to operate once a predator initiates an attack and function by increasing the chances of an individual surviving the prey capture process. Examples might therefore include deflection and startle (Chapter 13). Unfortunately, this dichotomy is not perfect and there are plenty of grey areas (warning signals can sometimes be apparent after pursuit begins). Other classifications have similarly run into difficulty: there has been a debate over whether masquerading as a twig, say, is an example of crypsis or mimicry. Our general philosophy, however, is not to get bogged down in semantics and classifications. We have decided on this course for two reasons: (a) most anti-predator adaptations are intuitively understandable without the need to classify at a high level of detail, and (b) as will be clear, anti-predator adaptations are not discrete and independent traits but form a continuum. For example, Müllerian mimicry (Chapter 9) can be seen as a case of adaptive resemblence, but its interpretation requires little more than an extension of the concept of warning signals.

Finally, we would like to convey our admiration for those researchers in the past who have attempted to review this huge and fascinating field of antipredator defence. Only on attempting a sythnesis ourselves have we come to truly appreciate the monumental works of A. R. Wallace (1889), Edward Poulton (1890), Hugh Cott (1940), and Malcolm Edmunds (1974). These authors have each set immensely high standards in terms of their clarity of writing and their depth of understanding. As will be seen, we have relied heavily on their collective works throughout this book, while attempting to give our own perspective and highlight the exciting new directions and initiatives. We cannot hope to emulate such fine books as those listed above, but we hope we can at least inspire the next generation of researchers to carry this research area forward.

## Acknowledgements

We are extremely grateful to a number of people who have generously commented on chapters at various stages of preparation. In particular, we wish to sincerely thank Chris Beatty, Kirsten Beirinckx, Carl Bergstrom, Tim Caro, William Cooper, Will Cresswell, Jennie Evans, Dan Franks, Francis Gilbert, Jean-Guy Godin, Mike Hansell, Peter Herring, Stuart Humphries, Sonke Johnsen, David Kelly, Leena Lindström, Anne Lyttinen, James Mallet, Johanna Mappes, Nicola Marples, Robbie McDonald, Sami Merilaita, David Montagnes, Arash Rashed, Candy Rowe, John Skelhorn, Bob Srygley, Martin Stevens, Birgitta Tullberg, Hans Van Gossum and Jayne Yack for their very helpful comments and advice. Dave Kelly and Chris Beatty gave us invaluable help constructing the index. As ever, all inaccuracies and misconceptions remain our own.

Fiona Burns, Liz Miller, Alexis Perry, and Helen Simmons very kindly made the excellent line drawings of various animals and plants for us, while Henri Goulet and Jim Kalisch provided the colour plates: we are very grateful to them all.

The smartest move we made in writing this book was asking Liz Denton to design and draw the figures: her imagination, attention to detail, accuracy, and industry are matched only by her patience.

Ian Sherman encouraged and advised very skillfully in the initial stages of the project while Anita Petrie at OUP was a very steadying and reassuring influence during the final editing of our manuscript.

PART I

# Avoiding detection

CHAPTER 1

# Background matching

It is common to describe organisms that come close to perfectly matching the environment so that detection of them is challenging as being *cryptic* (Endler, 1988). There are many reported cases of organisms that seem particularly adapted to match their environment (e.g. Norris and Lowe, 1964; Sweet, 1985; Armbruster and Page, 1996), that behaviourally select backgrounds that match their appearance (e.g. Papageorgis, 1975; Endler, 1984; Marshall, 2000), or that physiologically adapt their appearance to changes in their current background (e.g. McFall-Ngai and Morin, 1991; Messenger, 1997; Harper and Case, 1999; Chiao and Hanlon, 2001). There is also a small, but convincing, literature suggesting that such background matching leads to reduced predation risk (e.g. Feltmate and Williams, 1989; Cooper and Allen, 1994; Johannesson and Ekendahl, 2002). We will not catalogue this literature here (for a well-chosen review, see Edmunds, 1974), but instead discuss some general aspects of background matching.

In Section 1.1, we consider why crypsis might be selected over alternative forms of primary defence, such as the warning signals discussed in Chapter 7. In Section 1.2, we consider the most commonly cited example of background matching in textbooks: industrial melanism in moths and other insects. We discuss the importance of a careful definition of 'background' in Section 1.3. Realization that background is affected by properties of the viewer allows explanation of the possibility that organisms can signal to conspecifics without increasing the ease with which predators can detect them, as discussed in Section 1.4. In Section 1.5, we review the theory and empirical evidence behind the suggestion that the flicker fusion effect stemming from limitations of visual physiology may have a role to play in crypsis by background matching. In Section 1.6 we consider how polymorphism is selected and maintained in organisms that achieve crypsis by background matching. Section 1.7 considers the consequences of temporal and spatial heterogeneity of the backgrounds that an organism is viewed against. In particular, we consider whether an organism should specialize on maximizing its match to one particular background or be a jack-of-all-trades. We end the chapter with Section 1.8 by discussing disguise as a particular object in the environment (masquerade) in relation to background matching.

## 1.1 Why crypsis?

Before considering in detail how different forms of crypsis function, we ask a broader question, *why would an organism adopt crypsis rather than some other form of primary defence*? Given that it is in any prey's interests to avoid attack, prevention of detection seems the most obvious route. However defences, even crypsis, often incur costs and hence the question is more specifically about evaluating the factors that affect the cost–benefit trade-off. In order to consider this question in more detail we have adopted and modified a framework proposed by Tollrian and Harvell (1999). Although this framework was originally constructed to identify classes of costs to secondary defences such as chemical defences, it is equally applicable to primary as well as secondary defences. Tollrian and Harvell (1999) defined five general classes of fitness cost associated with defences. We modify them here and consider their application to background matching and crypsis in general.

**1.** *Allocation (or internal) costs*: Perhaps the most obvious form of cost is that in which the construction, maintenance, and operation of a defence requires allocation of limited resources at the expense of allocation to other fitness components. Evidence of allocation costs will be seen and measured most obviously as trade-offs between investment in defence and some other component of fitness such as growth, survivorship and fecundity. In cryptic prey, for example, there may be direct allocation of resources to costs of pigmentation; or of chemical energy to fuel movement between microhabitats within which chances of detection are especially low.

**2.** *(Indirect) Opportunity costs*: In essence this is a quantification of benefits from alternative activities that must be given up as a result of a decision to use a particular defence. Being cryptic may prevent prey from taking advantage of opportunities available in a habitat because these opportunities are incompatible with crypsis. For a prey of fixed appearance, there may be a range of microhabitats in which it cannot forage because its match to the background is too poor, and the risk of predation too great (for related models, see Sih, 1992). Similarly, basking in the sun for thermoregulatory reasons, for example, may be incompatible with minimisation of detection. Carrascal et al. (2001) recently showed that short-toed tree creepers (*Certhia brachydactyla*) select sunny or shaded substrates in accordance with ambient temperature, taking the sunny sites when the temperature is below 4 °C, but avoiding sunlit areas when the ambient temperature is above 9 °C. The authors interpret this behaviour as evidence that basking imposes increased risks of detection and that crypsis therefore imposes costs on thermoregulatory behaviours. In other instances, crypsis has been shown to reduce the efficiency of interspecific communication (e.g. in hermit crabs: Dunham and Tierney, 1983). It is easy to generate plausible, testable hypotheses that crypsis reduces opportunities for a range of important activities including sexual selection, foraging, and reduced exposure to parasites (Stamp and Wilkens, 1993).

The possibility that crypsis restricts opportunities for environmental exploitation suggests that the benefits that come from variable, facultative background matching, such as that shown by octopi, are more than simple anti-predator action. Animals that can rapidly change their colour patterns to match their backgrounds may do so both because this minimizes predation and because it removes or reduces opportunity costs associated with crypsis (Hanlon et al., 1999). The optimal state for many animals for whom transparency is not possible may be facultative background matching, but in many species anatomical constraints may prevent this.

**3.** *Environmental (or external) costs*: This class of cost applies when the generation or use of a defence incurs some costly interaction with the environment. For example, the operation of an anti-predator behaviour may move an animal into a suboptimal environment for some period and this may add additional costs beyond loss of opportunity. Thus, diel vertical migration moves zooplankton along the water column for the avoidance of predators. If it takes the zooplankton away from rich feeding areas then there is an opportunity cost; if it takes them into areas that are occasionally much colder, there could be a thermoregulatory cost *additional* to the lost opportunities from foraging.

**4.** *Design and self-damage costs:* Tollrian and Harvell proposed this form of cost especially in instances of chemical defence where there may be costs of autotoxicity or prevention of autotoxicity. However, more generally a defence may incur design costs, if the formation of a defence compromises the effectiveness of other organs and traits, this may well be particularly important for organisms that reduce their visual detectability by being at least partially transparent (see Chapter 4).

**5.** *Plasticity costs*: These are unique to prey that have inducible defences that are responsive to variation in predator environments. There may be costs incurred for instance in the development of necessary sense/response systems (chemosensors and hormones). This is especially true of animals with facultative background matching, such as squid, and amphibia.

We can now return to the question, *why crypsis*? Obviously, prey are likely to manifest some form of crypsis if risks of predation are high. However, of equal importance, prey are more likely to be cryptic if the opportunity costs of crypsis are relatively low. As we explain in more detail in Chapter 5, if the species life history is such that the opportunity costs of crypsis are very high, the species may be more likely to evolve effective secondary defences that enhance post-attack survival. The result is that such prey avoid the need for crypsis and therefore benefit from the opportunities present in a habitat, while paying relatively low costs of predation. The secondary defences that make this possible include any trait that sufficiently enhances survival during and after an attack; most obviously behavioural responses, morphological protections (such as spines) and chemical defences. Hence the answer to the question *why crypsis* is about both the benefits that crypsis confers and the costs it imposes.

## 1.2 Industrial melanism in *Biston betularia*

The most intensively studied species considered to display industrial melanism is the peppered moth (*Biston betularia*), so called for its pale, speckled morph, which is considered to provide effective background matching against trees with lichen-covered bark. This *typica* morph was the only form reported until around 1850, when an almost completely black morph *carbonaria* was increasingly observed. *Carbonaria* has generally grown in frequency to dominate many populations in heavily industrialised areas of the United Kingdom, but not rural ones. The explanation behind this (originally due to Tutt, 1896) has been that whilst *typica* should gain protection from predators in uncontaminated trees, pollution can both kill off lichens and darken the surfaces of trees with soot, selectively favouring the *carbonaria* morph over the *typica* morph, because it is less easily detected by predatory birds. The last 20–30 years have seen the reduction in the frequency of *carbonaria* in favour of *typica*. This has been explained by legislation leading to reductions in pollution, recovery of trees to a more natural state, and return of the selective advantage of *typica*, thanks to more effective background matching than *carbonaria*. Parallel phenomena have been reported both in other species and in continental Europe and North America. Hence industrial melanism in *B. betularia* has been a popular textbook example of evolution happening within human time-scales by easy-to-understand changes in selective pressures. However, recently this story has come in for criticism ranging from reasoned (Sargent et al., 1998) to emotional (Coyne, 1998) to sensationalistic (Hooper, 2002).

The main problem stems from our lack of understanding of the resting place during the day of these night-flying moths. It is during the day that moths will be vulnerable to detection and predation by birds. The problem is that we do not know for sure where these moths choose to rest during the day. There is growing evidence that they site themselves on the underside of horizontal branches near junctions with vertically inclined branches in the tree canopy (Majerus, 1988; Majerus et al., 2000), but it seems certain that the trunks of trees are not important resting places. This is important for a number of reasons. First, when measuring the effects of pollution on trees, it has generally been trunks that have been inspected, rather than the canopy, so our understanding of soot deposition and lichen cover may be quite misleading. Further, many experiments (starting with the classical studies of Kettlewell, 1955, 1956, 1973) have demonstrated birds taking living or dead moths from tree trunks. Any evidence that these studies produce about the relative predation rates of the two morphs against differently coloured tree trunks must be strongly tempered if trunks do not constitute the natural battleground between moths and birds. Kettlewell's experiments have been heavily criticized by Coyne (1998) and Hooper (2002) in particular. It is true that certain aspects of the design of his experiments do not meet with current standards, but they are certainly no worse as experiments than many influential works by his contemporaries. For example, it is unfortunate that Kettlewell scored the crypsis of moths against different backgrounds using human observers, and that recent research (Majerus et al., 2000) have shown that moth/substrate combinations that look cryptic to humans can be easily detected by birds and vice versa. However, it was common in the 1950s to consider that humans and birds had similar eyesight. We feel that Kettlewell should be judged in the context of his era, by which yardstick he performed a number of ingenious and highly influential experiments, which had flaws to be sure, but not sufficient to call for them to be discarded out of hand. The kindest thing we can say about Hooper's suggestion that Kettlewell was less than honest in the reporting of his experiments, is that her book falls short of providing the watertight case that ought to back up any such serious allegation.

The problem of tree trunks reappears in experiments (again starting with Kettlewell, 1955, 1956) using mark–recapture techniques that demonstrate apparent selection for *carbonaria* in polluted areas and for *typica* in pristine areas. Moths were released during the day, and generally settled on the nearest substrate including large numbers on tree trunks. However, substrate choice (and subsequent vulnerability to predation) in these experiments is likely to be quite misleading as a guide to substrate choice made under naturally occurring conditions, which likely happens at or before first light.

Where does this leave the textbook story of industrial melanism? Firstly, no one seriously disputes that the phenomenon is an example of natural selection acting over relatively short time-scales. However, the agent that has driven this selection is now not fully clear, although we agree with Majerus (1988), Grant (1999), and Cook (2000) that despite drawbacks to almost all studies, the overwhelming weight of evidence still suggests that differential predation by birds due to differential background matching is most likely to be the dominant selective agent. Sargent et al. (1998) are less convinced. They cite mark–recapture studies that did not find predicted differential survivorship as problematic for existing theory. We feel that this concern is exaggerated, since there is no reason to suppose that differential survivorship is sufficiently strong that small-scale experiments will always detect it, nor indeed that it must operate under all conditions. For example, if differential predation occurred only under certain weather conditions (e.g. bright sunlight) then some experiments will not detect it, but providing these conditions occur sufficiently regularly this intermittent selection could still drive the observed population-level trends. Sargent is also concerned that not all industrial areas showed the rise in frequency of the *carbonaria* morph whereas some apparently rural ones did. These cases are interesting, although an answer *may* be found in dispersal from neighbouring areas where selection is different; our lack of understanding of dispersal in this species, especially of pre-adult phases needs addressing. Also, no-one would suggest that all industrialised areas should be expected to produce identical selection pressures as the intensity and composition of pollution will differ between sites, and local topography and vegetation will affect the deposition of this pollution on trees. We have no argument with Sargent's contention that selection between the two morphs acting in pre-adult stages may be worth more close consideration, but this need not be an expense of reduced interest in predation as a selective agent. This textbook example still has a lot to attract practising scientists as well as students, and further work would be very welcome.

## 1.3 Background is a multivariate entity

In order to be cryptic, an organism must seek to match its background in all aspects of the visual signal that can be detected by the viewing organism (Endler, 1978). The viewer's eye will be sensitive to some or all of the following characteristics of a light source: intensity, wavelength and polarisation. The cryptic organism must match any spatio-temporal heterogeneity in these characteristics resulting from patterning and movement of the background. The background that must be matched depends on the viewer. If the viewer cannot detect polarisation of light, then there is no selective advantage for the cryptic organism to provide a good match to the polarisation characteristics of the background. Hence, different types of viewers can select for different background matching characteristics. Heiling et al. (2003), for example, argue that the crab spider *Thomisus spectabilis* is cryptic to humans but is highly conspicuous to the bees on which it preys. Intriguingly, these conspicuous markings make the flower that the spider rests on more attractive to bees, hence luring them into close proximity to the spider.

Furthermore, important characteristics of the background will differ according to the distances between the viewer and the cryptic organism and between the cryptic organism and physical features of the environment (Stamp and Wilkens, 1993). For example, a frog resting on a leaf improves its crypsis, when viewed by a human from a distance of around 2 m, if it matches subtle patterning on the leaf, say caused by beads of moisture. However, when viewed from 10 m away, such matching is unimportant, since the leaf appears to be a uniform block of colour (Stamp and Wilkens, 1993). The background will also depend on a number of qualities of illumination. Hence, background should not be thought of as simply as synonym for habitat. Background can be considered as the visual stimuli that the viewer obtains from sources other than the viewed organism. It is a function of all of the following: the physical habitat, illumination, the sensory physiology of the viewer, and the positions in the physical habitat of both the viewer and the viewed organism.

There has also been speculation by a number of authors that some colour patterns are cryptic when viewed from a distance but act as conspicuous warning displays when viewed close up (see review in Stamp and Wilkens, 1993). One possible example is the crab spider reported by Heiling et al. (2003) that is considered to be cryptic at a distance to bees as well as highly conspicuous close up. It is important to remember that crypsis does not necessarily imply dull coloration. In brightly lit environments, apparently garishly coloured reef fish may provide a good match to the background (Marshall, 2000).

## 1.4 Combining background matching with other functions

Since effective background matching depends on the visual abilities of the viewer, this can mean that an organism that does not appear cryptic to us is cryptic to its predators. One obvious example arises from interspecific differences in colour vision. If an organism's main predators are colour blind, then matching intensity of light is important to crypsis but spectral matching is not. Similarly, detail in the background that falls below the acuity threshold of predators need not be matched. This can present an opportunity for signalling to conspecifics without impairing crypsis, if the cryptic species has higher acuity than its predators or if signalling to conspecifics occurs under higher ambient light levels than predation events. For example, conspecific signalling can involve small multicoloured spots that do not match the pattern of the background but which blend over the coarser spatial frequency of predator acuity to provide a good match to the background (for fuller discussion, see Endler, 1991; Marshall, 2000). Similarly, it has been suggested that certain avian alarm calls decay rapidly over distance, so nearby conspecifics can be warned about an overflying bird of prey without alerting the predator to the caller's position (Klump and Shalter, 1984; Klump et al., 1986).

Some deepwater organisms may use red light for intraspecific signalling without giving their position away to their predators (a 'private communication channel'), since red light rapidly attenuates in water and (because of this) many deepwater organisms are not receptive to red light. Cummings et al. (2003) have suggested that ultraviolet (UV) light may have similar advantages to aquatic organisms as a short-range intraspecific signal, because of high signal degradation over longer distances due to UV light's high propensity to scattering by water molecules. In such situations, we must explain why the predator has not evolved its visual system to detect intraspecific signals between prey individuals. This is relatively easy for generalist predators, since a diversity of intraspecific signals combined with physical trade-offs in the visual system would prevent evolution of a system that could effectively detect the different signals of the predator's suite of prey types. Prey individuals, on the other hand, have no need to detect the signals of other species that share the same predator, and so do not have the same constraint on their visual system. Hence for these predators being a generalist presents sensory constraints that may prevent enhanced detection of 'private channels of communication'. In contrast, explaining why a specialist predator has not developed the ability to detect the intraspecific signals of its prey is more challenging, but not impossible.

One reason that specialist predators do not enhance their capacity to detect 'private channels of communication' is that prey generally have shorter generation times than their predators, producing different rates of evolutionary change. Furthermore, intraspecific signalling may be essential for reproduction, where a prey individual that does not signal does not reproduce. In contrast, predator individuals that can detect intraspecific signals would do better than those that could not, but those that cannot would still be able to contribute to the next generation (if the prey can also be detected by other means). Both this 'life–dinner' argument and the difference in generation times might allow the prey to evolve signals faster than their specialist predator can evolve to detect them. It is also worth remembering that specialist predators are uncommon, and generalization over a greater or lesser number of prey types is the norm for predators (Brodie and Brodie, 1999).

## 1.5 Flicker fusion

Movement by an organism can have an important bearing on its crypsis. When looking for fish in a stream from a bridge above, we spot them much more readily when they move than when they are still. The background of the stream bed is still, and so movement by the fish reduces it background matching. Generally, movement of an organism is not compatible with good matching of the background. However, an exception to this may come from the mechanism of flicker fusion. If a patterned object passes the viewer's visual plane faster than the critical flicker frequency, set by optical physiology, then the object appears to be monocoloured with a blend of the separate colours of the pattern. This effect can be demonstrated effectively with a child's spinning top, where patterns painted onto the top blur into a homogenous colour block as the top spins faster. It has been suggested that the banded pattern of snakes may produce this effect when the snake flees from a predator at speed, and that this confers a benefit to the snake in reduced predation risk (Endler, 1978; Plough, 1978). For example, flicker fusion may allow a monocoloured cryptic effect in the fleeing snake to be combined with an aposematic pattern when the snake is still or moving slowly. Alternatively, if the banded pattern is cryptic, then the eye may be deceived if the monocoloured fast moving object is suddenly transformed into a stationary cryptic one. It should be emphasized, however, that there is very little empirical support for these functions of snake colouration.

Lindell and Forsman (1996) suggested that they had evidence in support of the flicker fusion hypothesis. They compared survival in the field of males and females in a species of snake that has a monochromatic melanistic morph and a patterned morph. Melanistic males had lower survivorship than patterned males, but melanistic females had higher survivorship than patterned females. Lindell and Forsman (1996) proposed that this difference could be explained if males move about rapidly in search of sedentary females, and so benefit from flicker fusion. This is a very interesting suggestion, but in itself it provides at best weak support for the flicker fusion hypothesis. We agree with Lindell and Forsman that an experimental approach is required to evaluate the ecological importance of flicker fusion. Until then, the ecological relevance of this phenomenon should be considered as unproven.

## 1.6 Polymorphism of background matching forms

We now consider how the concept of background matching can be reconciled with the frequent observation of polymorphism in many cryptic species. Ford (1940) defined genetic polymorphism as the occurrence in the same locality of two or more discontinuous forms of a species in such proportions that the rarest of them cannot be maintained only by recurrent mutation from the other forms. This section will consider a number of different mechanisms by which polymorphism and background matching can be reconciled. The main mechanisms produce frequency-dependent *per capita* predation risk on prey morphs, and careful exposition of this requires a detour into the complex maze of terminology used to describe frequency-dependent selection. However, before starting on this, we remind the reader that predator–prey mechanisms are certainly not the only evolutionary factor involved in polymorphism, this appears to be true for perhaps the most frequently cited instance of polymorphism, snails of the genus *Cepaea*.

### 1.6.1 A case study: polymorphism in *Cepaea*

Although there are several species of land snails of the genus *Cepaea* that show genetic polymorphism the most intensively studied (and most variable) is *C. nemoralis*. This snail lives across Northern Europe in a wide range of habitats. It features polymorphism both of shell colour and in the presence and colour of up to five bands on the shell. This polymorphism is genetically determined. Polymorphism is extensive—with one study reporting only 20 monomorphic populations in a survey of 3000 (Jones et al., 1977)—and easy to score. This, combined with the commonness of the snails, has made the visible polymorphism in *Cepaea* a popular subject for population genetics studies. This body of work, in turn, has led to a reasonable understanding of the link between genotype and phenotype in this species. However, explaining the factors driving the evolution and maintenance of polymorphism has proved a much greater challenge.

Predation by birds (particularly song thrushes *Turdus philomelos*) is considered to be important in maintaining polymorphism (Cain and Sheppard, 1954). There is strong evidence of an association between shell colouration and background colour (e.g. Cain and Sheppard, 1954; Clarke, 1962) and furthermore that song thrushes are adept at detecting snails that provide a poor match to the background (Cain and Sheppard, 1954; Cain and Curry, 1963*a*,*b*). Hence, one possibility is that polymorphism is generated by disruptive selection pressure caused by background heterogeneity, either within the habitat of a single local population or between neighbouring populations with polymorphism being maintained by migration. Alternatively or additionally, it has been suggested that predation is frequency dependent with predators maintaining polymorphism by focusing on common morphs. There are numerous examples of such frequency-dependent predation in artificial systems (Greenwood, 1984; Allen, 1988*a*,*b*; Allen et al., 1988; Tucker, 1991). Hence, it is plausible that polymorphism in this species could be maintained by frequency-dependent predation although there is no direct evidence for this (Cook, 1998).

To complicate things, there are numerous other factors that could also influence the fitness of different morphs and hence the maintenance of polymorphism. One of these is the effect of shell colour on thermoregulation (reviewed by Jones et al. 1977). Further, it may be that the answer lies not purely in ecological mechanisms but in the combination of ecology and genetics, through such mechanisms as heterozygote advantage, drift, gene-flow, founder effects, population genetic bottlenecks, linkage disequilibrium, and co-adaptation (all reviewed by Cook, 1998). Jones et al. (1977) conclude that 'The most important lesson to be gained from intensive study of the *Cepaea* polymorphism is that many types of evolutionary force act upon it and the relative importance varies between different polymorphic loci, or even when the same locus is studied in different populations.' This general caution could equally as well be applied to the study of industrial melanism amongst others, the natural world is complex and heterogeneous, and the search for a simple explanation that explains all cases of a phenomenon may be fruitless.

### 1.6.2 Polymorphism through neutral selection

Endler (1978) gives a much quoted definition of crypsis which begins 'a colour pattern is cryptic if it resembles a random sample from the background . . . '. Taking random samples from the background would suggest that, providing the background is spatially variable, there can be a number of background-matching forms, each of equal crypsis, with neutral selection acting on this polymorphism. Hence, polymorphism could theoretically be maintained because there are equally cryptic visual forms. Although conceptually attractive, we know of no study designed to demonstrate that different morphs of a given species are equally as cryptic to their natural predators. Such a study would be useful. We now turn our attention to circumstances where there may be aspects of predation that may lead to positive selection pressure for polymorphism.

### 1.6.3 Positive selection for polymorphism

In order to examine the role of frequency dependence in the explanation of polymorphism in crypsis, Bond and Kamil (1998) designed an ingenious experiment using birds that were trained to search a computer screen and peck at images of moths in return for a food reward. The population of moth images stored in the computer from which individuals were randomly selected to appear on the screen was allowed to evolve over time, with offspring being slightly mutated versions of their parents. However, selection for crypsis was introduced by allowing greater representation of the least detected 'prey' in succeeding generations. Selection by the birds led to increased crypsis and greater phenotypic variance compared to control populations that were subjected to no selection. Bond and Kamil interpret their results as being suggestive that polymorphism is caused by the birds displaying pro-apostatic selection. That is, a given morph is less likely to be detected by the birds when it appears on the screen if the birds have little recent experience of that morph because it is rare in the population. This interpretation was strongly supported by the demonstration that both phenotypic variance and crypsis increased to higher levels in the experiments with avian predators than in controls with computer-controlled selection when any frequency-dependent effects were eliminated.

However, an earlier study using a similar experimental set-up (Bond and Kamil, 1998) demonstrated that polymorphism was not an inevitable outcome of this experiment. Thus, Bond and Kamil (1998) found that if all phenotypes are too easy for the birds to detect then there will be no selection, since all images are always found whenever presented to the birds. Conversely, if a morph is introduced that is impossible for the birds to detect, then this will come to dominate the population, reducing diversity. These experiments seem a convincing demonstration (albeit in an artificial system) of predator-induced selection for prey polymorphism of appearance. Further, this polymorphism seems to be driven by pro-apostatic selection. We know of no demonstrations that pro-apostatic selection maintains polymorphism in cryptic prey in a natural setting, but these laboratory experiments certainly suggest that such a situation is feasible. The terminology surrounding frequency-dependent selection is complex and we point the reader to the next section for definitions.

### 1.6.4 Definitions related to frequency-dependent predation

There is a strong body of experimental evidence for frequency-dependent preference: where predators prey more heavily on a prey type (and so per capita predation risk increases) when that type is more common relative to other prey (e.g. Reid, 1987; Allen, 1988*a*). Whilst there is general agreement that such frequency-dependent preferences often occur, identification of the mechanism behind them in any particular case can be challenging: because mechanisms may act together and because different mechanisms can produce very similar behaviours. There are several processes that occur in phases of the predation sequence subsequent to detection of prey that can lead to frequency-dependent preferences (Greenwood, 1984; Endler, 1988; Sherratt and Harvey, 1993); in this book we ignore the majority of these processes and focus on mechanisms that are related to detection of prey and in particular to polymorphic appearance within a group of similar prey. However, before we go any further, it is necessary to cover some of the terminology that is used to describe frequency-dependent predation (Table 1.1).

Let us consider a situation where a predator includes two prey morphs of the same species in its diet (labelled prey types 1 and 2).[1] Over a certain time period, the predator eats $E_1$ of type 1 and $E_2$ of type 2. The densities of the two prey types in the environment are $N_1$ and $N_2$. We follow Murdoch (1969) and Cock (1978) (amongst others) in defining the predator's preference between their two prey types ($p_{12}$) by

$$\frac{E_1}{E_2} = p_{12}\frac{N_1}{N_2}.$$

[1] It is straightforward to extend the following arguments and definitions to any number of prey types, but for simplicity we will stick to just two prey types.

**Table 1.1** A glossary of terms relating to frequency-dependent predation. Adapted from Allen (1989)

| Term | Examples of use | Meaning |
|---|---|---|
| Frequency-dependent selection (fds) | Fisher (1930), Ayala & Campbell (1974) | Selection that results in a positive or negative relationship between relative fitness and relative frequency |
| Positive fds | Levin (1988) | Fds where relative fitness of prey is positively related to relative frequency |
| Negative fds | Partridge (1988), Antovics & Kareiva (1988), O'Donald & Majerus (1988) | Fds where relative fitness of prey is negatively related to relative frequency |
| Apostate | Clarke (1962) | Rare morph maintained by apostatic selection |
| Apostatic selection | Clarke (1962) | Negative fds by predators in the absence of Batesian mimicry |
| Apostatic polymorphism | Clarke (1962) | Polymorphism maintained by apostatic selection |
| Pro-apostatic selection | Greenwood (1985) | = apostatic selection |
| Anti-apostatic selection | Greenwood (1985) | Positive fds by predators in the absence of Batesian mimicry |
| Potential apostatic selection | Greenwood (1984) | When the prey type taken is taken more often than expected by chance at all frequencies, and this selection increases with frequency |
| Switching | Murdoch (1969) | = apostatic selection (especially when prey are different species) |
| Matching selection | Bond (1983) | Apostatic selection (in matching backgrounds) |
| Oddity selection | Bond (1983) | = anti-apostatic selection |
| Reflexive selection | Moment (1962a) | = apostatic selection (on massive polymorphisms) |
| Reflexive polymorphism | Owen & Whitely (1989) | Massive polymorphism maintained by reflexive selection |

Of course, $p_{12}$ is a relatively uninformative measure of preference for one morph over another: in practice it is often difficult to separate a genuine 'liking' for one prey type over another from circumstantial details such as the true availability of alternatives. Nevertheless, if $p_{12}$ is greater than 1, then this indicates an overall 'preference' for prey type 1 over prey type 2; if it is less than 1 then the preference is reversed. In the boundary case, no preference is apparent. Most importantly, since $p_{12}$ is equivalent to the ratio of the per capita mortality of prey type 1 to the per capita mortality of prey type 2, it can be used as a simple measure of the relative fitness of these two prey types in the context of predation.

Say in a series of experiments, the ratio of prey availabilities ($N_1/N_2$) is varied, the ratio of consumptions ($E_1/E_2$) is noted, and $p_{12}$ calculated. If $p_{12}$

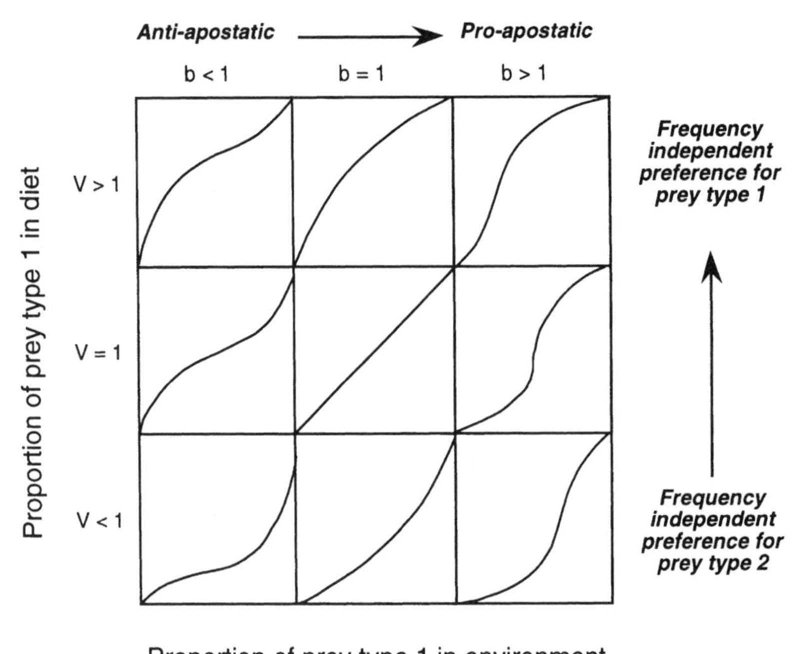


**Figure 1.1** Graphs showing how the frequency-independent and frequency-dependent components of predator preference can combine to influence the relationship between the proportion of a prey type in the environment and its proportion in a predator's diet. All nine graphs have limits 0.0–1.0. Here we have considered just two prey types (1 and 2) and used the Elton–Greenwood model (Greenwood and Elton, 1979) to generate the graphs. The parameter $b$ in the Elton–Greenwood model reflects the degree of frequency dependence in the preference ($b < 1$ anti-apostatic, $b > 1$ apostatic), while the parameter $V$ reflects the underlying frequency-independent selection ($V > 1$ preference for type 1, $V < 1$ preference for type 2).

is unaffected by the ratio of prey availabilities, then we have a *frequency-independent preference*. Otherwise, if $p_{12}$ increases as $N_1/N_2$ increases, then the predator displays a frequency-dependent preference, such that per capita predation rate on prey type 1 compared to the per capita predation rate on prey type 2 increases as prey type 1 becomes more common in the overall population. If $p_{12}$ (as a measure of relative fitness) rises from below 1 to above 1 as $(N_1/N_2)$ rises, then when prey type 1 is rare it will be selected for over prey type 2, and vice versa when it is common (see Fig. 1.1). We call this form of frequency dependence (*pro-*) *apostatic selection* (selection *for* the rare form). Conversely, when the preference for a prey type decreases from above 1 to below 1 as it becomes more common then such an outcome is referred to as *anti-apostatic selection* (see Fig. 1.1). Following Greenwood (1984), we reserve the term *potential pro-apostatic selection* for a subset of cases analogous to pro-apostatic selection where $p_{12}$ increases with increasing $N_1/N_2$ but one form is consistently at a selective advantage over the other (and *potential anti-apostatic selection* in a similar way). Naturally, you could invent more names for other outcomes (e.g. when $p_{12}$ falls over a range of values of $N_1/N_2$ and then rises[2]—as predicted in a theory paper of Sherratt and MacDougall (1995), but this is perhaps going too far.

Of course, it is possible to consider cases in which the two prey types are separate species. Once again, preference for one prey type may cross the value of one as a prey type gets more common. Here the analogous form of positive frequency-dependent predation has been called *switching* (Murdoch, 1969).

[2] If you find this form of selection in any of your experiments, we recommend that you call it something fancy.

Be warned when reading the literature that not everyone adopts the definitions given above. The term '*positive frequency dependence*' can be particularly confusing—it typically refers to cases where relative fitness of a prey type is positively related to relative frequency, but has sometimes been inappropriately interpreted in the converse: *attack rates* increasing on a prey type as it becomes more common (this is *negative frequency dependence*). Most authors do not differentiate between 'potentially pro-apostatic selection' and 'pro-apostatic selection', and refer to both as *apostatic selection* (e.g. Endler, 1988). Bond (1983) has called particular forms of apostatic selection *matching selection* and anti-apostatic selection *oddity selection*. Pro-apostatic selection has also been called *unifying selection* by Pielowski (1959). Reflexive selection (Moment, 1962; Owen and Whiteley, 1986) is also a synonym for pro-apostatic selection (Allen, 1988*a*,*b*). Bewildered? The abundance of terminology reflects the general importance of frequency-dependent predation, and the fact that it has been observed in many different contexts. Hopefully, Table 1.1 will help (modified from Allen, 1989*a*).

### 1.6.5 Search images

One of the most popular explanations for pro-apostatic selection is the formation of *search images* (Tinbergen, 1960). This term has been used to describe an enormous host of activities related to foraging. This has lead some (e.g. Dawkins, 1971) to suggest that the term is too imprecise to be useful. Whilst we have some sympathy with this viewpoint, we use the term here, but clearly define what we mean by it.

> We define a *search image* as a transitory enhancement of detection ability for particular cryptic prey types or characteristics.

One criticism of the term 'search image' is that it is imprecise about mechanism (Dawkins, 1971). This criticism could be applied to our definition; we do not specify the exact mechanism by which this 'enhancement of detection ability' comes about. However, as we define it (following Dukas and Kamil, 2001), search image is (1) a process of transitory attentional specialization that results in enhanced detection that (2) follows from repeated visual detection of an item over a relatively short time-scale. As such we can be explicit about the types of processes underlying our conception of a search image. Equally important, we are explicit about the types of processes that do not lead to search images under our definition. These include behavioural changes such as modification of search paths, and strategic decision-making about which detected prey items to accept or reject. Notice also we consider that a prey item must be to some extent cryptic for a search image to form for it. In the following discussion we consider detection of prey that vary in their levels of conspicuousness, but we are considering prey that are at the cryptic end of the conspicuousness continuum and are not therefore conspicuous outright.

A search image is formed after repeated detections of a particular prey type (Pietrewicz and Kamil, 1979), and requires further frequent detections of that prey type in order to be maintained (Plaisted and Mackintosh, 1995). Hence, a search image might form for a common prey morph, but not for other, less common forms. Search images should therefore lead to disproportionately high predation on common forms, and disproportionately low predation on rarer prey types. If rarity is beneficial, then search images may promote diversity of prey types leading to stable polymorphisms within cryptic species. This capacity to explain the evolution of diversity in cryptic prey species has made the existence of search images the focus of considerable attention and controversy. However, we should point out that although stable polymorphism is a likely outcome if predators use search images, it is not inevitable, especially if a rare morph is itself rather more conspicuous than a common form. Formation of a search image for a common prey type may reduce the per capita likelihood that a rare prey type is detected, but this reduction need not necessarily bring the likelihood of detection of a more conspicuous rare morph below that of the alternative, more common and more cryptic morph. Search images may therefore simply slow rather than reverse a decline in the relative frequency of a rare morph. Nonetheless, search image formation, and more generally pro-apostatic selection by predators on cryptic prey, can provide an evolutionary pressure for the development and maintenance of polymorphism in prey species, especially when any difference in the effectiveness of crypsis between alternative forms is not large.

### 1.6.6 Control of search rate

Frequency-dependent selection on cryptic prey can also arise from predators' control of their rate of searching the environment. There is likely to be a trade-off between how rapidly an area is searched for prey and the efficiency of detection (defined as the fraction of prey in the scanned area that is detected: for empirical support for this conjecture, see Gendron and Staddon, 1983). Further, cryptic prey may require a slower search rate to obtain a specified detection efficiency, compared to a more conspicuous prey type. Gendron and Staddon (1983) argued that it would be optimal for predators to reduce their rate of search when cryptic prey types are common in the environment. Conversely, when cryptic prey types are rare compared to more conspicuous prey, the optimal strategy is to increase search rate at the expense of detection efficiency. This will lead to a greater fraction of the more cryptic prey being missed but, the predator will be more than compensated by the larger number of the more conspicuous prey discovered. Hence, when a cryptic prey is rare, optimal control of search rate leads to a reduction in its per capita detection risk. In this way, strategies that control the rate at which predators search their environment can often explain empirical results, especially the disproportionate survival of some rare forms, that had originally been attributed to a search image effect (Allen, 1989*b*; Guilford and Dawkins, 1989*a*,*b*).

### 1.6.7 Comparing search image and search rate mechanisms

It is possible to identify a number of important differences in the behaviour of 'search rate' and 'search image' predators. First, with search rate predators the more conspicuous of a pair of cryptic morphs will always be at a disadvantage. Changes in the frequency of such a morph may affect the size of the fitness difference between a pair of morphs, but it will not change the fact that a more conspicuous prey will always be detected most often. Thus, when the more conspicuous prey is common, it will suffer disproportionately high predation because the optimal search rate is high and the predator's capacity to detect the more cryptic prey is low. Furthermore, when the more conspicuous morph is rare compared to a more cryptic form, it will cause a low optimal search rate, with a correspondingly high detection efficiency, and hence it may suffer even higher detection rates.

In contrast, a probability of detection for any given prey with a search image predator depends on that prey's conspicuousness and its frequency. One morph could be slightly more conspicuous than another, but if it is sufficiently rare the predator may have no search image for it and hence it may have better protection than a more common, but more cryptic alternative morph. Since both morphs lose fitness if their frequency increases, a stable equilibrium is likely even if one form is more conspicuous than another. One important consequence is that stable polymorphism may be maintained under 'search image' but not 'search rate' predators.

A second interesting difference between the search image and search rate behaviours is seen by comparing detection of prey in a monomorphic population to a polymorphic population where all morphs in both populations are equally cryptic. The search image mechanism suggests that frequency-dependent selection will occur in the polymorphic case, whereas the search rate hypothesis does not. The search rate model predicts entirely the same search rate in the two cases, and predicts that individuals of all prey morphs will be equally vulnerable to detection. In contrast, the search image hypothesis suggests that polymorphism will reduce the predator's capacity to detect prey; search images will either take longer to arise and will be less successfully maintained because the encounter rate of individual morphs decreases compared to the monomorphic case.

Hence, in this special case where the morphs have equal crypsis, polymorphism is stable for prey under predators operating the search image mechanism, but not under predators operating the optimal search rate mechanisms (Guilford and Dawkins, 1987; Knill and Allen, 1995). In support of the search image models Knill and Allen (1995) found that human 'predators' were less effective at detecting prey in the polymorphic case. Similarly, Glanville and Allen (1997) found that human subjects were slower to detect computer-generated prey displayed on a screen in trials where prey were polymorphic compared to monomorphic trials. The prey types were assumed to be equally cryptic, although this was not demonstrated explicitly. None the less, these results tend to support the suggestion that in these studies frequency dependent selection occurred through search image formation rather than search rate control. Furthermore, Plaisted and Mackintosh (1995) reported evidence from pigeons (*Columba livia*) strongly suggestive of search image processes. They found a temporary improvement in prey detection by pigeons following a run of encounters with the more common of a pair of equally cryptic prey types.

There is no reason why search image mechanisms and optimal control of search rate cannot occur simultaneously. Such a situation was explored theoretically by Dukas and Ellner (1993). Their model does not explicitly use the phrase 'search image', however, such a phenomenon is implicit in the assumptions of the model. Dukas and Ellner assumed that prey detection requires information processing, and the predator has a finite capacity for processing, which it must apportion to detecting different prey types. Increasing the processing ability devoted to a given prey type increases the chance that an encountered individual of that type will be detected. However, because of the finite processing ability available, increasing ability to detect one prey type can only be bought at the cost of reduced ability to detect other prey types. The other key assumption of the model is that the more cryptic a prey type is, the more information processing capacity is required to achieve a specified detection level. These assumptions are motivated by consideration of neurobiological experiments mostly on humans (see references in Dukas and Ellner, 1993; Bernays and Wcislo, 1994; Dall and Cuthill, 1997; Dukas and Kamil, 2001). This model predicts that when prey are very cryptic, then the predator should devote all its processing capacity to detection of one type. However, as the conspicuousness of prey increases, so the diet of the predator should broaden, as it becomes advantageous for it to spread its information processing capacity across a wider range of prey types. When the predator divides its 'attention' between several prey types, this division should not necessarily be even. If prey are cryptic, then the predator should give more attention to the least cryptic of the prey types; whereas if the prey are generally fairly conspicuous then the predator should attend most to the least detectable type.

In summary, much heated debate has surrounded the concept of search images over the last 30 years. This debate has arisen because many other mechanisms (most notably control of search rate) can also produce very similar behaviour to search image formation. This had lead some to overstate evidence of search image formation, without logically excluding plausible alternative explanations. However, more recent studies do seem to have demonstrated the existence of this mechanism, and it does appear that search image formation (perhaps working in concert with other mechanisms, such as control of search rate) could potentially select and maintain polymorphism in cryptic populations. This has been demonstrated in the laboratory, but not yet in the field.

### 1.6.8 Neutral selection again

Although pro-apostatic selection by predators may promote heterogeneity, Endler (1978) argued, 'pattern diversity among morphs or species subject to predation on the same backgrounds should decrease with increasing selection intensity'. His argument is that when predation pressure is weak, a morph that is a reasonable approximation to the background may be 'good enough', and hence a diversity of different forms could be equally 'good enough'. However, this would not be the case if the predation threat were greater. Despite its obvious importance, we know of no formal development or testing of this idea.

## 1.7 Coping with multiple backgrounds

A significant drawback to background matching as a strategy for crypsis is that almost all organisms will be viewed against a variety of backgrounds. Even if the habitat is physically homogenous, the temporal change in light conditions will change the nature of the background that the organism must attempt to match. Indeed, an organism can be viewed against two backgrounds simultaneously when viewed by two organisms with different visual sensory systems. This raises the question, should the organism specialize by maximizing its matching to one particular background, or should it seek a compromise that provides reasonable crypsis against more than one background but which is not maximally effective against any one of them. This was addressed using a simple model by Merilaita et al. (1999). We outline a very slight generalization of their analysis below.

We assume two backgrounds ($a$ and $b$). Let the probability of being viewed by a predator against background $a$ be $V_a$. The corresponding probability that the potential prey is not detected whilst in the predator's view is $C_a$. The prey is always viewed against background $a$ or $b$, so

$$V_a + V_b = 1.$$

Hence, the overall probability of being detected by a predator is given by

$$D = V_a(1 - C_a) + V_b(1 - C_b).$$

The probability of escaping detection is given by

$$E = 1 - D = 1 - V_a(1 - C_a) + V_b(1 - C_b).$$

In some cases, there is likely to be a trade-off between crypsis against the two backgrounds, and improved crypsis against one background can only be bought with reduced crypsis against the other. Mathematically, $C_b$ is a declining function of $C_a$.

$$C_b = f(C_a),$$

$$\frac{\mathrm{d}f(C_a)}{\mathrm{d}C_a} < 0.$$

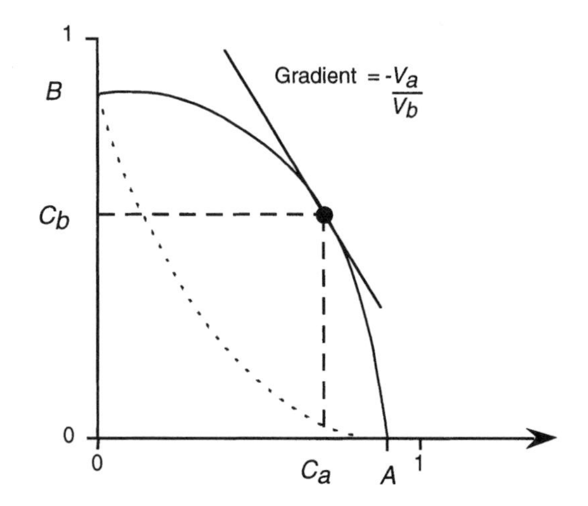


**Figure 1.2** The probability of not being detected when viewed against background $b$ ($C_b$) as a function of the equivalent probability against background $a$ ($C_a$). If the function is convex (like the solid line) and has a point where its gradient is $-V_a/V_b$, then a compromise level of crypsis is optimal, otherwise the organism should maximize crypsis against one of the backgrounds.

It is trivial to show that $E$ has a turning point (i.e. either a maximum or a minimum) if there is a value of $C_a$ that satisfies

$$\frac{\mathrm{d}f(C_a)}{\mathrm{d}C_a} = \frac{-V_a}{V_b}.$$

Further, this point is a maximum if, at that point,

$$\frac{\mathrm{d}^2f(C_a)}{\mathrm{d}C_a^2} < 0.$$

Figure 1.2 describes this situation graphically. We assume that the prey individual is free to adopt any value of $C_a$ from 0 (which maximizes crypsis against $b$ but provides no crypsis against $a$), to a value $A$ (which maximises crypsis against $a$ but provides no crypsis against $b$). If the shape of the trade-off curve is 'convex' like the solid line in the figure, and if there is a point where that line has gradient $-V_a/V_b$, then the optimal strategy for minimizing predator detection is a compromise value that provides some crypsis in both environments. Increasing $V_a$ (or decreasing $V_b$) moves this compromise towards improving crypsis against background $a$, as we would expect. However, if no such point can be found, or the trade-off line is concave, like the broken line in the figure, then the optimal strategy is to

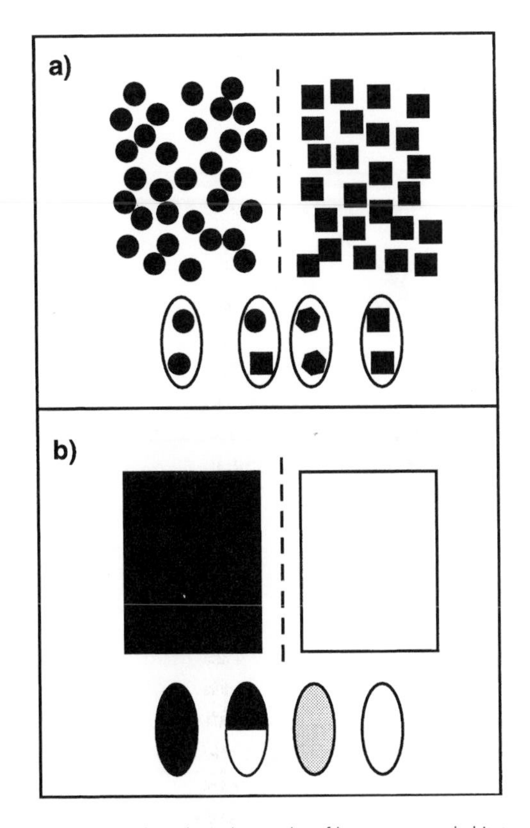


**Figure 1.3** Two hypothetical examples of heterogeneous habitats and animals relying on crypsis through background matching. In (a) the habitat consists of two different microhabitats, one with circular and the other with square elements. The two outermost of the four animals have adapted to the microhabitats with respective patterns only. The two animals in the middle, one with a circle and a square to the left and one with two hexagons to the right, represent compromised adaptations for crypsis in both microhabitats. Successfully compromised colourations give the trade-off between crypsis in these microhabitats a convex form. In (b) one microhabitat is black and the other is white. Again the outermost animals represent adaptations to one microhabitat only. However, this time the compromised colourations in the middle are apparently very poor, making the trade-off between crypsis in the two habitats concave.

specialize and maximize crypsis against one of the backgrounds. In the figure, the optimal strategy is to maximize crypsis against *a* providing

$$AV_a > BV_b,$$

where $B$ is the value of $C_b$ corresponding to $C_a = 0$, that is, the maximal probability of not being detected whilst in the predator's view against background *b*. Otherwise crypsis against *b* should be maximized. That is, crypsis should be maximized in the environment where the product of the probability of being viewed by a predator and the probability of that view not leading to detection is maximized. Hence, both background specialization and compromise can be predicted, a key determining factor being the shape of the trade-off between crypsis in one environment and another.

We now look to give a more ecological description of this shape. Imagine a prey organism that specializes in crypsis against background *b*, and so adopts strategy $C_a = 0$. If it can increase its crypsis against *a* by losing a relatively smaller amount of its crypsis against *b*, then we have the type of convex shape of trade-off curve than can lead to evolution of an intermediate level of crypsis that provides some protection in both environments. However, if a little crypsis against *a* can only be bought with a relatively large decrease in crypsis against *b*, then we have a concave shape and background specialization is favoured. Merilaita et al. give simple examples of abstract background combinations that might lead to these two different types of situation, these are reproduced in Fig. 1.3.

Of course, the situation can become much more complex than the simple example considered here. The costs associated with being detected need not be the same for detection against both backgrounds (since, e.g. the predator may be more effective at capturing prey against one background). The trade-off curve can have both convex and concave segments, or can incorporate straight line segment and/or discontinuities, and many more backgrounds than two can be considered. In such circumstances, the adaptive landscape will be much more complex and there will be considerable potential for evolutionarily stable polymorphisms. However, the essential points of our analysis above will remain unchanged, specifically:

**1.** The optimal cryptic strategy for an organism may be one that does not maximize their crypsis against any one of the backgrounds against which it is viewed, but rather provides some measure of crypsis against a suite of the backgrounds that it is viewed against.

**2.** In some cases, specialization against one background will be favoured, in which case this will be the background against which the organism can maximize the rate of occurrence of viewings of it by a predator that do not lead to detection.

The case where specialization is predicted warrants further scrutiny. Let us return to our simple case of the two environments $a$ and $b$. Let us imagine that $AV_a$ is only slightly bigger than $BV_b$. The optimal strategy is to maximize crypsis against background $a$. However, if there is a small change in one of the parameters, such that $AV_a$ becomes slightly smaller than $BV_b$ then this has dramatic consequences and now the optimal strategy is to maximize crypsis against $b$. This has important ramifications:

**1.** Two types of organisms could have very similar ecologies but adopt very different appearances because the difference in their ecologies means that they lie on different sides of the knife-edge described above and so adopt colorations that maximize their crypsis against different backgrounds.
**2.** If the environment inhabited by a population changes (perhaps only slightly) such that it moves to the other side of this knife-edge, then individuals in that population that previously had the optimal choice of colouration can now find themselves with a colouration that is far from the optimal. Evolution towards the new optimum may be particularly challenging as colourations only slightly different from the 'old' optimal may still be selectively worse than the current situation. An example of this is shown in Fig. 1.4. Here we have a concave trade-off function shown in Fig 1.4(a). This means that the turning point in $D$ is a maximum, and so the optimal strategy is one of the extremes. Imagine first that we are in a situation represented by the solid line in Fig. 1.4(b), $C_a = A$ is the optimal strategy. However, imagine now that $V_a$ is changed such that we move to the broken line: now $C_a = 0$ is the best strategy. However, a small mutation away from $C_a = A$ produces an increase in detection rate $D$. Only a macromutation producing a $C_a$ value less than $\beta$ in the diagram would be selected over the strategy $C_a = A$.
**3.** Polymorphism within a population may be maintained by even slight fluctuations in the proportions of different backgrounds that individuals experience, if those fluctuations continually move the system across the knife-edge.

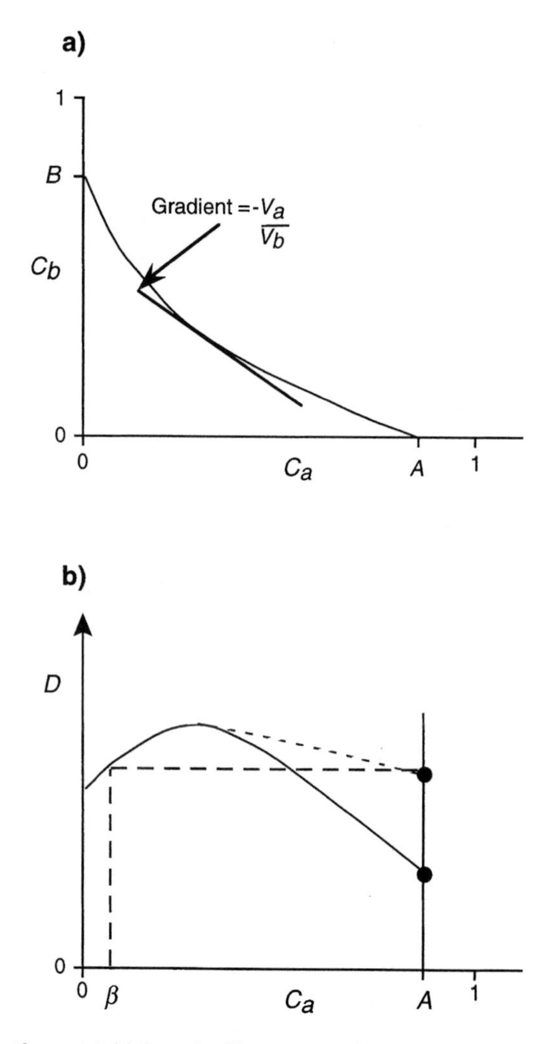


**Figure 1.4** (a) The trade-off between $C_a$ and $C_b$ is concave so specialization is predicted. (b) When $V_a$ is such that detectability ($D$) is given by the solid line then $C_a = A$ is the optimal choice, but if $V_a$ changes slightly so we move to the broken line, then $C_a = 0$ is optimal.

In a follow up paper, Merilaita et al. (2001) reported the results of experiments with captive birds searching for artificial prey. Birds were faced with one of two background types (differing in the size of patterning, see Fig. 1.5). There were three types of prey, one which matched the small background pattern, one which matched the larger patterned background, and an intermediate-sized pattern (see Fig. 1.5). On the small patterned background, small patterned prey were most cryptic (measured by mean time between prey discoveries), followed by

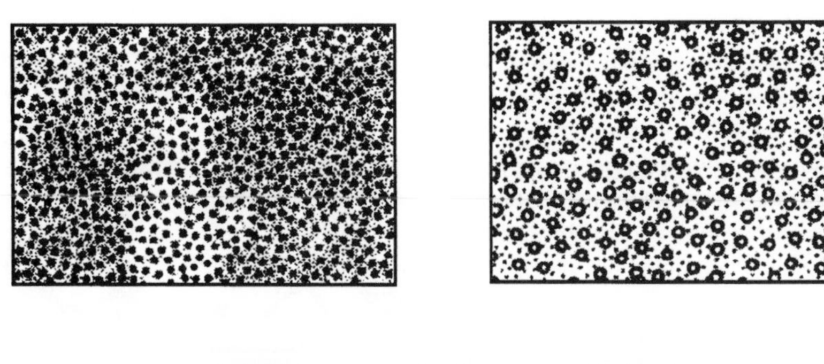

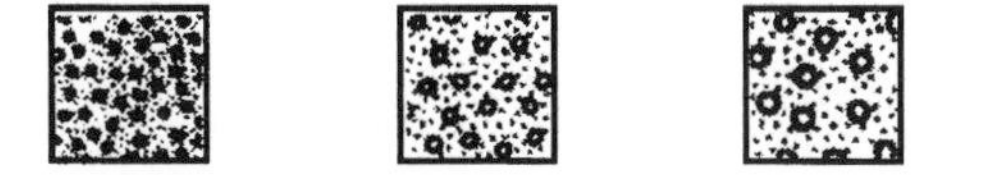

**Figure 1.5** The two large rectangles show the small and large background patterns and the three small squares show the small, intermediate (compromised) and large prey patterns used in the experiment.

the intermediate pattern, which in turn was more cryptic than the large pattern. On the large patterned background, the small patterned prey was least cryptic, but there was no difference between intermediate and large patterned prey. This means that in a situation where both backgrounds were encountered with equal frequency, the intermediately patterned prey would be the best protected of the three, and generally that circumstances exist where some intermediate type between the two extremes of background matching would be optimal. These results can be seen as supportive of the theory presented in this section, however, further empirical study of this theory would be very welcome.

The arguments in this section have many parallels with the arguments of Edmunds (2000) on why some mimicry systems display poor mimicry such that humans can easily tell model and mimic apart. He suggests that under some circumstances evolution will drive a mimic species to specialize on one particular model species, whereas in other circumstances generalized mimicry of a number of different models (necessitating imperfect mimicry of each one) will evolve (see Barnard, 1984 for a similar idea). Further discussion of the evolutionary pressures on 'quality' of mimicry can be found in Section 10.5.5.

## 1.8 Masquerade

Many organisms mimic inedible objects such as leaves, thorns, sticks and bird-droppings, that is, objects of no inherent interest to the potential predator. Remarkable examples include the sea dragon (*Phyllopteryx eques*), a sea-horse found off the coast of Australia that has numerous outgrowths that together create the impression of a sea weed (Fig. 1.6). The northern potoo (*Nyctibius jamaicensis*) from central America rests motionless on trees in the day, and closely resembles a broken branch (Edmunds, 1974). Naturalists have long been impressed with the way in which certain leaf mimics not only have a leaf shape, but also exhibit patterns resembling the leaf's veins and even their blemishes. Although conceptually similar, this general form of mimicry of inedible objects is often distinguished from the mimicry of whole organisms, such as that involved in Batesian

**Figure 1.6** The leafy sea dragon, a sea-horse with outgrowths that may cause it to be misidentified as a sea weed.

and Müllerian mimicry (Chapters 9 and 10), and is therefore sometimes called *masquerade* (sometimes called 'special resemblance to inedible objects': (Cott, 1940; Endler, 1991). As the value of masquerade rests on not being recognized as potential prey (or a predator), yet may also include aspects of the abiotic background such as bird droppings and pebbles, then many cases of masquerade fall somewhat in between crypsis and mimicry.

One charming way to think of the difference between masquerade and crypsis was proposed by Allen and Cooper (1985). In the first chapter of Milne (1926) we are introduced to Winnie-the-Pooh who was attempting to reach honey that was high in the trees and guarded by bees. Pooh first thought of floating up on a blue balloon which would be cryptic against the blue sky (the option of having a green balloon like a tree was also considered, and indeed Pooh was unsure which form of crypsis would provide the most effective concealment—yes, Pooh tackled the problem of coping with multiple backgrounds long before Meritalia et al. 1999—see Section 1.7). However, Christopher Robin remarked that the bees would identify him hanging below. As a solution, Pooh rolled himself in mud so that he would resemble a small black cloud. It is this difference between resemblance of background (balloon), and resemblance of an uninteresting object (cloud), that is often used to distinguish crypsis from masquerade. We will not dwell on the various classifications proposed (see Vane-Wright, 1980; Edmunds, 1981; Endler, 1981; Robinson, 1981; Pasteur, 1982), but note that some forms of masquerade (such as Pooh's disguise, or the remarkable resemblance of the Amazon fish *Monocirrhus polycacanthus* to a dead leaf, see Fig. 1.7) may facilitate access to resources rather than protection from predators.

Scientific research on masquerade has been somewhat limited, perhaps in part because this adaptation appears relatively straightforward to understand. De Ruiter (1952) found that captive jays (*Garrulus glandarius*) could not discriminate twig-like caterpillars from the twigs on which the caterpillars normally fed, but could distinguish these caterpillars from the twigs of other trees, suggesting that masquerade sometimes involves a close evolutionary relationship. It is also noteworthy that one

**Figure 1.7** The Amazon fish (*Monocirrhus polycanthus*) that strongly resembles a dead leaf.

characteristic that appears to help predators identify masquerading moths is their bilateral symmetry. Hence, many moths that resemble leaves also hold their wings in an asymmetric posture (Preston-Mafham and Preston-Mafham, 1993). There is also a methodological challenge to studying masquerade. Say we observe a herbivore passing over a plant that mimics a stone. Is this background matching (i.e. failure to detect the plant as an entity) or masquerade (i.e. detecting it as an entity but misclassifying it as a stone)? Unless you can read the herbivore's mind, it is hard to know. As Getty (1987) notes 'we do not have good operational definitions that allow us to recognize and count encounters and rejections unambiguously'.

One final question is whether masquerade shows any form of frequency-dependent advantage, such that it is more beneficial to the individuals that possess it when they are rare. Intuitively, one might expect that as the number of prey that masqueraded as twigs increased compared to twigs, then it would pay potential predators to pay more attention to twig-like objects. This is precisely what Staddon and Gendron (1983) predicted in an early application of signal detection theory. However, it is perhaps not entirely by chance that masquerading prey items have evolved to resemble extremely common objects in their environment such as bird-droppings, twigs, and leaves. Hence we feel that other factors may well regulate the population size of masquerading prey long before they reach densities at which it pays predators to attack both of them and the things they resemble.

We note in passing that Winnie-the-Pooh was ahead of the game in recognizing the frequency-dependent advantages of a disguise. Black clouds only tend to appear when it is about to rain so he asked Christopher Robin to open an umberella and pretend it was raining: 'I think, if you did that, it would help the deception which we are practising on these bees.' As Pooh said, 'you never can tell with bees.'

## 1.9 Conclusion

Background matching (and crypsis in general) has not been as intensively studied in recent years as some of the other topics in this book. This neglect is unwarranted since there are important issues where our understanding is far from complete. In particular, we emphasized in Section 1.1 that there are costs as well as benefits to adopting crypsis. The benefits are rarely quantified for any given system, but the costs are rarely even considered. This is a significant omission, since ultimately we want to develop an understanding of why some species have evolved crypsis and others have evolved say Batesian mimicry or investment in defences coupled with warning signals. Further, we want to understand why a given species has evolved to one form of crypsis rather than plausible alternatives (e.g. background matching rather than transparency). This requires us to be able to quantify and compare the costs and benefits of these alternatives for a given ecological system. There is no doubt that developing a quantitative cost–benefit framework to address such questions will be challenging, but progress in similar endeavours in fields such as life-history theory and mating systems suggests that it is not be beyond us.

CHAPTER 2

# Disruptive colouration

## 2.1 Introduction

Most sources quote Cott (1940) as the first to formalize the idea of disruptive colouration. Essentially, an organism's body will present characteristic shapes to a potential detector due to the shapes of different body parts and their organization within the organism. These characteristic shapes can be used to enhance the ease with which the organism is detected in the environment and identified for what it is. Edges and boundaries between adjacent body segments, or between the edge of the organism and the environment, play important roles in visual recognition; the idea behind disruptive colouration is to make the detection of edges and boundaries more difficult.

This can be done in two ways. First, false boundaries can be created by abutting contrasting colours in places where no real boundary occurs (see Fig. 2.1a). Cott calls this technique *constructive relief*. He suggests that this is best achieved in a pattern on a smooth surface 'when two contiguous patches in the pattern—one being light and one dark—are both shaded in such a way that the pale tint becomes lighter and the dark deeper as each leads up to the line of contact with its fellow'. This shading leads to the optical illusion that the smooth surface actually contains a boundary between body segments at the juncture of the lightly and darkly shaded regions. Essentially, this use of shading to create the illusion of physical structure that does not exist is the mirror image of the use of shading to disguise physical structure discussed in the next chapter. Cott's principle of *maximum disruptive contrast* suggests that this illusion works best when the two abutting areas of colour are very different in tone.

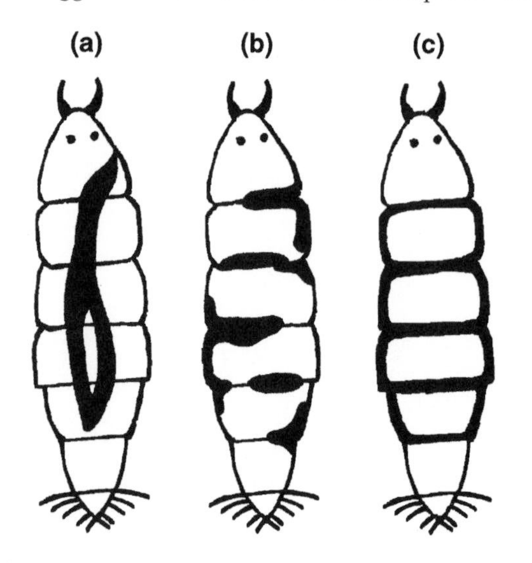


**Figure 2.1** Illustrations of disruptive colouration.

Second, real boundaries (again either between adjacent body segments or between the body of an organism and its background) can be made more difficult to detect by being broken up so that there is not a constant colouration (and so a constant contrast) along the boundary (compare Fig. 2.1b and c). Cott calls this *differential blending*: 'by the contrast of some tones and the blending of others, certain portions of the object fade out completely while others stand out emphatically'. This technique distracts the viewer from appreciating the full extent and shape of the real boundary, since some parts of it are easier to discern than others.

Both of these mechanisms may give the impression that the organism is quite different in shape (and even in number) than it actually is, and so reduce its chance of being detected in the environment and correctly recognized.

Disruptive colouration may also be used to reduce the ease with which a specific body part (rather than a whole organism) is detected and recognized. A particularly common recipient of such protection may be the eye, which in many taxa seems to be embedded in a contrastingly coloured stripe or other pattern (e.g. see Barlow, 1972; Neudecker, 1989). There may be other functions for such marking, such as an aid to accurate striking at the individual's own prey or as a means of reducing the effect of sun-glare (see Ortolani, 1999), but disruptive colouration also seems a plausible explanation for such head markings. This seems particularly likely given that eyes often appear to be specifically targeted in attacks (see Section 13.2.3) and can be physically fragile (and less expendable) compared to other body parts. We know of no experiments with real predators that test the concealing nature of such eye markings. However, Gavish and Gavish (1981) demonstrated that eye markings can reduce the ability of human subjects to detect eye-like patterns. Markings were particularly effective if the eye was positioned close to or on the line of contrasting tone generated by the boundary of the eye marking. Gavish and Gavish explain this with reference to physiological theories that the sensory system of the detector is distracted by the dominant signal of the nearby boundary between the contrasting colours of the eye marking and the background colour of adjacent head areas, and so is less likely to notice the pattern generated by the eye.

## 2.2 Separating disruptive colouration from background matching

We would expect that organisms using disruptive principles would often also seek to gain reduced detection by background matching. It is challenging, but not impossible, to explore whether the cryptic nature of an organism relies on disruptive mechanisms over and above benefits from background matching. Merilaita (1998) suggests three ways that disruptive colouration may conflict with the optimal appearance for background matching.

**1.** The elements of a background matching pattern should be distributed so as to match the spatial arrangement of appropriate elements in the environment. However, if elements are more concentrated on the edges of the organism than would be predicted from the distribution of similar elements in the background, then this could be suggestive that the patterning of the organism is influenced by disruptive mechanisms over and above background matching.

**2.** For background matching, the size and shape of pattern elements on the organism should match exactly similar elements in the background. If these background elements are relatively similar to each other, then disruptive considerations could drive such elements on the animal to be more variable (and so more complex) than those in the background, so as to give the impression of separate objects rather than a repeating pattern.

**3.** For background matching, colours used should be a random sample of those of the background. This may be modified by the need for strongly contrasting colours to produce a disruptive effect.

## 2.3 Empirical evidence

There have been so few studies of disruptive colouration, that we consider each in turn.

Merilaita (1998) applied the principles of the last section to the white-spotted morph of the marine isopod *Idotea baltica* (see Fig. 2.2), which appears cryptic to humans against its normal background of brown alga *Fucus vesiculosis* that has white-coloured epizoites (*Electra crustulenta* and *Balanus improvisus*).

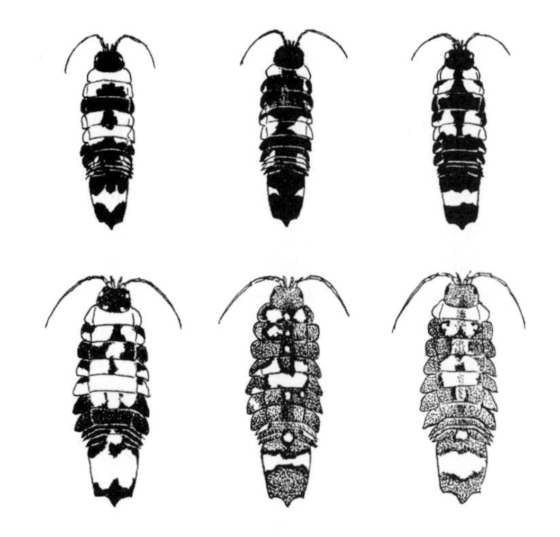

**Figure 2.2** *Idotea baltica* females (upper row) and males (lower row), not to same scale, redrawn from Merilaita (1998).

He found that white spots on the isopod touched the body outline more often than would be expected if they were distributed solely to match the background. In addition, the spots on the isopod were more variable in area and in shape than those of the background. Merilaita argued that these observations are suggestive of a role for disruptive colouration in the evolution of the appearance of this isopod. As an empirical study of disruption specifically, this paper is a very important step forward in our understanding of disruptive colouration. However, further work could be done on this species in order to really convince that disruption provides a benefit over and above any crypsis obtained by background matching. Specifically, there is a need to demonstrate that the isopod would be less cryptic to its natural predators if it had fewer marginal white spots and/or if its spots were less variable. Such a test should be possible using artificial prey items whose colouration could be easily manipulated.

There may be instances where markings that function as background matching also function disruptively without any compromise between the two functions. An example may be the saddle pattern of alternate light and dark stripes of several benthic freshwater fishes (see Armbruster and Page, 1996). These stripes may help fish match the background of large dark stones on a lighter, finer substrate, as well as visually breaking up the single fish into smaller units, consistent with disruptive colouration. In such cases, where disruptive patterning requires no compromise with background matching, identifying the relative contributions of these two mechanisms to crypsis would be very challenging.

Silberglied et al. (1980) suggested that many butterflies are considered to have an apparently disruptive wing stripe. Utilizing a small, isolated, and spatially contained population of one such species the Fatima butterfly *Anartia fatima*, they demonstrated that experimentally obliterating the white wing stripe with ink had no effect on either survival or wing damage in the wild, relative to a control group which also had the same ink added to their wings but in such a way as to leave them visually unchanged. Certainly, this result does not encourage the suggestion that the wing stripe has a disruptive function. Endler (1984), however, postulates a straightforward explanation for this result. He suggests that obliteration of the white stripe may simply convert the butterfly's pattern to one of equal crypsis. 'There are a large number of equally cryptic random samples of the same background; some may contain a white mark and others not'. This interpretation still suggests that the original striped pattern provided no added crypsis through disruptive colouration on top of the protection afforded by background matching. An alternative interpretation is given by Waldbauer and Sternburg (1983), who suggests that obliteration of the white stripe may have induced unintended mimicry of several other species that are locally abundant and distasteful, and this mimicry is equally as protective as the original disruptive patterning. We agree with Endler that a useful way to explore whether the stripe has a disruptive effect would be to use an experimental manipulation that left the butterfly with an equivalent area of the same light colour but along the wing-edge rather than as a stripe through the wing. If the wing stripe provides a disruptive function, we would expect that individuals with a coloured wing-edge rather than a stripe would suffer higher predation.

Production of downward directed light that matches incident levels is discussed as a means of crypsis in deepwater organisms in Chapter 4. Two cases have been reported of shallow-water fish that produce downward directed light but of insufficient intensity to match the incident downwelling radiation (McFall-Ngai and Morin, 1991; Harper and Case, 1999). In both cases, the intensity of produced light increased with increasing intensity of incident light, such that physical limitation does not seem a wholly convincing explanation for failure to match incident levels. Further, light production was heterogeneously distributed on the underside of the fish, such that some patches of the underside produced light intensities close to ambient and others much less. Both papers speculate that such incomplete matching of the background light may function as a type of disruptive camouflage, although neither includes any direct test of this. Alternatively, it may be that surface ripples tend to produce heterogeneous light fields in the top layers of water, and so heterogeneous light production is consistent with background matching. This alternative hypothesis should be empirically testable under laboratory conditions where the background can be manipulated.

We could find only one other study that provides empirical evidence relating to disruptive colouration: Götmark and Hohlfält (1995) compared the detectability (to humans) of bright black-and-white male and dull brown female pied flycatcher (*Ficedula hypoleuca*) models hidden in trees. No difference in detectability between the sexes was found. The authors interpret this as being suggestive that male plumage produces disruptive colouration equivalent in effectiveness to the crypsis provided by background matching in the female. However, they suggest that 'the leaves and twigs of the trees formed a contrasting mosaic of light and dark patches, which made the males difficult to detect'. To us, this suggests background matching, rather than disruptive colouration. However, another quote from the same paragraph does suggest disruption: 'Several observers noted the black and white patterns of male mounts, but at first did not believe they belonged to a bird.' As with the studies previously discussed, careful measurement of the background and experiments with artificial prey (and realistic predators) would be required to explore whether such avian plumage produces crypsis by background matching, disruption or a combination of the two.

## 2.4 Conclusion

In conclusion, we cannot resist quoting the concluding paragraph of Silberglied et al. (1980) in full

> 'Few concepts in the theory of adaptive colouration are as well accepted, but as poorly documented, as that of disruptive colouration. No direct experimental tests demonstrating its efficiency have yet been performed. We are reluctant to cast doubt on this logical general concept from the results of a single experiment, but consider it important to point to the need for caution in interpreting animal colouration from untested hypotheses.'

This warning holds as true now as it did then, although the work of Merilaita (1998) suggests that things need not always be this way. Until further progress is made, we urge that you should take statements like 'disruptive colouration is common in the animal kingdom' (Chiao and Hanlon, 2001) and 'disruptive patterning is, of course, a widespread and powerful camouflage technique among animals' (Messenger, 1997) with a strong dose of caution.

# CHAPTER 3

# Countershading and counterillumination

## 3.1 Introduction

Many animals have a systematic gradation in shading and colouration that is neither obviously a case of background matching nor disruptive colouration. Most often, such animals are darker on their dorsal than their ventral regions. This trait is known as *countershading* (Poulton, 1888; Thayer, 1896). Sometimes the transition from dark to light areas in countershaded animals is very sharp, in penguins, for example; in other cases, such as squirrels, the transition is gradual. Since the existence of countershading is generally limited to animals that are thought to be cryptic, it is easy to draw the conclusion there may be some special mechanism by which countershading renders animals difficult to detect.

In this chapter we describe the original explanation for the cryptic properties of countershading, also known as *obliterative shading* (Cott, 1940), but now known as *self-shadow concealment* (Kiltie, 1988). In essence the theory of self-shadow concealment suggests that the transition between dark and light regions acts to obliterate the visible effects of differential illumination, when illumination comes from above and the object is viewed from the side. If shading is obliterated, the argument goes, the capacity of onlookers to recognize and detect a three-dimensional object is greatly diminished. However, though this hypothesis is widely believed, there are alternative explanations for countershading. For example, aquatic animals will usually be seen by predators that are above or below them, but rarely to their sides. In a situation where light comes from above, an aquatic individual viewed from below must try and match the bright downwelling light in order to reduce the ease with which it can be detected. Conversely, when viewed from above, it should match the backdrop of the dark deeper waters. This too should generate countershading and relies only on the well-understood mechanism of background matching discussed in Chapter 1.

To evaluate these and other explanations for countershading we critically review the literature, focusing on (1) whether the results of direct experiments and other analyses lead to the general conclusion that there is indeed an enhancement of crypsis through countershading, and (2) whether any demonstrable advantage is gained through either the 'classical' mechanism of self-shadow concealment, the alternative of background matching mechanism or indeed some other mechanism(s). This chapter is based on a recent review by Ruxton et al. (2004).

## 3.2 Self-shadow concealment and countershading

Although Poulton (1888) arguably originated the idea of self-shadow concealment, it is more usually attributed to the painter and naturalist Abbot H. Thayer (1896). Thayer proposed a 'beautiful law of nature' which he described as 'the law of gradation in the coloring of animals . . . responsible for most of the phenomena of protective colouration except those properly called mimicry'. Thayer noted that 'animals are painted by nature, darkest on those parts which tend to be most lighted by the sky's light, and vice versa'; and, according to Thayer, this pattern of countershading 'makes an animal appear not to exist at all'. Thayer used painted animals and models to demonstrate this vanishing trick (see reproductions in Ruxton et al. 2004). Reporting a demonstration to the Oxford University Museum, Poulton (1902) wrote that 'the model which is the same shade of colour all over appears to be a different shade everywhere because of the difference in illumination: while the model which is a different shade at every level appears to be the same shade all over because the differences of shade exactly counterbalance differences in illumination'.

This mechanism of self-shadow concealment is supposed to reduce the capacity of a viewer to recognise the animal as a three-dimensional solid object, and hence to reduce the chances of it being detected. In support of the theory, the presence of shading is known in humans to make solid objects stand out from their backgrounds (Ramachandran, 1988). Furthermore, Thayer was certainly correct that countershading *is* a common trait in many animal species, and this view has been shared over the decades by a number of authors (notably Stephen J. Gould, (1991), who wrote that 'light bellies [are] perhaps the most universal feature of animal colouration'). Gould (1991) identified shadow self-concealment as the mechanism underlying countershading, writing that: 'Thayer correctly identified the primary method of concealment—a device that makes creatures look flat.' The idea gained currency from an early stage despite the fact that supporting evidence is at best patchy. Authorities such as E. B. Ford (1957) used the idea to explain paler undersides in larvae of the purple emperor and the brimstone butterflies, also see discussions in for example Cott (1940) and Edmunds (1974). The self-shadow concealing mechanism is now a textbook explanation for countershading, often presented as an established fact (Encyclopaedia Brittannica, 2001). However, as Kiltie (1988) argued, the frequency of the phenomenon does not prove the existence of a particular cause; countershading may be common, but for a plurality of reasons of which self-shadow concealment may, or indeed may not, be one.

In order to distinguish the trait in question from Thayer's causal explanation for it, we refer to a pattern of dark/white dorsoventral patterning as countershading, and Thayer's mechanism as self-shadow concealment (see Kiltie, 1988). Note that this is a very general definition of countershading, that does not specify whether the transition from dark to light colouration is abrupt or gradual, a point we return to in Section 3.6.

## 3.3 Direct empirical tests of the advantages of countershading

One way to examine the hypothesis of defensive countershading by self-shadow concealment is to test the prediction that, when illuminated from above, a countershaded animal does indeed appear to be uniformly shaded when viewed from the side. Since the grey squirrel was a species identified by Thayer as an example of countershading, Kiltie (1989) used it as a case study to test this prediction. Taxidermic mounts of the squirrel were illuminated from above and photographs were then taken when the mounts were placed horizontally (as if running along a branch parallel to the ground) and vertically (as if running up the trunk of a tree, perpendicular to the ground). In both cases, the specimen was placed in a natural position (with its feet in contact with the substrate) and a photograph taken of the animal's flank. The specimens were also laid on their flank, so that photographs of their dorsa could be taken. The pictures were examined for a statistically significant gradient in shading across transects taken over the animal's flank. Evidence of a significant gradient was assumed to indicate a pronounced shadow and therefore high visibility. When placed horizontally, the flank views showed lower gradients than the dorsal views, consistent with the hypothesis that countershading enhances crypsis of animals viewed from the side. However, for vertically orientated specimens, flank views produced higher gradients than dorsal views. This suggested to Kiltie that countershading may work when squirrels are horizontally but not vertically orientated. In addition, Kiltie (1989) noted that on horizontal substrates, 'the degree of shadow obliteration is imperfect and hence of questionable value in deterring predators'.

Kiltie's investigation rests very much on the assumption that steep gradients of dark to light colouration indicate high levels of visibility, and vice versa. A number of more direct studies have attempted to test this prediction with caged or free living predators. In an attempt to test whether countershading enhanced crypsis, de Ruiter (1956) used freshly killed caterpillars of the species *Endromis versicolora* and three captive jays (*Garrulus glandarius*) as predators. These caterpillars are reverse-shaded with a dark ventra and light dorsa. Some caterpillars were tied in their naturally occurring position beneath twigs (such that their dark side was in contact with the lower side of the twigs) and in the reverse position above twigs (such that their dark side was again in contact with the twigs). Hence, in their normal position, the caterpillars were countershaded, with their dark parts being uppermost; whereas in the abnormal position on top of leaves their lightest regions were uppermost (and so most strongly illuminated). Consistent with the prediction that countershading enhances crypsis, de Ruiter found that normally positioned countershaded prey were taken by jays less frequently than those put in reverse positions. However, since it can be argued that the jays simply preferred prey positioned above rather than below the twigs, the results of this experiment do not represent strong evidence that countershading itself enhanced crypsis.

A second, direct test was reported by E. R. A. Turner (1961) who used a now much copied experimental set-up in which garden birds were predators and dyed pastry baits were presented as artificial prey. Turner (1961) found that wild birds took uniform green artificial pastry prey more than countershaded prey when simultaneously presented. However, Edmunds and Dewhirst (1994) criticized Turner's experimental design on the grounds that since the countershaded prey were a slightly brighter hue than the uniform green controls, an aversion to this brighter hue might explain the results. In a subsequent experiment Edmunds and Dewhirst (1994) rectified this problem by making two types of pastry: light and dark green. From these they fashioned four types of prey: uniform dark, uniform light, countershaded (dark on the top half, light on the bottom) and reverse-shaded (countershaded prey turned upside down).

Equal numbers of these four types were presented to wild garden birds. Uniform light prey and reverse-shaded prey where taken approximately equally; these two groups were taken significantly more often than uniform dark prey; in turn, dark prey were taken significantly more than countershaded prey. This is, in our view, one of the best and most important direct experimental attempts to demonstrate that countershading, by whatever mechanism, can enhance crypsis in terrestrial habitats. However a demonstration that countershaded prey were attacked less often than the other prey forms is necessary, but it is not in itself sufficient to conclusively establish an effect of countershading in enhancing crypsis. We need, in addition, to be able to rule out the possibility that all four prey types were equally easy to detect, but the birds expressed post-detection preferences. Eliminating this alternative could be achieved by presenting the four prey types on a contrasting, white background as well as on one, such as a lawn, against which the colours match. Against the white background all prey types should be equally visible and differences in attack rates would be indicative of post-detection preferences. If one assumes that such preferences remain unchanged in the lawn trial, then they can be statistically controlled for in the results of that trial, and any remaining difference in attack rates between morphs might more safely be interpreted as resulting from differences in ease of detection.

In acknowledging that the birds may have shown an aversion to countershaded prey after detecting them, Edmunds and Dewhirst proposed two lines of refutation. First, they suggested that an aversion by predators to edible prey with horizontal stripes would not benefit a predator and hence would not be stable over an evolutionary timescale. An inhibition of attacks on the two-toned prey would then soon be wiped out. Second, they argued that an aversion to two toned prey *per se* is an unlikely explanation given that the predators attacked reverse-shaded prey at a high rate not significantly different to light prey.

Though sound, neither of these refutations, in our view, entirely and conclusively removes the problem. The first would not be true if countershading were used by some prey as an honest signal of some repellent defence such as toxicity in prey. The second assumes that the aversion is away from two-toned prey in general, rather than specifically away from countershaded prey. It does not rule out the *possibility* that the avian predators show special aversions to countershaded prey but not reverse-shaded prey. These reservations may be merely devil's advocacy since we know of no evidence for an association between countershading and noxiousness in prey.

However, a more challenging interpretation is that the birds were unfamiliar with countershaded, two-toned food items and preferred uniformly light and dark items because these were generally more familiar. The fact that the birds took reverse-shaded prey at a rate not significantly different to light prey may indicate that birds are not averse to two-toned prey, as Edmunds and Dewhirst suggest. Equally though, it may indicate that the birds did not discriminate a uniformly light prey from a prey that is light on top and dark on the bottom, especially since the darkest surface is positioned away from the source of day light and is therefore relatively inconspicuous (field observations indicate that reverse-shaded baits are to human observers visually very similar to light prey when presented on lawns: Speed, unpublished). If this latter explanation were true, then simple unfamiliarity, rather than enhanced crypsis could explain Edmunds and Dewhirst's results. Hence, the possibility of alternative explanations that are at least plausible makes further replication with prey on a colour matching and a contrasting background essential.

The experiments of Kiltie, de Ruiter, Turner, Edmunds, and Dewhirst represent, to our knowledge, the sum total of published manipulative experimental tests of the adaptive value of countershading. The experiment of Edmunds and Dewhirst does suggest a protective mechanism for countershading, but in our view none of these experiments provides clear and unambiguous evidence that countershading actually reduces detectability. Furthermore, only Kiltie's experiment probes the specific mechanism that might lead to such a reduction and even here there is no test of alternative hypotheses such as background matching. In the absence of clear direct tests, we next turn to indirect evidence from non-manipulative studies to evaluate evidence that countershading contributes to crypsis.

## 3.4 Indirect evidence

### 3.4.1 The naked mole-rat

Braude et al. (2001) studied the prevalence and development of countershading in the naked mole rat (*Heterocephalus glaber*). Many naked mole rats are countershaded, with a purple-grey dorsa and pale-pink ventral area often separated by a jagged line. In laboratory litters countershading develops between 18 days and 3 months. However, over a period of years pigmentation of the dorsa fades and countershading disappears when the dorsa eventually matches the pink ventral areas. Countershading is common in younger workers, but is completely absent in newborn pups, most queens, breeding males and animals older than 7 years of age.

Braude et al. (2001) considered that countershading could have one or more causes in these animals. Alternative, non-exclusive explanations for countershading include:

**1.** That countershading seen in modern mole-rats is *a vestige of adaptive countershading present in the fur of surface dwelling ancestors*. This possibility was not supported by Braude et al. because deposition of skin pigmentation seems to have a different mechanism to deposition of fur pigmentation. If the dark brown areas of a furred bathgyrid is shaved, its skin colour is the same as under the areas of unpigmented fur, (Braude et al., 2001: 355).
**2.** *Protection from UV light*; this possibility was also not supported because most surface activity takes place in darkness. Even when daytime exposure does takes place pigmentation does not match the areas exposed to sunlight.
**3.** *Thermoregulation*: also not supported because (i) most thermoregulation is controlled behaviourally, and (ii) this explanation could not easily explain why reproductive males, queens, and older animals are not pigmented.
**4.** *Protection from abrasion*: again, not supported because it does not fit with the observation that queens and dominant males are subject to abrasion, but have no pigmentation; or with the observation that tails are fully pigmented and suffer no abrasion, while the soles of the feet suffer most abrasion but have no pigmentation at all.

5. *Camouflage*: pigmentation is mainly limited to young workers who are most likely to disperse from their colonies. Braude et al. note that there is a high turnover of workers suggesting that many disperse between colonies. Although dispersal is a nocturnal activity, pink bodies may be especially conspicuous in moonlight. Hence, Braude et al. suggest that pigmentation on the dorsa makes dispersing animals less visually conspicuous to nocturnal predators.

However, rather than providing evidence of a self-shadow concealing mechanism, countershading may simply, as Braude et al. suggest, be a by-product of the ventral side remaining pink because it is not visible to predators. This is especially likely given the short legs of the mole-rat. In addition, there may be non-trivial costs associated with the production of pigment. Hence, countershading may well simply be a by-product of the optimization of the distribution of costly pigmentation.

### 3.4.2 Countershading in ungulates

In a major study of colouration in even-toed ungulates Stoner et al. (2003) gathered together a very large dataset that enabled a complex comparative study of the prevalence and function of different character states. Characteristics of colour, behaviour, and ecology were collected for 200 species and coded in a present/absent format. After controlling for shared ancestry, the dataset was analysed to see if certain character states were shared by animals likely to be subjected to similar selection pressures. Lightening of coat colour in winter was, for example, common in animals that are exposed to bright winter conditions, such as in the arctic latitudes. This suggests adaptive background matching to avoid detection by predators. Striped coats were associated with hiding young and in adults that live in light forests. However, the evidence that countershading enhanced crypsis is more equivocal. Stoner et al. predicted that countershading should be common in diurnal species, especially those that live in well lit, relatively uniform visual environments. Consistent with this, countershading was associated with desert dwelling species, but there were no other associations with open habitats. Furthermore, when a similar analysis was performed on lagomorphs there were no associations of countershading with any open habitats. Stoner et al. concluded that 'the widespread importance of countershading in helping to conceal mammals is still questionable'.

### 3.4.3 Countershading in aquatic environments

Korner (1982) found three examples of a species of fish louse that showed countershading. The three examples of this louse species (*Anilocra physodes*) were each attached to the flank of a fish that itself showed countershading. One particular point of interest is that two lice were attached to the right flank of the fish with their head facing forward, but the other was attached in the same position on the left side. Despite this, all the lice showed darkening on their upper side, even though morphologically this was a different half of the louse in the two cases. Korner suggested that the colour pattern emerges after the louse attaches. This seems perhaps a more plausible explanation than postulation of the existence of two flank-specialist colour morphs.

Furthermore, Korner proposed two benefits that the louse might receive from countershading: (1) avoiding being detected by cleaner-fish, and (2) avoiding making its host more conspicuous so reducing the risk of being eaten along with its host by a predator. The mechanism of crypsis for the louse was, in Korner's view (1982: 250), that countershading 'increases the optical illusion of flattening in the attached fish louse'. Kiltie (1988) later agreed with this interpretation. However, whilst a fascinating piece of natural history, this study provides no evidence for the adaptive value of countershading in this louse. Indeed, countershading could be interpreted more simply (a) as a means to background matching when viewed from above or below; and perhaps (b) background matching against the countershading of the fish's flank when the animals are viewed from the side.

Some catfish of the family Mochokidae swim with their ventral side uppermost whilst feeding from (or breathing at) the water's surface at night (e.g. Chapman et al., 1994; Stauffer et al., 1999). These fish show reverse countershading with a light dorsa and dark ventra. Particularly interesting is the report by Nagaishi et al. (1989) that one of these species (*Synodontis nigriventris*) is uniformly coloured by day (when it avoids the surface) but changes to reverse countershading at night. However, we note that there is again no evidence available to differentiate between self-shadow concealment and background matching as potential drivers of this colouration.

Reports of countershading in other aquatic animals are similarly ambiguous. Seabirds that feed on fish at medium depths seem to be countershaded with greater frequency than either bottom or surface feeders (Bretagnolle, 1986; Cairns, 1993). Cairns, in particular, concluded that this colouration provides self-shadow concealment. However, no evidence is presented to support this. Furthermore, self-shadow concealment works only when the organism is viewed from the side. It seems to us that such predators would particularly value being hidden to prey beneath them in the water column, rather than to the side. There is good experimental evidence that a white underside confers such cryptic benefits during flight for seabirds foraging for fish in surface waters: Phillips (1962), Cowan (1972), and Götmark (1987). Hence, we have evidence for background matching from below (if not from above), but no evidence for self-shadow concealment.

Many mid-water fish are predominantly silver on their lateral and ventral aspects but darker on their dorsa. This is also likely to be an adaptation to provide background matching (Denton, 1971) with the dark dorsa matching the dark deep waters below, while the reflective ventral/lateral fish scales may thus match the intensity and wavelength of downwelling background light (so-called 'radiance matching'). Recently Shashar et al. (2000) demonstrated that cuttlefish predators are able to discriminate light reflected from fish scales from background light using the partial linear polarisation that is characteristic of reflected but not scattering background light. Shashar et al. propose that sensitivity to linearly polarized light in the cuttlefish may therefore function as a means of breaking the background-matching countershading camouflage of light-reflecting silvery fish. Nonetheless, a role for self-shadow concealment cannot at this point be either strongly supported or ruled out for this or other aquatic predators. Despite the considerable research that has been devoted to understanding the physiology of reflective surfaces (Herring, 1994), the key experiments needed to explain functionality are presently lacking.

### 3.4.4 Counterillumination in marine animals

Several aquatic animals use bioluminescence apparently to produce light that matches the downwelling ambient light so as to make detection from below more difficult; this is often called *counterillumination* (for a wider review, see Widder, 1999). Some species have been demonstrated to alter the intensity and wavelength distribution of the light that they produce appropriately as they change depth, so as to match the changing downwelling light (Young and Roper, 1976; Young and Arnold, 1982; Young, 1983).

Animals have also been demonstrated to produce light that matches the angular distribution of the ambient light (Latz and Case, 1982). When some countershaded cephalopods change from their normal orientation they can often rapidly change their chromatophore use over the body so as to retain countershading in their changed orientation (Ferguson et al., 1994). Although there is no direct evidence, we agree with Kiltie (1988) that only the most cynical would argue that these adaptations are not fine-tuned to reduce the conspicuousness of the animals involved. Also, to the extent that light production is biased to the ventra of animals, this tends to generate countershading. However, the primary mechanism seems once again to be background matching rather than self-shadow concealment, since in the three-dimensional world of the open ocean, attack can come from any direction—not simply from the side. Indeed prey can generally be detected most easily from below (against the strong downwelling light).

In summary, a growing number of studies have used indirect evidence from observation of unmanipulated systems and through comparative analysis to infer a concealment mechanism for observed countershading. However, none of these studies satisfactorily demonstrates that self-shadow concealment is a more plausible candidate explanation than any alternatives. We find that in most cases, background matching is at least as plausible (and often a more plausible) explanation than self-shadow concealment. However, many of these unmanipulated studies, by their very nature, do not lend themselves to definitive evaluation of the relative plausibility of different mechanisms.

## 3.5 Countershading in aerial, aquatic, and terrestrial systems

In this section we take a more general overview, considering how different selection pressures might lead to the same countershaded phenotypes. In theory, background-matching could act to produce countershading on flying animals as well aquatic ones, both to avoid detection by predators and to avoid being seen by prey. However, this mechanism is likely to be less important in air than in water. One reason for this is that flying animals will seldom need to hide both from eyes above and below. For example, owls would benefit from hiding from their mammalian prey beneath them, but have little to fear from above. Most attacks on flying animals are mounted from above, where the attacker can work with gravity rather than against it. At least in terrestrial ecosystems the background against which flying animals would be seen from above is less uniformly dark than in aquatic habitats. These arguments suggest that background matching will be a less potent mechanism for generating countershading in flying animals than in swimming ones. However, it is important to remember that flying animals also spend a significant part of their time at rest, and it may be that this terrestrial part of their existence can exert selection pressure for countershading.

It is also true that the classical explanation of countershading (self-shadow concealment) may also work best in an aquatic environment. The reason for this is that scattering of light in water tends to lead to an unchanging angular distribution of light direction, whereas in terrestrial ecosystems the direction of the strongest light source, the sun, changes markedly throughout the day, and is strongly affected by cloud cover, and so the position and intensity of shadowing changes throughout the day. However, as noted above, there are no experiments that conclusively determine whether countershading in aquatic animals results from self-shadow concealment or from selection for matching of different backgrounds on the ventral and the dorsal areas.

What about terrestrial animals? An alternative explanation for countershading is that the dorsal surface of the animal is pigmented to give some adaptive advantage (crypsis or protection from UV light or abrasion; see discussion in Kiltie, 1988; Braude et al., 2001) and the ventral side is unpigmented because there would be no similar benefit to pigmenting this area. This applies particularly if pigmentation is itself costly. As with aquatic background matching, this explanation argues that the phenomenon of countershading is an epiphenomenon; the result of different selection pressures operating on the dorsal and ventral surfaces of the animal. Hence, one general class of explanation could be that if epidermal darkening brings benefits (e.g. crypsis or UV protection) but also incurs costs (e.g. pigment production and deposition or reduced efficacy of intraspecific signalling) then we would expect animals not to be shaded in areas that make no difference to these traits. Countershading may not be specifically advantageous in itself, but may reflect different trade-offs between the costs and benefits of dark colouration experienced in the upper and lower portions of the body.

Attractive though this is as a general explanation for countershading, we stop some way short of rejecting the entire notion that countershading itself reduces detection. We have three reasons for this position. First, and most obviously the necessary experiments have not been performed. Second, it is possible to envisage more or less plausible hypotheses by which countershading could reduce detection rates. It could be that the Poulton–Thayer hypothesis of self-shadow concealment really does function at least in some habitats. However, we are inclined to agree with Kiltie that for many countershaded animals frequent variation in direction and intensity of ambient light means that a countershaded phenotype will rarely find itself in optimal conditions so that the shading of ventral and dorsal areas are balanced. Alternatively, countershading may provide protection because it subverts a predators 'search image' formed by the normal shading patterns of objects in an environment. Most objects lit from above are lighter on top than below, whatever the intensity of ambient light. Predatory observers may be expected to detect three-dimensional objects based on this general pattern of shading, particularly if there are time constraints in searching. However, some countershaded objects subvert this pattern and may escape detection and recognition, especially if their colourations are similar to their backgrounds. If countershading can function in this manner, it would presumably be subject to density dependent recognition, like many other forms of crypsis. In contrast, self-shadow concealment does not rely on subversion of a search image, and may therefore not be subject to strong density dependence. Whether or not countershading does work by subverting the search image for 'normally illuminated' objects, our main point is that it is possible to generate plausible hypotheses that could explain countershading as a means of facilitating crypsis, and hence that well designed manipulative experiments are well overdue.

A third reason that we do not reject the possibility that countershading really does decrease detection rates comes from the observation articulated at the start of this chapter, that countershading is closely associated with crypsis. If countershading in terrestrial animals is simply and only a result of distributing expensive pigment on those areas that are likely to be visible to predators, we would not expect it to be limited to cryptic animals. It should perhaps be present in species that are not cryptic, but instead display their unprofitability with conspicuous warning signals. However, countershading is not widely known in animals with bright warning displays. This problem could be easily resolved if it were established whether animals with warning displays are visually detected from a greater range of angles than those that are cryptic. However, such data have not been systematically collated. One constructive answer to this problem is a comparative review of the distribution of pigmentation in cryptic and noncryptic animals, factoring in the range of angles from which the animals will be seen and key components of the visual environments they inhabit. Until we have really systematic studies like this, we cannot discount the possibility that there may be some aspect of countershading that does indeed decrease the detectability of prey items.

## 3.6 Conclusion

In order to include as many lines of evidence as possible, we have adopted a very general definition of countershading, without specifying how gradually or abruptly dark colouration changes to light. It is clear that there is much variation between countershaded species in this spatial gradient. It does seem that (except for oddly shaped objects) self-shadow concealment requires a gradual change from dark to light. However, we do not consider that an abrupt change should necessarily be predicted if background matching is the mechanism underlying countershading. Such a prediction would only hold if countershaded prey were observed by predators whose orientation towards the prey (in three dimensions) was highly repeatable between encounters. This is unlikely to be the norm in either terrestrial or aquatic ecosystems. Hence, the abruptness of change from dark to light is not a perfect discriminator between candidate explanations for the mechanism underlying countershading in a given species.

In summary, there appears to be no conclusive evidence that countershading *per se* provides any enhancement of crypsis in terrestrial or aerial environments. The highly refined adaptations of some marine organisms to match the different background light conditions against which they are set when viewed from different aspects strongly suggests an adaptive advantage to countershading in these environments. Experimental quantification of the effectiveness of these adaptations would clearly be very welcome.

There is no experimental evidence that conclusively tests and supports the explanation of self-shadow concealment. The alternative explanation, that the dorsa and ventra face different selection pressures (often associated with background matching) is often more plausible, and indeed there may be a plurality of explanations for countershading. Current understanding of the selection pressures that drive countershading is very patchy indeed. If possessing contrasting dorsa and ventra is really 'the most universal feature of animal colouration' (Gould, 1991) then we must strive to explain this pattern.

CHAPTER 4

# Transparency and silvering

Superficially, *transparent* may seem like a synonym for *visually undetectable*. However, in Sections 4.1 and 4.2, we will discuss how transparent organisms can still be visually detected by their predators or prey. That said, section 4.3 will demonstrate that there are circumstances where a little transparency can go a long way to reducing an organism's visibility. Some body parts cannot be made transparent, but Section 4.4 argues that opaque body parts need not always significantly increase the detectability of a generally transparent organism. Section 4.5 will consider the distribution of transparency among natural organisms. The distributional observations we seek to explain are the greater prevalence of transparency among aquatic than terrestrial organisms, and the particular prevalence among mid-water species. An alternative but related strategy to transparency, adopted by some mid-water fish, and considered in Section 4.6, is silvering of the body to provide crypsis by broadband reflection.

## 4.1 Transparent objects still reflect and refract

A perfectly transparent object is one that does not absorb or scatter incident light. However, perfect transparency does not make an object perfectly invisible. Glass can be very close to perfectly transparent yet we can see glass window panes. To be invisible, an object must have no effect on any light that strikes it. Perfect transparency removes two physical mechanisms, absorption and scattering, that could affect any incident light; however, that still leaves several others. Whenever light passes from one medium into another, then that light is both reflected and refracted even if both media are transparent. It is this reflected and refracted light that allows us to see even the purest glass. Every material that is not perfectly opaque has a physical quantity called its refractive index. When light moves between media of different refractive indices, some of the light is refracted and some reflected. The magnitude of both of these effects increases with the difference between the refractive indices of the two media.

We will deal with reflection first. At normal incidence (when the light beam is perpendicular to the interface between the media), the proportion of light that is reflected is given by the simple formula

$$R = \left(\frac{n_1 - n_2}{n_1 + n_2}\right)^2, \qquad (1)$$

where $n_1$ and $n_2$ are the refractive indices of the two media that form the interface (Denny, 1993). These values vary with the wavelength of light but representative values for different materials are given in Table 4.1.

Biological tissues have refractive indices between 1.34 and 1.55 (Johnsen, 2001). Using these values in eqn. (1) suggests that the proportion of the light striking some biological tissue that is reflected is

**Table 4.1** Refractive indices for a selection of media. Modified from Denny (1993) and Johnsen (2001)

| Medium | Refractive index |
|---|---|
| Air | 1.00 |
| Water | 1.33 |
| Cytoplasm | 1.34 |
| Densely packed protein | 1.55 |
| Glass | 1.56 |

much greater when the animal is in air (2–5 per cent) than when it is in water (0.001–0.6 per cent). These simple calculations go a long way towards explaining why transparency is much more commonly observed among aquatic organisms than terrestrial ones. A transparent organism in air reflects a fraction of incident light many times greater than that of the same organism in water. Hence, a transparent organism has a greater optical effect on the light that ultimately reaches the eye of its predators or prey when in air than in water. This conclusion holds regardless of the angle between the light ray and the interface between the organism and the surrounding media, although the formulae get more complex than for normal incidence.

Refraction is manifested by the bending of the path of a light ray as it passes between media with different refractive indices. The greater the difference between these indices (and the less orthogonal the angle of incidence), the greater the distortion of the path. Hence, for biological tissues with refractive indices between 1.34 and 1.55, refraction is greater in air than in water. The practical consequence of refraction is that the image seen through a transparent object is not the same as the image that would be seen if the object where not there. This can be verified by holding a piece of glass at arm's length at a slight angle to the perpendicular. The view that you see through the glass does not perfectly match up with the view that you get from light that does not pass through the glass; the piece of the view that you get through the glass seems slightly shifted. This shift is due to refraction of light travelling through the glass. The magnitude of refraction would be less if we were viewing under water, because the refractive index of water is closer to that of glass. This is easily observed whilst washing up drinking glasses—unless you wash glasses by switching on a machine!

Not only is the extent of refraction physically less extreme for an organism in an aquatic environment than in a terrestrial one, the apparent effect of any refraction is likely to be less easy to detect in aquatic environments. As a generalization, aquatic environments are relatively featureless compared to terrestrial ones (especially in midwater) and so the effect of image distortion will be harder to detect in such environments. This may help explain why transparency is particularly common among midwater aquatic species (a theme that we will return to in Section 4.5).

In summary, transparent individuals can still be detected because, although they do not absorb or scatter visible light, they still reflect and refract it. In order to minimize these potentially crypsis-breaking mechanisms, the refractive indices of transparent tissues must conform as closely as possible to that of the surrounding media. This is much easier to achieve if the medium is water rather than air. Reflection and refraction are two physical mechanisms that can allow a perfectly transparent organism to be visually detected. The next section considers several more.

## 4.2 More reasons why perfect transparency need not translate to perfect crypsis

### 4.2.1 Polarization

Light waves oscillate in the plane perpendicular to the direction of travel of a light beam. Where on this plane a specific light wave oscillates is described by the polarization of that light. This matters because ambient light is expected to become partially polarized as it passes through water (Mobley, 1994). In addition, the polarization of light can be changed when it passes through so-called polarization-active materials, of which muscle fibres seem to be an example (Shashar et al., 1998). Several fish, crustaceans and cephalopods have been shown to be sensitive to the polarization of light, and there is some evidence to suggest that this may be used to enhance detection of transparent prey. Novales-Flamarique and Browman (2001) found that juvenile rainbow trout appeared better able to find zooplankton prey in a laboratory aquarium under polarized light compared to unpolarized light. This suggests that light is changed in polarization as it passes through the zooplankton in a way that the trout can detect and exploit. Shashar et al. (1998) obtained similar results for juvenile squid preying on live zooplankton in an aquarium. Shashar et al. also demonstrated that adult squid showed a marked preference for preying on transparent glass beads that were polarization-active, compared to beads that looked identical—at least to humans—but that had no effect on polarization. Hence, change in polarization may be another physical process that can allow a transparent organism to be visually detected.

### 4.2.2 Other wavelengths of light

An organism can be transparent to visible light but still absorb and/or scatter significant amounts of UV radiation. The ability to detect UV light is common in fish (Losey et al., 1999) and other aquatic predators (Johnsen, 2001). Browman et al. (1994) found that two species of zooplanktivorous fish appeared to find prey less easily in a laboratory aquarium when the UV part of the ambient light spectrum was removed. However, Rocco et al. (2002) were unable to replicate these results. Rocco et al. suggest two reasons for this difference between their study and that of Browman et al. Firstly, Rocco et al. use natural sunlight as opposed to Browman's artificial illumination; secondly Rocco et al. used direct counts of rate of prey consumption rather than indirect estimates of detection distances. Hence, at the present time, there is no unequivocal evidence that UV vision is important in predator–prey interactions involving visibly transparent prey, but it certainly seems physically possible.

UV radiation is capable of damaging living cells (Paul and Gwynn-Jones, 2003) and can cause increased mortality among aquatic larvae (Morgan and Christy, 1996), suggesting that there may be a trade-off for an organism between transparency to UV radiation for crypsis reasons and having a UV-absorbing outer surface to protect internal organs from damage (Morgan and Christy, 1996; Johnsen and Widder, 2001). UV is most intense at the surface of the water column and is rapidly absorbed by water and dissolved organic matter (Peterson et al., 2002), which explains why UV damage has seldom been reported at depths greater than 25 m (El-Sayed et al., 1996). In contrast, it has been estimated that sufficient UV light for vision can be found at depths of 100–200 m (Losey et al., 1999). Hence, it has been predicted that if a trade-off between radiative protection and crypsis influences the distribution of transparency to UV radiation among aquatic species, then UV transparency should be more common among deeper living species (Johnsen and Widder, 2001), or that the pigmentation of an individual may decrease, and transparency increase, as organisms migrate vertically into deeper waters (Morgan and Christy, 1996).

In order to test the hypothesised trade-off between crypsis and avoiding radiative damage, Johnsen and Widder (2001) measured the transparency of zooplankton collected from epipelagic (0–20 m) and mesopelagic depths (150–790 m). They report that the two groups did not differ in their transparency to visible light but those from the surface layers were less transparent to UV, in line with their expectation from the trade-off argument of the last paragraph. However, these authors argue that the greater UV absorbency of surface dwelling zooplankton need not necessarily substantially increase their detectability to predators. This is based on two arguments.

First, absorbency in the UV-A range (320–400 nm) was generally low in their study compared to that in the UV-B (280–320 nm). This is important because only UV-A has been implicated in vision (Losey et al., 1999) and UV-B seems much more damaging to living tissues than UV-A (Johnsen and Widder, 2001).

Second, Johnsen and Widder also report that species with low UV transparency also had low transparency to visible light. UV radiation attenuates more rapidly in water than blue visible light, hence if an individual has low transparency to both blue visible light and UV radiation, then (all other things being equal) the visible light will permit its detection by organisms at a greater distance, compared to UV. Consider a predator moving closer and closer to an as-yet-undetected prey item. If the prey item has low transparency to both blue light and UV light, then it will be detected initially because of its effect on the visible light reaching the predator. After its effect on the visible light has caused the prey to be detected, the effect of UV light, when the two organisms have drawn even closer to each other will matter little. Conversely, if the prey is highly transparent to visible light, then a predator could move very close to it without detecting it on the basis of its effect on the visible spectrum. In this case, even small changes in UV transparency may substantially affect the ease with which the prey organism can be detected.

These caveats suggest that the situation may be rather complex, and in order to understand the selection pressures on UV transparency we need a quantitative understanding of the effects of depth and degree of transparency on both risk of UV damage and risk of predation. We also need an understanding of the fitness consequences of UV damage. Such understanding is not currently available and should be the focus of future research.

### 4.2.3 Snell's window

Another method by which visual predators could detect perfectly transparent prey near the surface in an aquatic medium is through using Snell's window. Figure 4.1(a) illustrates the light reaching a fish's eye. All light reaching the eye from angles from the perpendicular less than 49° is light from above the water's surface. Indeed the fish's whole view of the world above the water's surface is compressed into

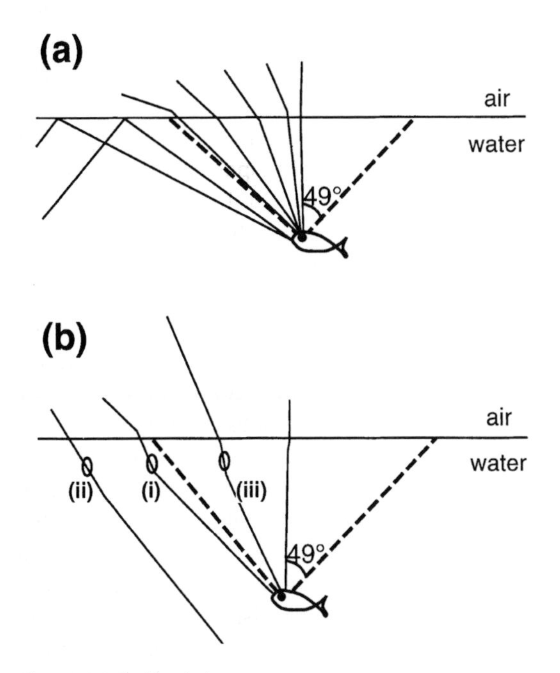


**Figure 4.1** Snell's window.

this so-called Snell's window by the properties of refraction discussed earlier. All light entering the fish's eye at angles greater than 49° is light that was travelling upward through the water column but was reflected at the surface of the water. The intensity of light entering the water column from above is generally much greater than the light from below that has been reflected back from the surface. Hence, a fish sees the world above the surface as a bright image in Snell's window surrounded by a much dimmer (reflected) image of the depths below. This could be a problem for a zooplankter situated just outside Snell's window. Light that would pass through that space if the zooplankter was not there can be refracted by the zooplankter so that its course is changed and it reaches the fish's eye (zooplankter *i* in Fig. 4.1b). If the zooplankter were not there, that light would not reach the fish's eye. This light will show up as a bright spot against the dark background of the image of the depths, and so the zooplankter will be very visible to the fish. This will be less of a problem for a zooplankter well away from Snell's window (zooplankter *ii*), which will not refract light sufficiently dramatically to enter the fish's eye. Nor is there a substantial problem for a zooplankter in Snell's window, as light refracted through it will be seen against the rest of the bright light in the window, and so will not provide a strong contrast.

This theory was supported by the observations of captive blueback herring by Janssen (1981). These fish are visual feeding planktivores, and characteristically swim upwards towards their prey. This trajectory will cause the ring of visibility just outside Snell's window to sweep inwards as the fish swims towards the surface, allowing the fish a chance to detect prey across a wide area of the surface waters. The median angle of attack by the fish in Janssen's observations was 54°. This is only slightly greater than the 49° that defines Snell's window and is consistent with attacking individuals detected just outside Snell's window. Such calculations can only be considered approximate since the position of the edge of the window will be affected by any surface waves; S. Johnsen, personal communication. Zooplankton can do little about their relative positioning with respect to predators; their best defence to this vulnerability is to minimise refraction by keeping their refractive index as close to that of water as possible.

We indicated earlier that refractive index is different for different wavelengths of light; this too may have ecological implications. Johnsen and Widder (1998) studied the effect of the wavelength of incident light on the transparency of 29 species of gelatinous zooplankton. In the overwhelming majority of cases, they found that transparency increased linearly with wavelength, albeit only very gradually. This means that the spectrum of light passing through such a zooplankton will be different from that of the background against which it is seen, potentially allowing detection in the same way that a green coloured glass bottle is easier for us to see than a clear one. However, light of different wavelengths attenuate differentially in water such that the spectrum of ambient light is much broader in the surface layers than at depth (where blue–green visible light predominates). Hence, Johnsen and Widder argue that this potential weakness to crypsis by transparency is more of a problem when hiding near the surface. Even in the surface waters, we currently lack a demonstration that the visual systems of appropriate aquatic organisms can detect the relatively modest changes in light spectrum predicted by this mechanism.

Section 4.1 suggested that transparent organisms could still be detected through the light that they may reflect and refract. This general section has identified several other mechanisms that *might* allow an organism that is perfectly or partially transparent to visible light to be detected by either its predators or its prey: change in light polarization, Snell's window, and differential absorption of different wavelengths (especially combined with UV vision). We say 'might' because definitive experimental work is still lacking. Rocco et al. (2002) cast doubt on previous evidence that UV radiation helps predators capture visually transparent prey. Further work using systems as similar as possible to natural field conditions are required to resolve this issue. Similarly, further experiments on the possible role of polarization in increasing the detectability of visually transparent prey under natural light conditions would be useful. Another related topic is the extent to which changes in hue caused by differential transparency of organisms to different wavelengths of visible light can be detected by potential predators. Janssen's fascinating idea and experiments on Snell's window are certainly worthy of further work to explore the ecological relevance of his results. We have now demonstrated that transparency is not the perfect form of crypsis and that *transparent* is certainly not a synonym for *undetectable*. However, the next section argues that despite these caveats, even less than perfect transparency can sometimes be an effective form of crypsis.

## 4.3 Imperfect transparency can be effective at low light levels

The visibility of an underwater object generally depends more upon its contrast than its size (Lythgoe, 1979). Contrast is the difference in intensity of light reaching an organism's eye that originates from the viewed object compared with the intensity of light reaching the eye from other sources. A predator's ability to detect prey (and vice versa) depends on the minimum contrast that it can detect: the contrast threshold. Specifically, prey whose contrast falls below the contrast threshold will not be visually detected. Contrast threshold decreases with increasing ambient light intensity. That is, an object can be seen at a greater distance as the ambient light intensity against which it is viewed increases. This means that aquatic prey will be more easily detected from below, against the bright (downwelling) light from the sky, than from above, against the much reduced scattered or reflected light directed upwards. However, downwelling light intensity decreases with depth, and so this should lead to an increase in contrast threshold with depth. Anthony (1981) estimated the contrast threshold for cod and suggested that this changed from around 0.02 at light intensities equivalent to those in the first 20 m of clear water on a bright sunny day, up to 0.5 in light intensities more representative of depths of around 650 m (Johnsen and Widder, 1998). Consideration of Fig. 4.2, based on theoretical modelling of Johnsen and Widder (1998), suggests

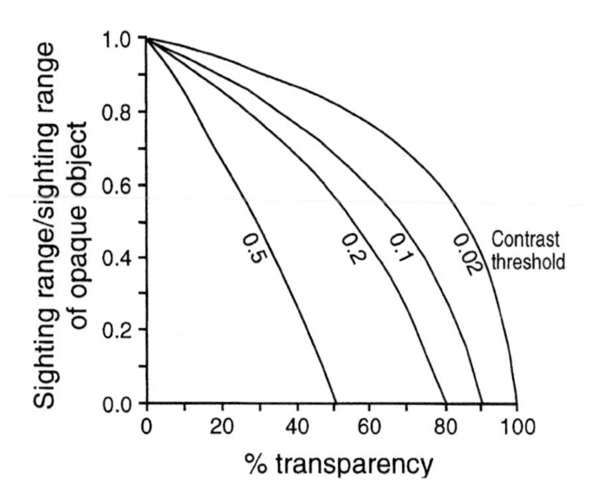


**Figure 4.2** Sighting distance of a prey item versus its percent transparency when viewed from below by predators with visual systems that have different minimum contrast thresholds for object detection. The sighting distance is divided by the sighting distance of an opaque object to control for water quality, prey shape, size etc. The ratio gives an estimate of the advantage of transparency for crypsis in a given situation.

that this change should have a substantial ecological effect. At the top of the water column when the contrast threshold is 0.02, transparency has to be greater than 85 per cent to halve the sighting distance of an object compared to a fully opaque counterpart. In contrast, at depth, when the threshold has increased to 0.5, transparency need be only greater than 30 per cent to more than halve the sighting distance. Indeed at this contrast threshold, individuals with a transparency of greater than 50 per cent remain visually undetected no matter how close the predator or prey may approach. Johnsen and Widder (1998) measured the transparencies of 29 species of zooplankton as being between 50 and 90 per cent.

Utne-Palm (1999) demonstrated that the sighting distance of predacious gobies was higher for opaque copepods than transparent ones. This is in line with our expectation that transparent organisms are actually more challenging to detect than opaque ones. Experiments to measure the contrast threshold and how it is affected by properties of the eye, the target, the medium and the light are technically challenging. However, the arguments in this section urgently require a stronger empirical foundation.

## 4.4 Some parts of an organism cannot be made transparent

Eyes must contain light-absorbing pigments and so cannot be wholly transparent. Food is usually opaque, with the result that the stomach cannot generally be transparent after the organism has fed. Feeding on transparent prey does not seem to be a solution to this, as transparent organisms rapidly turn opaque after death (Herring, 2002). This change suggested to Herring and to others (e.g. McFall-Ngai, 1990; Hamner, 1995) that transparency of cellular tissues can only be achieved with a continual investment in active maintenance. However, direct evidence of such a continuous metabolic cost to crypsis through transparency is currently lacking, and investigation of this would be very useful.

There has been considerable empirical investigation into whether the size of opaque parts increases predation risk for otherwise transparent organisms. As these studies take different approaches, we will try to provide a synthesis of their results. Giguère and Northcote (1987) found that the predation rate of juvenile coho salmon on transparent phantom midges (Diptera : Chaoboridae) increased with increasing stomach fullness of these insect larvae. Considering the relative sizes and mobilities of predator and prey, this is more likely to result from increased prey visibility following a meal, rather than decreased mobility. Tsuda et al. (1998) obtained similar results for herring larvae and salmon fry feeding on copepods. Similarly, Wright and O'Brien (1982) concluded that the detectability of mainly transparent midge larvae to predatory fish was based on the size of high-contrast structures within the larvae rather than the overall size of the prey individuals. Essentially identical conclusions were drawn by Hessen (1985) after experiments with roach feeding on zooplankton. Bohl (1982) offered predatory fish a combination of generally transparent daphnia with and without eggs in their dorsal pouch. Eggs are opaque and should act to increase the vulnerability of egg-bearing individuals. He found no preferences for egg bearers under complete darkness and under bright light conditions, but a strong preference for egg-bearers under low light conditions. This seems to represent strong evidence that the egg-bearing individuals were more easily detected by predators than non-egg bearers when visual detection was challenging, but not impossible. Brownell (1985) performed laboratory experiments to explore cannibalism in the (relatively transparent) larvae of cape anchovy. He concluded that the formation of pigmented eyes, at a certain development stage, triggered a substantial increase in larval detectability and so vulnerability to cannibalism.

Zaret (1972) reported laboratory experiments where predation by fish on a cladoceran appeared to be related to the amounts of pigmentation in their black compound eyes. These experiments used a species with two morphs that differ considerably in eye pigmentation; the morph with the higher pigmentation (i.e. the larger eye) was attacked more frequently than the morph with the smaller eye. Zaret took advantage of the fact that the stomach of these organisms lies near the eye to produce a 'super eye spot' by allowing individuals of the 'small eye' morph to pre-feed in water to which India ink had been added. This led to a reversal in relative predation rates in subsequent experiments where the small eyed morph with the 'super eye spot' created by black pigment in their stomach was preyed on more heavily than the larger eyed morph without the India ink modification. Whilst alternative explanations based on behaviour or chemosensory changes are possible, the most likely explanation for these results is that the size of the opaque parts of the prey influence detectability by predators. A follow-up set of laboratory trials (Zaret and Kerfoot, 1975) demonstrated that variation in predation rate on a largely transparent cladoceran was better explained by variation in the size of its large compound eye than variation in the size of the whole body. However, a similar study by O'Brien et al. (1979) did not support these conclusions, finding actual size to be a better predictor than eye size. Similarly, studies have found no effect of ink pre-feeding by daphnia on detectability by bluegills (Vinyard and O'Brien, 1975) or lake trout (Confer et al., 1978). Finally, it may also be adaptive to have an opaque stomach if feeding on bioluminescent (or brightly coloured) prey, in order to reduce the visibility of stomach contents to potential predators or prey.

In conclusion, there is evidence that opaque parts to a generally transparent organism can have a substantial effect on predation risk, although this effect is not shown universally in empirical studies. However, we should not expect there always to be such an effect. If the organism can be detected by other means, and we have considered a number of methods by which a perfectly transparent organism can still be detected visually, then small opaque parts need not necessarily increase predation risk.

## 4.5 The distribution of transparency across habitats

In Section 4.1, we explained why arguments based on the refractive indices of air and water suggest that transparency is a much more attractive option for aquatic than terrestrial organisms (although the transparent wings of many flying insects may help to reduce the ease with which they are detected). We have also presented arguments (based on UV damage, low contrast thresholds and the spectra of ambient light) for why transparency will be less attractive (in comparison to alternative means of crypsis) very close to the water's surface.

In the very deep ocean (the aphotic zone), beyond 1000 m, solar illumination is absent. In these waters, animals are mostly red or black in colouration (Marshall, 1971; McFall-Ngai, 1990). The only source of illumination at these depths is bioluminescent light produced by organisms. Under these circumstances transparency seems less attractive than pigmentation designed to minimize the light reflected back by the organism. The reason for this is not completely clear to us, although it may be related to generally lower encounters between predators and prey at these depths making continuous metabolic investment in transparency and/or the structural compromises required by this whole-body style of crypsis less attractive than crypsis based on pigmentation. However, it may be that transparency is more common in the deep sea than we currently believe. Our knowledge of deep-sea organisms tends to be greater for species with larger body sizes, and large body size tends not to be associated with transparency.

One remaining challenge is to explain the rarity of transparency in benthic organisms. It may be that the backdrop of the substrate provides opportunity for other types of crypsis using pigments that avoid the structural compromises and/or continual maintenance requirements of transparency. Transparency is also rare among those animals living on the surface film (the neuston). It may be that the complex light environment produced by the ever-changing surface of the water makes crypsis by pigmentation easier, although this hypothesis is currently untested (Hamner, 1995). Additionally or alternatively, for such animals, the conflict between allowing UV radiation to pass through the body to achieve crypsis by transparency and minimizing cell damage by UV radiation may favour alternative methods of crypsis, based on pigmentation of the organism's surface.

Hence, the distribution of transparency predominantly where sunlight is present (but not in the surface layer) in aquatic systems seems understandable, especially since alternatives such as hiding in physical structures are generally not available in open water. Although, note that several of our arguments above rest on the suggestion that tissues can only be kept transparent by active maintenance; this conjecture urgently needs critical testing.

Unlike other forms of crypsis, transparency involves the entire body, not just the exterior (McFall-Ngai, 1990). It has been argued that transparent organisms will gain an advantage by having the optical properties of their body (such as refractive index) as similar to those of the surrounding medium as possible. Many aquatic organisms achieve this by being gelatinous, where thick layers of extracellular watery material separate very thin layers of cellular material. Transparency of muscle and other dense tissues is more easily achieved if one dimension is very thin (McFall-Ngai, 1990; Herring, 2002). Clearly a gelatinous body plan leads to a huge increase in size. This increased size will have costs and benefits other than improved crypsis. One added benefit of a gelatinous body plan will be decreased body density and so increased buoyancy. For suspension feeders, a gelatinous body plan can bring increased capture rates without the increase in metabolism that would normally result from growing bigger (Acuña, 2001). Costs are likely to include increased drag and vulnerability to mechanical damage. These non-crypsis costs and benefits may have important implications for the evolution of crypsis by transparency through adoption of a gelatinous bodyplan.

An organism may well experience very different types of illumination even on a daily time-scale. This can be brought about either by the daily cycle of sunlight and/or by vertical migration of the organism. Hamner (1995) contends that 'at sea, predation drives most, if not all, vertical migratory behaviour'. From the discussion above, we can see that at different depths (or, more generally, under different lighting conditions) different mechanisms of crypsis will be favoured. Hence, it is no surprise to find that individuals can adapt their external appearance on behavioural timescales so as to adopt the most effective mechanism for their current situation (Morgan and Christy, 1996). An alternative behavioural mechanism employed by some species is to adjust their depth so as to remain within a range of preferred irradiances (Frank and Widder, 2002).

## 4.6 Silvering as a form of crypsis

In terrestrial ecosystems, there is generally strong directionality in the ambient lighting, since the sun takes up only a small part of the visible sky, and this part of the sky produces much more light than other equivalent sized parts. This directionality can be seen clearly from the existence of shadows. As light passes through water, scattering by water molecules and suspended particles has the effect of destroying this strong directionality. Denton et al. (1972) suggests that 300 m depth of clear water is required before direct sunlight is scattered sufficiently to become vertically symmetrical, reducing considerably (to perhaps around 50 m) if the water is turbid or the sky cloudy. Thus, in the midwaters of the oceans and deep lakes, the light environment is symmetrical about the vertical axis. In these waters, there is also generally a complete absence of background features. Under such conditions, a vertical mirror will be difficult to detect from all angles *except directly above and*

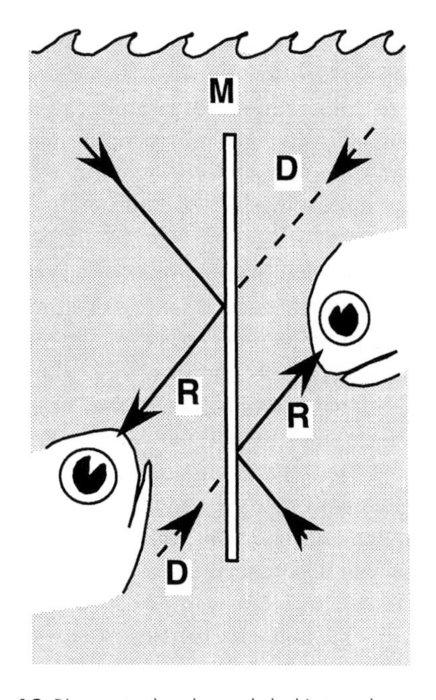


**Figure 4.3** Diagram to show how a dark object can be camouflaged in the radially symmetric radiance distribution of the ocean by making it reflective, vertical mirror (M) cannot distinguish between reflected rays (R) and direct rays (D), so the mirror (or silvered) fish is invisible (from Herring, 2002).

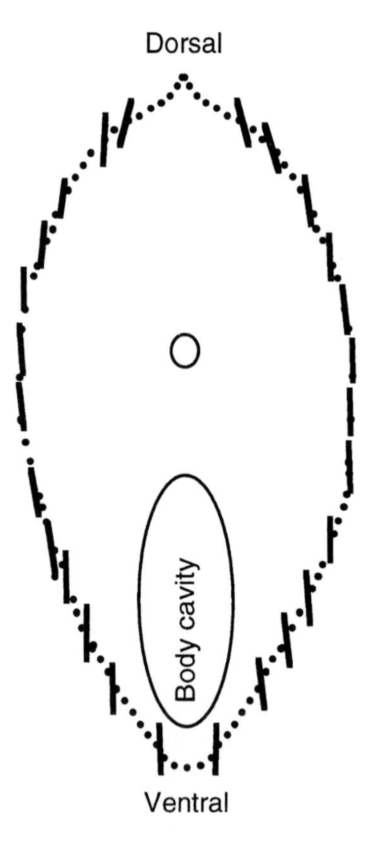


**Figure 4.4** Reflectors aligned to the curve of the body surface would not be an effective camouflage for the flanks of a muscular fish of elliptical cross-section like a herring. Hence, the individual reflectors are aligned vertically, providing an effective vertical mirror surface (from Herring, 2002).

*below* (Fig. 4.3). Hence, in these environments an animal with a body plan like a vertical mirror can make itself invisible to detectors viewing from side-on. This is not an option for terrestrial organisms or organisms near the substrate, where background features mean than reflection is no longer an effective disguise. We can plainly see the mirrors hanging on our walls, because they produce a reflection that does not match the background of the wall.

One strategy to reduce the vulnerability of disguise by silvering to detection from above and below is to be laterally flattened so as to restrict the vulnerable cross-sectional area. This is adopted by some silvered fish, such as hatchetfishes. However, this bodyform restricts locomotive performance, and so muscular fish, such as herring, mackerel, and salmon, adopt a bodyform with an elliptical cross-section where the reflective scales act as tiny mirrors and each is aligned vertically rather than parallel to the curved body surface (Fig. 4.4). However, the problem of providing reduced detection from above and below still remains.

One feature of mid-water environments is that the downwelling light is orders of magnitude more intense than the upwelling light (Widder, 2002). Thus, dorsally placed mirrors reflecting downwelling light would make an animal very visible against the dark background of the depths. Hence, mirrors are not the solution to reducing detection from directly above. Instead, silvered animals tend to have a dark pigmented upper surface, so as to provide a match to the background. The reverse problem applies when being viewed from below, where reflected upwelling light provides a large contrast to the strong downwelling light. One solution is for the animal to produce its own downward-directed light to match that of the ambient downwelling illumination (for an overview see Widder, 1999). Some species have been demonstrated to alter the intensity and wavelength distribution of the light they produce appropriately as they change depth, so as to match the changing downwelling light (Young and Roper, 1976; Case et al., 1977; Latz and Case, 1982; Young and Arnold, 1982; Young, 1983). Organisms have also been demonstrated to produce light that matches the angular distribution of the ambient light (Denton et al., 1972; Latz and Case, 1982). Such counterillumination is only feasible at considerable depths when the intensity of downwelling light is sufficiently reduced that self-generated light can significantly affect the contrast (for further discussion, see McFall-Ngai, 1990).

Land (2000) argues that less than perfect matching of the background illumination can still substantially reduce the ease with which the organism can be detected. An opaque organism has a contrast of 1.0 against the downwelling light, as defined by Land and Nilsson (2002), whereas a perfectly transparent organism has a contrast of zero. If contrast can be reduced from 1.0 to 0.5, then an eye must capture four times as many photons in order to detect the object. The light requirement changes as the inverse of contrast squared, so reducing the contrast from 1.0 to 0.1, requires the detector to collect 100 times as much light to provide the same detection ability. Hence, any reduction in contrast reduces the maximum depth in the water column at which the organism can be detected. Land suggests that a 100-fold change in downwelling illumination occurs over a distance of around 140 m of clear water.

Both Denton (1970) and Herring (2002) suggest that the tapered keel of many silvered fish act to reduce visibility from below, although the mechanism for this is not specified. Reflective scales could be angled so as to reflect light partially downward, and whilst this would reduce visibility from below, it would do so only at the expense of increasing visibility from the side. Alternatively, it could be that this shape promotes diffraction of light passing close to the body so as to reduce visibility from below, but further exploration of this is needed.

When light is reflected, its polarization changes. This may provide a mechanism by which the crypsis provided by silvering can be reduced. As noted in Section 4.2.1, cephalopods are sensitive to the polarization of light, and Shashar et al. (2000) demonstrated in a laboratory experiment that cuttlefish targeted fish with normal polarization reflection over fish that, by means of filters, were made not to reflect the linearly polarized component of the light. Further experiments in a more natural environment would be very interesting.

Silvering seems largely confined to adult stages of fish. It may be that for juvenile stages and smaller plankton, transparency is a more effective form of crypsis, but that this 'whole body' disguise becomes impossible for large fish that need to move quickly, and must have large amounts of (relatively opaque) respiratory and connective tissues associated with their substantial muscle mass. The different depth distributions of larvae and adults may also have a role. Giske et al. (1990) suggests than juveniles tend to be distributed more in shallower waters than adult stages, and we have already discussed why silvering is more effective at greater depths where the light field is more homogeneous.

Salmonid fish characteristically change from a dark and disruptive body patterning to a classical silvered appearance before migrating from stream and river to the ocean. This can be understood in terms of our argument above about the inefficiency of silvering in shallow water with visible background features, whereas their juvenile colouration would make salmonids highly visible when viewed from the side against the featureless ocean. Silvering can also be a solution to crypsis of parts of an organism that cannot be made transparent in an otherwise transparent animal. Effectively, the eye or digestive organ can be thought of as a mini-organism and disguised by silvering in exactly the same way as described in this section for whole organisms.

Badcock (1970) and Denton et al. (1972) report that many silvery mesopelagic fish darken their silvery reflective surfaces at night. Recall that silvering is only an effective means of crypsis if the light field around the object is symmetrical. This will not be true at these depths at night, where the main light will not be downwelling star and moonlight, but rather the light generated by luminescent organisms. Under such circumstances, a dark light-absorbing surface may be a more effective disguise than a silvery reflective one.

## 4.7 Conclusion

Visual detection of an organism requires that the light reaching a detector's eye has been modified by the organism. Perfect transparency means that light can pass through an organism without any modification through the processes of absorption and scattering. This does make transparent organisms challenging to detect. However, other physical processes (such as reflection and refraction) can also alter light as it passes between an organism and the surrounding medium (air or water). Hence, transparent organisms are challenging but not impossible to visually detect. The environmental conditions where transparency are likely to be particular effective in reducing detectability occur in mid-depths of open water regions of lakes and seas. These regions also provide the conditions where a highly reflective skin can make an organism very challenging to detect. This chapter provides an overview of the current state of knowledge of crypsis by transparency or reflection. It seems that the fundamental physical processes are relatively well understood, but quantification of the importance of these processes to natural predator–prey systems is currently lacking. This may well reflect the bias of terrestrial-living scientists towards the processes they see in the world around them. However, this field does seem to have experienced renewed interest recently (as evidenced by the recent papers cited in the preceding sections and recent reviews by Johnsen, 2001; Johnsen and Widder, 2001), and we can expect further developments in the near future.

In particular, study of the costs of transparency would be particularly welcome. In the language of Section 1.1, there may be opportunity costs associated with crypsis through transparency—since transparency is much more effective in some habitats (and light fields) than others. Much of the arguments in Section 4.5 rested on the assumption that there was a recurrent allocation cost to maintaining transparency of body tissues and fluids. Lastly, in contrast to the other forms of crypsis discussed in this section, the effectiveness of transparency depends on the whole body of an organism, not just its exterior surface. Hence, there are likely to be very significant design costs associated with adoption of transparency as a primary defence against predators. Quantification of these costs would provide a very useful step forward for this area of study.

PART II

# Avoiding attack after detection

CHAPTER 5

# Secondary defences

In this chapter we take a necessary diversion from the main theme of this book and consider the evolution of secondary defences in prey. Secondary defences are essentially post-detection defences that serve to make an encounter unprofitable for a predator (see Edmunds, 1974). Aggressive retaliation, rapid escape movements, morphological defences such as spines, and chemical defences such as toxins all commonly function as secondary defences. If the unprofitability of a prey is recognized quickly by a predator it may break off an attack before serious injury is inflicted. Furthermore, if predators can learn about prey unprofitability and reduce attacks in future then the inclusive fitness of a prey's alleles that confer defence might subsequently be raised.

One major reason for looking at secondary defences here is that much anti-predator signalling involves making predators aware that potential prey are unprofitable to predators. Furthermore, the form a defence takes is likely to have an affect on the way in which it is signalled to predators. As we explain in Chapter 7, visible external defences such as the spines of a porcupine may, for instance, require a different sort of warning display compared to 'invisible', internally stored chemical defences in toxic prey species. Hence, before we consider the evolution and design of warning signals and displays in Chapters 6–8, it is important to consider selected aspects of the evolutionary ecology of secondary defences.

In the first part of the chapter we consider two important and related factors: diversity and costs of secondary defences. We examine in outline how secondary defences incur costs and provide benefits to prey. We place a particular emphasis on chemical defences because they are strongly associated with the aposematic warning displays we examine in Chapters 7–11. In the second part of this chapter we consider how secondary defences might have evolved and consider theoretical issues that await resolution.

## 5.1 The diversity of secondary defences

Diversity in secondary defences has at least two important components. First, most obviously, is form. Forms of secondary defences can be classified into: behavioural (escape, retaliation, etc.), morphological (more generally mechanical: spines, sticky secretions, modified organs, etc.) and chemical (toxins, venoms, irritants, etc). A second component of diversity is timing: some secondary defences are constantly present and active (constituitive), others are to greater or lesser extents induced according to threat. Inducibility of a defence can occur in two ways. First, a defence may be completely absent in a prey organism unless it is induced by appropriate environmental cues that generally predict threat; the production of defensive spines of rotifers and daphnia and the variable tail morphologies of tadpoles are examples (see Section 5.2.2). Second, the defence may be present but only deployed when a prey animal is threatened; a common example is the behavioural induction of chemical defences such as the selective deployment of defensive secretions by many arthropods.

For simplicity in this chapter we reserve the term 'inducible defence' to refer to secondary defences that can be entirely absent until an appropriate, but not necessarily imminent, threat is registered by the prey. We use the term 'behavioural response' for those secondary defences characterized by selective deployment of existing defensive capability.

What kind of factors affect the form that a defence takes? Consider a hypothetical terrestrial insect subject to sufficient predation pressure that secondary defences might evolve. In outline the following factors will strongly influence the form of an animal's defence.

First, the appropriate form that secondary defences take will be determined by the diversity and intensity of forms of attack by relevant predators and consequently by the range of prey characteristics that effectively defend prey against these threats. Thus, our hypothetical insect may be attacked by both vertebrates such as birds and invertebrates such as ants. Since different defences may be effective against these different predators, the prey may evolve some combination of defensive components such as mechanical protection from a tough cuticle to defend itself against predators and rapid escape behaviours to defend itself against birds.

The form of a prey's defences will also be affected by its evolutionary history. Though some defences may be highly effective against predators, they may require excessively demanding rearrangements of a prey animal's physiology and anatomy. Even if a particular poison may be effective against ants and birds, our hypothetical insect prey may lack the capacity to acquire or biosynthesise it. The existence of historical constraints means that most prey can evolve defences only from a limited subset of all possible mechanisms of effective defence.

Further constraints on the form of a defence may be imposed by the prey's environment, limiting the range of resources or developmental conditions and again further limiting the subset of possible defences. Thus, for instance, our hypothetical insect prey may be able in some circumstances, but not others, to acquire a food material that it sequesters in its cuticle in order to make it unprofitable to its predators. If the food source is too scarce and unpredictable, the prey may have to evolve other more reliable lines of defence that are less constrained by environmental conditions.

Finally, the precise form of a prey's defences will obviously depend on the functions that define the benefit/cost ratio of each possible defensive component. As in the preceding section of this book on crypsis, we assume that costs of secondary defences can generally be classified as one or more of: allocation, opportunity, environment, design (self-damage), and plasticity (Tollrian and Harvell, 1999). It may be that two secondary defences generate, for instance, equal benefits but incur different levels of allocation cost. Hence, the cost/benefit ratios of alternative forms of defence will be pivotal in determining the precise form that defences take. Although the benefits of any particular defence will be largely determined by predatory risks, secondary defences may bring additional benefits, such as removing opportunity costs imposed by crypsis, and we discuss this idea in relation to warning displays in Chapter 7. If defences are costly, then optimization of the form of a secondary defence may lead to localization of defensive traits to parts of an organism that are most dispensable, so that vital organs are not injured in an attack (see optimal defence theory in Rhoades, 1979). A number of prey, ranging from nudibranch molluscs (Penney, 2002) to Lepidoptera (Brower and Glazier, 1975) do indeed localize chemical defences in body parts that are least vital and most vulnerable to attack (Poulton, 1890).

Although the question of form is multifaceted, the question of inducibility adds a further layer of complexity. A defence of any form is most likely to be inducible if, as Harvell and Tollrian (1999) suggest, the predatory risk is variable and unpredictable, predators present a reliable cue that indicates threat and if defences are sufficiently costly so that constituitive expression is not cost effective. A similar argument may apply to the timing of secondary defences that are deployed by behavioural response, such as defensive secretion.

Overall then, factors that affect the diversity of secondary defences in form and timing are complex, but can be reduced to a number of key determinants: form, intensity, and predictability of predatory threat; evolutionary and ecological constraints and determinants of the cost/benefit ratios of different defensive phenotypes. As we can see from the above discussion relative costs and benefits are important not only to the question of whether secondary defences evolve, but also to the form that they take. In the next section we therefore examine examples of selected secondary defences for which costs and benefits have been determined. We pay attention to defences that are commonly associated with warning displays, but do consider other defences where they add to our understanding of the complexity of secondary defences.

## 5.2 Costs and benefits of some behavioural and morphological secondary defences

### 5.2.1 Behavioural defences

Rapid escape velocities, as well as erratic, 'protean' evasive movements are well known in terrestrial as well as aquatic systems. Indeed, rapid escape itself may be so effective as a deterrent that predators advertise this capacity with conspicuous displays (van Someren, 1959; Lindroth, 1971; Gibson, 1974; Baker and Parker 1979; Hancox and Allen, 1991). It has been repeatedly shown that escape behaviours enhance the survival of an individual prey and incur allocation costs large enough to have fitness consequences (e.g. Gilbert's study of the ciliate *Strobilidium velox*; Gilbert, 1994). However, escape and other evasive behaviours can impose costs in a much more complex manner than simple reallocation of resources.

A fascinating example is the freshwater snail, *Lymnaea stagnalis*, that retreats deeply within its shell to escape attacks by predatory crayfish (Rigby and Jokela, 2000). There are potential allocation costs in shell growth and in the muscular energy required to draw back into the shell. However, to retreat in this manner the snail must also expel some blood. When the attack is over the lost fluid is replaced with freshwater from the surroundings. Replacing lost blood in this manner may introduce microbial pathogens into the snail and therefore incur further costs of immune activity. The extent of the cost of antimicrobial defence is dependent on the microbial contents of the environment at the time of the avoidance response; hence the snail may have to pay both direct allocation costs and environmentally imposed costs according to the levels of pathogens in the surrounding water. By manipulating frequency of bleeding and microbial content of the water, Rigby and Jokela (2000) found that increases in the quantity of blood expelled per day greatly reduced the survival value of the snails, as well as probability of reproduction, reproductive output and storage in fat bodies. This study represents a remarkable demonstration that in some prey organisms an anti-predator defence system is directly traded off against anti-parasite defence.

In order to make rapid evasive manoeuvres possible, investment in muscles and other body components may be necessary and such investment may incur allocation, design and other costs. One important case study with direct relevance to warning displays and mimicry is the flight mechanics of neotropical Lepidoptera. Butterflies and diurnal moths can reduce risks of predation while in flight by fast, evasive movements that may confuse predators (Humphries and Driver, 1970) and/or by the possession of chemical defences and warning displays which advertise the unprofitability of prey to predators. Quick, evasive flights are facilitated in Lepidoptera by investment in thoracic flight muscles and, indeed, one measure of the power available for flight is the ratio of flight muscle to body mass (Kingsolver and Srygley, 2000).

The relationship between investment in flight muscles and edibility has been examined extensively in neotropical butterflies. As expected from a trade-off model, edible butterflies have higher thoracic mass than chemically defended (and mimetic) butterflies and moths; conversely edible butterflies and moths have a lower abdominal mass than chemically defended individuals (Marden and Chai, 1991). Srygley and Chai (1990*a*) found similar relationships between the edibility of butterflies and investment in thoracic and abdominal mass (though see discussions on the relationship between thoracic mass and survival, Kingsolver and Srygley, 2000). Since the abdomen contains organs of digestion, food storage, and reproduction, it appears that powerful flight muscles are bought at a cost of both nutritional and reproductive functions. This is especially true of females since ovaries and their contents in Lepidoptera are relatively large and undoubtedly expensive organs. Hence, the possession of chemical defences directly affects the form and investment in other defences, often relieving the organism of the need to invest heavily in rapid escape behaviours. It appears that for a number of Lepidopteran species the return on investment in chemical defences is much more cost-effective than the return on investment in behavioural secondary defences. Furthermore, in Marden and Chai's study (1991), Müllerian mimicry (i.e. a shared warning signal: see Chapter 9) between defended species seems to significantly reduce the investment necessary in flight muscle. This would follow if the risks to defended co-mimics are lower than for defended species not involved in mimicry (Marden and Chai, 1991).

In order to estimate the efficacy of flight as a secondary defence, Chai and Srygley (1990*b*) and Srygley and Dudley (1993) took a set of body measurements of neotropical butterflies and moths, and found that the ratio of body length to thoracic width was the 'best single variable for predicting flight speed'. Marden and Chai (1991) calculated the acceleration that thoracic muscles could supply in each butterfly species and then estimated the maximum acceleration of their avian predators; palatable butterflies had acceleration that was almost always higher than that for the birds, whereas acceleration for defended species was generally lower. Hence, this example shows in a clear and informative manner how costs and benefits of secondary defences can be estimated and can be related to the form that an individual's defence actually takes.

### 5.2.2 Morphological and other mechanical defences

If prey are likely to be attacked after they have been detected they may utilise morphological protection to render themselves unprofitable to predators. Sharp spines (and horns, bills, claws, etc.), withdrawal into shells or other forms of armour and the possession of tough, impenetrable integuments are common across phyla. As Edmunds (1974) points out, many morphological structures that can be used in anti-predator defence may well have other uses in intraspecific interactions (such as male–male competition for mates) making the cost/benefit ratio more favourable.

Inducible morphological defences have been well-studied particularly in aquatic systems. Chemical cues from predators, called kairomones, are used by many prey species as interspecific cues responsible for induced morphological changes (though see discussion of the term in Pasteels, 1982). In addition, alarm signals from distressed or damaged conspecifics are important in both morphological and behavioural defensive responses in some aquatic invertebrate prey. Stabell et al. (2003), for example, recently showed that induced morphological change in *Daphnia galeata* results from alarm signals of conspecifics rather than from predator-derived cues.

Induced morphological change for mechanical or other defences occurs in many species and has been well studied in amphibia (McCollum and Van Buskirk, 1996; Lardner, 1998; Van Buskirk and Relyea, 1998; Van Buskirk, 2000; Benard and Fordyce, 2003), fish (Pettersson and Brönmark, 1997), gastropods (Trussell and Nicklin, 2002), bivalves (Leonard et al., 1999), bryozoa (Iyengar and Harvell, 2002), and odonate larvae (Johansson, 2002). Induced change is also very well studied in plants (e.g. Agrawal and Karban, 1999). The very fact that these morphological and behavioural changes are inducible rather than constituitive strongly suggests that their production incurs fitness costs. However, it remains important to evaluate how inducible defences incur costs and benefits and again, since the literature is large and growing (reviewed in Tollrian and Harvell, 1999) we limit ourselves to describing some illustrative examples here.

In some systems, it is possible to understand how external conditions cause morphological change, to measure the fitness costs and to identify the proximate mechanisms that incur ultimate costs. A great deal of work on inducible defences in planktonic life-forms reveals much about the cost/benefit ratios of induced defences. One of many examples are ciliates in the genus *Euplotes*. In *Euplotes daidaleos*, morphological changes are induced by specific predators and the extent of morphological change increases with predator density (Kusch, 1995). In *Euplotes octocarinatus* the presence of predators induces development of extended 'wings' and other appendages by a reorganization of the exoskeleton, a process that does not require cell division. In experiments reported by Kuhlmann et al. (1999), such rearrangements increased rejection rates by predators between 2 and 20 fold). That the defence incurs allocation costs is indicated by an 18 per cent increase in generation time during the development of a defensive morphology, after which generation time remained 16 per cent higher than in the uninduced state (Kusch and Kuhlmann, 1994). A proximate source of such costs is likely to be protein synthesis during the growth phase of the ciliate's life cycle; 'winged' forms had a protein content 55 per cent higher than non-winged forms.

Induced change has also been very well studied in aquatic animals such as *Daphina* (review in Tollrian and Dodson, 1999) and rotifers (Stemberger, 1988). In an instructive example in tadpoles, McCollum and Leimburger (1997) examined how the grey treefrog (*Hyla chrysoscelis*) changes its morphology and appearance in the presence of a dragonfly larva. When the predator was caged in the same container as free-swimming tadpoles, the tadpoles grew deeper tails which, unlike the uninduced form, were coloured yellow/red with distinct black spots.

McCollum and Leimburger found that changes to the tail structure were likely to increase swimming speeds and therefore enhance the evasiveness of the animal. Induced tadpoles are therefore less vulnerable to predation than the uninduced form. However, they survive less well when predators are absent, indicating that some form of cost is incurred by morphological change (McCollum and Van Buskirk, 1996). Similar benefits and costs have been reported in other amphibian species (e.g. the wood frog *Rana sylvatica*; Van Buskirk and Relyea, 1998) but the causal basis of the fitness costs is not necessarily clear.

Similarly, the reason for the change in tail colour in *H. chrysoscelis* is also not certain. McCollum and Leimburger consider that the induced colour change could direct predatory strikes toward the tadpole's tail, thereby protecting the body (see Chapter 13), or instead that it may have some visually disruptive, cryptic effect (see Chapter 2). In either case, it is not clear why such colour changes should be induced rather than constituitive, unless the pigments themselves are expensive. Perhaps a more consistent explanation is the tail colours make the tail more conspicuous, drawing a predator's attention to the rapid evasive movement that only the induced form can achieve. If this results in fewer net attacks, the bright colours of the tail may function as a rare example of an inducible warning display (see also Sword, 2002).

As these few examples demonstrate, inducible morphological defences present excellent opportunities for the evaluation of costs and benefits of defences. In outline, experiments are relatively straightforward (measure responses of the animal to the presence or absence of the inducing stimulus and evaluate its survival in the presence or absence of predators). However, in detail the experiments are often complex and appropriate design may be a highly complex matter (see discussion in Gilbert, 1999). A growing number of cases clearly document and measure benefits and demonstrable costs; however, it should be stressed that when they are identified they have diverse, sometimes surprising origins. Finally, a detailed consideration of the question of inducibility vs. noninducibility is important but beyond the scope of our discussion (see e.g. Tollrian and Harvell 1999). We must be aware that examples of inducible secondary defence represent a specific subset of organisms with anti-predator defences—perhaps their defences are inducible simply because they carry much greater costs than is typical for other secondary defences?

## 5.3 Chemical defences

The chemical ecology of prey defences is perhaps the most diverse and complex of the secondary defences. Chemical defences such as toxins, irritants, and venoms can act before, during, and after attacks to raise the inclusive fitness of a prey animal. When there is a significant threat of an attack or when an attack has commenced, many animals combine behavioural responses with secretion of noxious chemicals, often utilizing morphological specialisations to do so. Because of the complexity of chemical defences, and their frequent association with warning displays and mimicry we place particular emphasis on this sort of defence in the remainder of the chapter. There are a number of excellent reviews to which readers are directed for more detail on particular systems (for terrestrial arthropods see, e.g. Benson, 1971; Blum, 1981;

Pasteels et al., 1983; Brower, 1984; Bowers, 1990, 1992; Whitman et al., 1990; for marine invertebrates, see e.g. Lindquist et al., 1992; Pawlik, 1993; Hay, 1996; Lindquist and Hay, 1996; McClintock and Baker, 1997*a*,*b*; Baker et al., 1999; Amsler et al., 2001; Dobler, 2001; McClintock et al., 2001; Lindquist, 2002).

Before considering whether chemical defences are generally costly, it is important to discuss a number of points that characteristise this sort of secondary defence. In particular, chemical defences: (1) are widespread and diverse across species and phyla; (2) often contain complex mixtures of chemicals; (3) can be acquired from a prey's environment or a symbiont or alternatively (4) may be self-generated by biosynthesis or reduced nutritional quality; finally, (5) chemical defences commonly show high levels of intraspecific variation. We consider these points briefly below. Readers already familiar with these ideas but interested in evidence about the potential costs of chemical defences may wish to go directly to Section 5.3.2.

### 5.3.1 Some characteristics of chemical defences

*Chemical defences are widespread and highly diverse across species and phyla*

Chemical defences are widely known in invertebrate animals (Whitman et al., 1990). In addition, there is growing evidence that noxious secondary metabolites may defend many marine organisms from predation by fish and invertebrates (Lindquist et al., 1992; Pawlik, 1993; Hay, 1996; Lindquist and Hay, 1996; McClintock and Baker, 1997*a*,*b*; Baker et al., 1999; Amsler et al., 2001; McClintock et al., 2001; Lindquist, 2002). A large number of adult, soft-bodied, sessile, and other benthic and pelagic invertebrates (sponges, molluscs, including the nudibranchs, ascidians, etc.) are known to synthesize and/or sequester secondary metabolites from food materials that may be important in defences against predation (see discussions in, e.g. Hay and Fenical, 1988; Lindquist et al., 1992; Pawlik, 1993; Hay, 1996; Lindquist and Hay, 1996; McClintock and Baker, 1997*a*,*b*; Baker et al., 1999; Amsler et al., 2001; Stachowicz, 2001; Lindquist, 2002). These secondary metabolites may also have other functions such as allelopathy and inhibition of settlement of animals that may compete for space, and also protection from UV radiation (Pawlik, 1993; Amsler et al., 2001).

Chemical defence is also well known in vertebrates, for example, snakes (biosynthesis of neurotoxins Mebs, 2001), fish (acquired through biosynthesis or symbiosis, e.g. the pufferfish, Yotsu et al., 1987; Mebs, 2001); amphibia (acquired through food materials, possibly symbionts and biosynthesis, e.g. Andersen et al., 1982; Brodie et al., 1991; Brodie and Formanowicz, 1991; Daly et al., 1997; Brodie, 1999; D'Heursel and Haddad, 1999; Vences et al., 2000; Summers and Clough, 2001; Pires et al., 2002); birds (in some instances acquired through food materials, Cott, 1940; Dumbacher et al., 1992, 2000; Weldon and Rappole, 1997; Dumbacher, 1999; Archetti, 2000); and mammals (e.g. the notoriously malodorous defensive spray of skunks, Andersen et al., 1982; Lariviere and Messier, 1996). As with other examples, the chemical defences described in this chapter may have uses beyond predator deterrence; they may be used to restrain and kill prey; they may be used in aggressive intraspecific interaction and to prevent the actions of parasitoid, parasites and other enemies.

*Chemical defences of an individual often contain highly complex mixtures*

Defensive secretions are very often complex mixtures of compounds (e.g. Blum, 1981; Pasteels et al., 1983; Eggenberger and Rowell-Rahier, 1991, 1993) and it seems likely that at least some components act synergistically to deter predators (e.g. Muhtasib and Evans, 1987). Pasteels et al. (1983) suggest that such complexity and variability may reflect a strategy to reduce counteradaptation by predators. Alternatively however, several components of a secretion could be the precursors of active defensive compounds. Blum (1981) discusses the complexities of defensive secretions in detail, pointing out that minor components of secretions may play a vital role in determining the effectiveness of a secretion as a secondary defence. One pertinent example is the defensive secretion of the whip scorpion *Mastigoproctus giganteus*, which contains 84 per cent acetic acid and 5 per cent octanoic acid (Eisner et al., 1961).

The octanoic acid acts as a spreading agent and as a penetration-promoting agent, ensuring that the acid penetrates the cuticle of a predatory arthropod; it thus adds substantially to the effectiveness of the acetic acid as an aversive agent.

*Chemical defences can be acquired from a prey's environment*

Some organisms feed on elements that contain noxious materials and gain a degree of protection from predation by the presence of defensive compounds in the gut. A good example is the grasshopper *Schistocerca emarginata*, whose toxicity is gut-content-mediated (Sword, 1999, 2002; Sword et al., 2000). Many other prey, however, actively sequester defensive substances from food items and accumulate them in their bodies (Bowers, 1992). Sequestration may involve movement of a chemical substance to some localized storage organ, and may perhaps involve biotransformation of the molecule. Sequestration is well studied in a number of arthropod systems (e.g. Pasteels et al., 1983; Bowers, 1984, 1986, 1990, 1992, 1999; Pasteels, 1993; Trigo, 2000; Nishida, 2002) and is often associated with dietary specialisation by prey. However, it is not always clear whether dietary specialization is a cause or a consequence of the defensive qualities that a food can confer (Bernays and Graham, 1988). It should be noted in addition that biosynthesis and sequestration are not mutually exclusive; many prey use both means of generating a defence (see e.g. Dobler et al., 1996).

Sequestration has been particularly well studied in a number of terrestrial arthropods, including Lepidoptera and Coleoptera. Nishida recently identified a number of substances sequestered by Lepidoptera (and see also Trigo, 2000); we summarize these in Table 5.1, adding some information

**Table 5.1** Description of some plant chemicals sequestered by Lepidoptera, based on text in Nishida (2002) (with some consideration to Coleoptera). The table describes some examples of organisms that sequester the substance and evidence that it can deter a range of predators

| Class of substance | Found in | Evidence of predator deterrence |
|---|---|---|
| Cardenolides (often as cardiac glycosides) | Lepidoptera (Nishida, 1994,1995) Coleoptera (Pasteels *et al*, 1986) | Strong evidence of anti-predator effects e.g. birds, spiders (Brower,1969; 1984 Pasteels *et al.*, 1990; Theodoratus & Bowers, 1999) |
| Grayanoids | Species of diurnal moth in the genus *Arichanna* | Birds and reptiles (Nishida, 1994,1995; Nishida, 2002) |
| Iridoid glycosides | Lepidoptera (eg checkerspot, *Euphydryas* spp, and buckeye butterflies: *Precis (Juonia coenia)* | Birds (Bowers, 1990), spiders (Theodoratus and Bowers, 1999), ants (Dyer & Bowers, 1996) |
| Pyrrolizidine alkaloids | Widespread use in Lepidoptera esp. danaine, ithomiines and arctiid moths (Nishida,1994,1995), Coleoptera (Pasteels *et al.*,1986; Pasteels *et al.*,1990) | Invertebrate predators (Nishida, 2002). Inconsistent evidence with vertebrates (Rowell-Rahier *et al.*, 1995; Cardoso, 1997), May be ineffective with birds (Ritland, 1991; Yosef, *et al.*,1996) |
| Aristolochic acids | Found in pipevine butterflies (Aristolochicacae) | Can act as feeding deterrent (Nishida, 2002) |
| Cyanoglycosides | Widespread (may be biosynthesis) (Blum, 1981). Lepidoptera, esp. zygaenids and nymphalids | Toxicity and capacity for predator deterrence is well established (Muhtasib & Evans, 1987; Nahrstedt, 1988) |
| Cycasin | Lycaenid butterflies (*Eumaeus* and, *Taenaris* spp.) and an arctiid moth (*Seirarctia echo*) | Deters ants and unpalatable to the Gray Jay (Nishida, 2002) |
| Glucosinolates | Some lepidoptera, e.g. cabbage whites *Pieris* spp. | Deters ants (Müller *et al*, 2002) |

about sequestration in Coleoptera. As Whitman et al. (1990) point out, merely because a substance is present in chemically defended animals does not mean that the substance necessarily functions as a secondary defence. There is therefore a large and growing literature that evaluates and demonstrates the effectiveness of sequestered substances of predator deterrence (Table 5.1). Many marine organisms are similarly known to sequester defensive substances (see especially Pawlik, 1993; Hay, 1996; McClintock et al., 2001). Secondary metabolites from seaweeds and benthic invertebrates are utilised in this manner. For instance, the tube-building amphipod *Ampithoe longimana* selectively lives on and consumes a brown algae (*Dictyota menstrualis*). This seaweed produces diterpene alcohols and renders the invertebrates unacceptable to fish predators (Duffy and Hay, 1994; Hay, 1996). Similarly, many marine molluscs such as nematocysts utilize food items for the sequestration of defensive substances (Pawlik, 1993) and organelles.

In other marine examples chemical defence can be acquired without ingesting the defensive substance. Thus the Carribbean amphipod (*Pseudamphithoides incurvaria*) builds a bivalved domicile from a chemically defended *Dictyota* seaweed. This domicile protects the amphipod from fish predators and without it the amphipods are quickly eaten (Hay, 1996). Even more exotically, the edible amphipod *Hyperiella dilatata* captures the noxious sea butterfly (the pteropod *Clione antarctica*; Bryan et al., 1995) attaches it to its back and, in a form of 'shared doom', ties its avoidance from predation to that of the pteropod (McClintock and Janssen, 1990).

*Chemical defences can be biosynthesized*

In contrast, many prey synthesize their own chemical defences *de novo*. Glandular biosynthesis is, as Whitman et al. (1990) describe, the most common form of defence in terrestrial arthropods. It is especially well developed in the Hymenoptera, Opiliones, Diplopoda, Isoptera, Hemiptera, and Coleoptera (and see extensive reviews in Blum, 1981; Pasteels et al., 1983). The uses of chemical defences are highly diverse within groups such as arthropods, ranging from explosive secretion (such as the bombardier beetle, e.g. Eisner and Aneshansley, 1999) to enteric discharge, as when grasshoppers 'spit tobacco' regurgitating partly digested food, salivary and digestive secretions, perhaps fortified with defensive substances (Blum, 1981; Whitman et al., 1990). Alternatively, faeces can itself be used as a defensive substance forcibly discharged at prey, perhaps with fortifying repellents, or fixed to some part of the body; some chrysomelid larvae attach their faecal products to their backs using them as shields (Whitman et al., 1990; Morton and Vencl, 1998; Vencl et al., 1999). Prey can autohaemmorhage (or reflex-bleed), in which blood perhaps with additional repellent substances is exuded, often from intersegmental membranes (e.g. Holloway et al., 1991; Grill and Moore, 1998; Müller et al., 2001). However, the defensive mechanism by which reflex-bleeding functions is still unclear in groups such as froghoppers (Homoptera: Cercopidae; Peck, 2000). Spines with noxious chemicals and stinging, irritating (urticating) hairs are common, and well studied in lepidoptera larvae, as are noxious volatiles (e.g. osmertia in Papilionidae caterpillars; Bowers, 1993) and defensively functioning salivary and mandibular glands (e.g. whip scorpion, Eisner et al., 1961). In addition, biosynthesis is known and well studied in some marine invertebrates (Pawlik, 1993).

*Symbiosis explains some cases of chemical defence*

An alternative method by which chemical defence is procured is by playing host to an appropriate symbiont (Mebs, 2001). Tetrodo-toxin (TTX), for example, is a potent neurotoxin that is present in marine and terrestrial animals and can be synthesized by bacteria (Mebs, 2001). There is some evidence that the presence of TTX in some marine organisms is conferred by a bacterial symbiont. For instance, Mebs (2001: 93) reports that 'pufferfish become toxic after "infection" with TTX producing bacteria' (see Yotsu, 1987). However, the mechanism by which TTX is produced in newts and other amphibia remains obscure (though see Daly et al., 1997). In marine systems it has been suggested that microbial symbionts are important in a number of systems including sponges, bryozoans and colonial tunicates (see discussions in Lindquist, 2002). Species in these groups harbour microbial symbionts and produce secondary metabolites similar to known microbial metabolites (Kobayashi and Ishibashi, 1993; Lindquist and Hay, 1996; Mebs, 2001). However, the function of symbionts is not always certain; it may be to prevent pathogenic microbes rather than to deter predation (Gil-Turnes et al., 1989). In a fascinating example, Lindquist (2002) describes evidence that a marine isopod that develops in maternal brood pouches gains a microbial cyanobaterial symbiont from its mother, which colours it bright red and deters predators.

*Reduced nutritional value is a further chemical defence*

Reduced nutritional quality may be common as a component of defence in organisms that live permanently in the plankton (the macro-holotplankton). Bullard and Hay (2002) examining cnidarians, ctenophores, and other members of the macroplankton found that many may utilise low nutritional quality (measured as mass of soluble protein per unit volume) and/or nematocyst cells that discharge coiled venomous threads at predators (and prey) to defend themselves. In many cases, prey were apparently defended by the possession of low nutritional status. When organisms surveyed had higher protein levels they were accompanied by additional defences such as higher densities of nematocysts, floating on the sea surface away from predators, or nocturnal activity. Similarly, Penney (2002) found that nudibranchs, which lack protective shells, contained less than half the organic mass of sympatric shelled snails, making up the difference with water and/or raised levels of ash. In feeding trials, crabs took considerably more artificial pellets with 'prosobranch' nutritional characteristics than with 'nudibranch' nutritional characteristics. Hence, though we may normally think of chemical defence as synonymous with the presence of toxins, some chemical defences include the absence of nutrients and the presence of other non-nutrients.

*Chemical defences can exhibit high levels of intraspecific variation*

A final point about chemical defences is that they frequently exhibit a high level of intraspecific variation in marine (e.g. Bryan et al., 1995; Hay, 1996; Lindquist, 2002) and terrestrial organisms (Brower et al., 1968; Cohen, 1985; Holloway et al., 1991; Bowers, 1992; Eggenberger et al., 1992). Variation within a species means that some individuals are less well defended than others (see Chapter 12) and Bowers (1992) considers this situation to be 'probably very common, if not ubiquitous, in unpalatable insects'.

In animals that sequester their defensive chemicals or otherwise acquire them from the environment, intraspecific variation in defensive chemicals can be explained by variation in food materials or some other external component. However, even when chemical defences are acquired from external sources, genetic variation may explain some of the differences between defended individuals (Müller et al., 2003).Variation in defensive levels of insects that biosynthesize toxins has, likewise, been shown to have a genetic component in some species. Eggenberger and Rowell-Rahier (1992) demonstrated that in the Chrysomelid beetle, *Oreina gloriosa*, there is a substantial genetic component to variation in the proportions of components in the animal's defensive secretions (average heritability circa 0.5–0.58) and in the concentrations of secretions (average heritability, 0.45). In addition, chemical defences can vary according to sex, stage of life cycle, and age (sometimes increasing with age, e.g. Eggenberger and Rowell-Rahier, 1993), sometimes decreasing (Brower and Moffitt, 1974; Alonso-Mejia and Brower, 1994).

### 5.3.2 Are chemical defences costly?

Having very briefly reviewed a number of characteristics of chemical defences, we now turn to the question of costs. It is relatively easy to identify sources of proximate physiological costs of a defence that could reduce an animal's fitness. In biosynthesis, there are obvious costs of (1) direct gene expression of a toxin, or (2) of enzyme production necessary for complex metabolic pathways that generate defensive secondary metabolites. Furthermore, costs are incurred in (3) generating and (4) maintaining facilities for storing toxins (e.g. in glands). Finally, (5) prevention of, and susceptibility to, autotoxicity may incur further costs. Mebs (2001) points out that the prevention of autotoxicity may 'include drastic changes at the molecular level at receptor or ion-channel sites', which may affect the effectiveness of other physiological systems.

In sequestration, the cost of biosynthesis itself is replaced by one or more of the costs of: (a) finding appropriate food materials; (b) transportation of a molecule across the gut, which may require synthesis of carrier molecules (see review in Weller et al., 1999); (c) concentration and perhaps biotransformation of defensive compounds, for example, in the blood (or haemolymph, perhaps against osmotic gradients; Bowers, 1992); (d) in some, especially insect, species there may be further costs from a need to protect pupal or adult forms from defensive molecules acquired by larvae; and finally, (e) sequestration may, additionally, incur environmental costs if, for example, it exposes prey to variable levels of intraspecific competition for materials essential for generation of the defence. Thus, it is often not clear that acquisition by sequestration or symbiosis is necessarily cheaper than direct biosynthesis. As Dobler (2001) recently pointed out, the rules of whether to sequester or not, on economic and other grounds, are still poorly understood.

Other means of acquiring unprofitability to predators may also be costly. Thus, collection of defensive substances from symbionts may incur costs of accommodating the symbiotic organisms (Mebs, 2001) and there may be physiological or design costs to a prey associated with reducing nutritional components of its anatomy (Penney, 2002).

*Empirical evidence for costs of chemical defences*

Though they are easy to describe in principle, costs are often much more difficult to detect and measure. Most obviously, and perhaps most easily, researchers have evaluated the trade-offs associated with allocation costs; looking, for instance, for a reduction in growth, survivorship, or fecundity in relation to the possession or induction of a defence. Alternatively, some authors demonstrate opportunity or design costs associated with chemical defence. We discuss examples of these studies below (and see Table 5.2 for a selection of studies that evaluate the costs of defences in insects).

If defences are secreted in response to a threat, then repeated triggering of such defences may incur costs measurable as changes to other phenotypes. Grill and Moore (1998) investigated this possibility in the Asian ladybeetle *Harmonia axyridis*, which produces reflex-blood when provoked. Repeated bleeding in larvae retarded subsequent development and reduced the final size of the adult, indicating that reflex-bleeding is a costly activity.

Chrysomelid leaf beetles have a diverse range of chemical defences (Pasteels, 1993). In some species of the genus *Oreina* specialized glands on the pronotum of the adult secrete defensive cardenolides (biosynthesized *de novo*) and/or pyrrolizidine alkaloids (sequestered from host plants; Dobler and Rowell-Rahier, 1994). However, the larvae do not have defensive glands and store defensive substances in their bodies and their haemolymph (Dobler and Rowell-Rahier, 1994) . In order to investigate the effects of sequestration on growth, Dobler and Rowell-Rahier (1994) raised larvae of *Oreina elongata* on one of two species of natural host plants that did or did not contain pyrrolizidine alkaloids. They found that those raised on the alkaloid-free hostplant gained more weight than those raised on the pyrrolizidine alkaloid-containing plant, indicating a trade-off between chemical defence and growth. An interesting aspect of the results in this study is that the individuals of *Oreina elongata* were selected from two populations, one of which is accustomed to feeding on the pyrrolizidine alkaloid-containing host, the other not. When compared, animals from the former (alkaloid familiar) population had growth rates that were less severely retarded than animals from populations that were unfamiliar with the alkaloid. This suggests perhaps that when acquisition is costly, selection can act to lower reductions in fitness (Leimar et al., 1986).

Although Dobler and Rowell-Rahier's study provides clear evidence for a trade-off between acquisition of a defensive chemical and growth, the authors point out that the alkaloid itself may not have been the direct cause of retarded growth; other differences between the host plants, such as

**Table 5.2** A selection of insect species investigated for the cost of acquiring a chemical defence, manifest through a trade-off with some other components of fitness, most often growth. Developed from Bowers (1992)

| Insect species | Life Stage | Chemical defence compounds | Evidence of a cost | Reference |
|---|---|---|---|---|
| ■ *Danaus plexippus* (Lepidoptera: Nymphalidae) | Adult | Cardenolides | Yes | Cohen, 1985 |
| □ *Danaus plexippus* (Lepidoptera: Nymphalidae) | Larva | Cardenolides | Yes | Zalucki *et al*, 2001 |
| □ *Danaus gilippus* (Lepidoptera: Nymphalidae) | Adult | Cardenolides | No | Cohen, 1985 |
| □ *Euphydryas anicia* Lepidoptera: Nymphalidae) | Adult | Iridoid glycosides | No | Bowers, 1988 |
| □ *Junonia coenia* (Lepidoptera: Nymphalidae) | Larva-pupa | Iridoid glycosides | Yes | Bowers & Collinge, 1992 |
| □ *Junonia coenia* (Lepidoptera: Nymphalidae) | Larva | Iridoid glycosides | Yes | Camara, 1997 |
| O *Phratora tibialis* (Coleoptera: Chrysomelidae) | Larva-adult | Methylcyclopentanoid monoterpenes | Yes | Rowell-Rahier & Pasteels, 1986 |
| O *Phratora vitellinae* (Coleoptera: Chrysomelidae) | Larva-adult | Salicyaldehyde | No | Rowell-Rahier & Pasteels, 1986 |
| O *Plagiodera versicolora* (Coleoptera: Chrysomelidae) | Larva-adult | Methylcyclopentanoid monoterpenes | Yes | Rowell-Rahier & Pasteels, 1986 |
| O *Chrysomela 20-punctata* (Coleoptera: Chrysomelidae) | Larva-adult | Salicyaldehyde and unknown compounds | Varies* | Rowell-Rahier & Pasteels, 1986 |
| O Oreina elongata (Coleoptera: Chrysomelidae) | Larva | Pyrrolizidine alkaloids | Yes | Dobler & Rowell-Rahier 1994 |
| O *Chrysomela confluens* (Coleoptera: Chrysomelidae) | Larva | Salicyaldehyde | No | Kearsley & Whitham, 1992 |
| O *Harmonia axyridis* (Coleoptera: Coccinellidae) | Larva-adult | Unknown compounds | Yes | Grill & Moore, 1998 |
| □ *Neodiprion sertifer* (Hymenoptera: Diprionidae) | Larva | Pine resin acids | Yes | Bjorkman & Larsson, 1991 |

*Costs in *Chrysomela 20-punctata* were shown when the anima was fed on *Salix caprea* but not *S. nigricans*

O Defensive secretions were stimulated during larval stages and growth rates were monitored and compared to those of an unstimulated control group.
□ Animals were fed on hostplants or equivalent food that did/did not contain a defensive metabolite that the animal sequesters and their growth was then monitored.
■ The relationship between adult size and levels of defensive substances were evaluated in wild-caught animals.

water, nitrogen, carbohydrates and other alleleochemicals may cause a difference in growth rates. This point applies equally in the converse situation, in which larvae grow more quickly on defended than on undefended host plants. Smith (1977) found that larvae of the butterfly *Danaus chrysippus* were significantly larger when grown on milkweed hostplants that contained defensive glycosides that this species sequesters as a defensive agent, than when reared on an asclepiad plant without toxins. Smith concluded that his results demonstrated that the use of host plants with defensive secondary metabolites conferred a major benefit on the larvae and seemed to incur no extra cost. However, it could well be the case that the glycosides did incur some level of metabolic costs, but that these were more than compensated for by other qualities present in the cardenolide-containing plants.

Such is the difficulty in designing and performing experiments that evaluate whether the defensive agent in a host plant itself reduces growth in herbivorous insects that it has only very recently been shown that in the monarch butterfly, *Danaus plexippus*, cardiac glycosides in a milkweed host plant actually incur a physiological cost that makes a causal contribution to diminished larval growth rates (Zalucki et al., 2001*a*). This is despite the fact that this butterfly has been examined as a paradigm case study in chemical ecology for decades (e.g. Brower et al., 1968; Brower, 1969; Brower and Moffitt, 1974; Brower and Glazier, 1975; Cohen, 1985). Even here the picture is complex; Zalucki et al. (2001*a*) concluded that cardiac glycosides were not the only components of the hostplant that retarded larval growth (see also Zalucki et al., 2001*b*). One of the problems in demonstrating costs of chemical defences may be that it is much more difficult to use neonate larvae than late-instar larvae in this kind of experiment. Yet, as Camara (1997) pointed out, very young animals may be most susceptible to these toxins and, in addition, detrimental effects to young animals may have the biggest effects on overall growth. The complexities of establishing the existence of physiological costs of sequestration are discussed in an excellent review by Bowers (1992) and we recommend this article for a detailed discussion of the evaluation of costs of chemical defences.

*Cost-free chemical defences?*

It remains possible that in a number of situations, chemical defences can be actually, or almost, cost-free. Thus, Table 5.2 shows a number of cases in which researchers could find no evidence that a chemical defence reduced growth or the quality of other traits. Net costs may be avoided particularly if the production of a defensive substance is associated with the release of energy-rich molecules (Rowell-Rahier and Pasteels, 1986). Notably, Kearsley and Whithan (1992) examined the economics of defence in larvae of a Chrysomelid beetle (*Chrysomela confluens*), which extracts salicin (a phenol glycoside) from its host plant and transports it for storage in two dorsal glands. Salicin is metabolized in these glands, producing glucose and salicylaldehyde. The deterrent effects of salicylaldehyde were shown by Kearsley and Whithan. However, of equal interest, the glucose product can supply significant quantities of chemical energy; up to one-third of the animal's calorific needs if its glands are emptied and replenished daily (Rowell-Rahier et al., 1995).

When Kearsley and Whithan examined the economics of defence in *Chrysomela confluens*, no evidence for net costs were found: there were no significant differences in developmental rate, adult dry mass or feeding rate between animals that were 'milked' daily and control sibs that were not milked. Kearsley and Whithan (1992) concluded that the beetle gains a 'cost-free' defence because the source of the chemical defence is also an energy source. However, thorough and ambitious though this study is, the authors themselves point out that they could not evaluate the costs of creating, maintaining and using glandular structures and musculature necessary for the defence to be effective. Nor could they evaluate whether adult survival is affected by high levels of defensive larval secretion. However, the possibility that prey gain energy from their sequestrates as well as chemical defence (see, e.g. Blum, 1981) does mean that defences themselves could be very cheap, if not actually 'for free'.

*Indirect evidence of costs*

An alternative, less direct strategy is to examine the energetic or other physiological costs of the defence and to then evaluate whether they can be sufficiently large to make fitness costs likely. Hetz and Slobodchikoff (1982) directly quantified the proximate cost of defence, measuring the energy content of defensive secretions in an aposematic beetle (*Eleodes obscura*) and estimated the costs of its defence relative to a female's annual investment in her eggs. The beetle stores a toxic mixture (of benzoquinones, 1-alkenes, and octanoic acid in an aqueous solution of glucose) in a pair of reservoirs. An 'average' beetle was found to carry circa 21.6 J of chemical energy in the fluid in its reservoirs at any one time and Hetz and Slobodchikoff estimate that one-fourth of the contents of the reservoir would be used during an attack by a predator. It was estimated that over a year chemical defence costs the beetle circa 18 per cent of the energy that is put into eggs by females over the same time period. Consistent with the idea that chemical defences are costly, Holloway et al. (1991) demonstrated that reflex-bleeding in the seven-spot ladybird (*Coccinella septumpunctata*) can utilize as much as 20 per cent of the animal's fresh body weight, albeit in exceptional circumstances. Conservation of secretion also suggests costs of chemical defence. Here animals are very reluctant to secrete defensive substances and only do so after repeated threat from a predator. Secretion may be localized so that only the region of the animal that is under direct threat produces the defensive chemical (Blum, 1981; Eisner and Aneshansley, 1999) and prey may attempt to reclaim part or all of their secretions when danger has passed (Blum, 1981; Whitman et al., 1990).

## 5.4 Costs, benefits, and forms of defence

We have highlighted a number of important characteristics of secondary defences and evaluated the possibility that defences bring costs as well as benefits. Because of their particular association with conspicuous warning displays we placed special emphasis on chemical defences. In many, but by no means all cases, costs and benefits of chemical defences have clearly been established. One reason for the prevalence of chemical defences may be that they present a favourable cost/benefit ratio; highly effective predator deterrence for a small expenditure relative to other forms of defence. If this were generally true we would expect to see prey organisms favouring chemical defence over other forms such as escape behaviours or the creation of sharp spines, because the ratio of protection per unit invested in a defence is larger for chemical than some other defences. In tropical Lepidoptera, as we saw in Section 5.2.1, chemical defences alongside common warning displays appear to reduce the need for investment in flight muscles, thereby allowing more resources to be allocated to feeding and reproduction.

A further benefit from investment in toxins, rather than some other form of defence is that the gain in protection may be markedly increased if there are other chemically defended individuals present, even if these individuals belong to different species (see discussion of synergistic traits in Leimar and Tuomi, 1998). Predators may learn aversions to prey more quickly if they have higher toxin loads, especially if toxins from different prey species act synergistically to harm their predators. Furthermore, if individuals from one prey species start to poison a predator, it may lead to higher levels of aversion to other species that may otherwise be used as a food source. The gain from such mutualistic interactions may be higher than for other forms of defence, such as rapid escape, or tough cuticles, because toxins can cause higher and more persistent levels of harm to predators. The cost/benefit ratio from investment in chemical defence therefore becomes more favourable for a prey if it is surrounded by other defended individuals of its own and other species. Increasing the benefits of toxicity without extra cost in this way has been termed a Müllerian mutualism (Turner and Speed, 2001) because it describes a mutualism implied in Müllerian mimicry (see Chapter 9 and Leimar and Tuomi, 1998) without necessitating resemblance in warning display between defended species. One important prediction that follows from a recognition of Müllerian mutualism is that cost/benefit ratios of chemical defences will be quite variable between localities within populations according to the presence of other chemically defended species.

Variation in cost/benefit ratios may therefore be important in explaining why high levels of intraspecific variation are observed within populations of chemically defended species (see Chapter 12). If the cost/benefit ratios vary between individuals then we might expect to see variation in investment in chemical defences that could help explain why such variation is common. Cost/benefit ratios probably vary with the abundance of defended prey and with predator pressure within localities. Similarly, individuals within a population may encounter wide variations in resources needed for a variety of fitness components and may therefore have subtly different optimal investment strategies, one result of which is variation in secondary defence. Thus, as we postulated in Section 5.1, costs and benefits of defences probably do play an important role in determining the form that defences take both within and between species.

## 5.5 The evolution of defences

### 5.5.1 Evolutionary pathways

Recent advances in techniques of molecular phylogeny means that we are beginning to learn about evolutionary pathways that may be followed in the evolution of defences, especially chemical defences. An interesting feature of biosynthesis in marine and terrestrial environments is that some animals appear to synthesise defensive molecules very like those they ingest from their food materials. This is known at least in Lepidoptera (Brown et al., 1991) and marine molluscs (Gavagnin et al., 2001*a*,*b*).

Why should a prey animal biosynthesize what it has already? One explanation is that the prey animal must first evolve the capacity to break down a toxic substance before it can utilize some material as a source of nutrition; having achieved that, a pre-adaptive capacity for biosynthesis reconstructs the defensive molecule or something very similar to it. The benefits to the prey animal are likely to include reduced reliance on particular food items for defence, in other words *emancipation from a food source*, especially if the availability of that food source is erratic or otherwise unreliable. This area has been well reviewed and discussed by a number of authors in relation to Lepidoptera (Brown et al., 1991; Nishida, 2002); and in nudibranch molluscs (Cimino and Ghiselin, 1998, 1999; Gavagnin et al., 2001*b*). With the nudibranchs, Cimino and Ghiselin (1999) discuss this question in some depth, suggesting that certain animals may be pre-adapted to evolve biosynthesis of molecules that exist in their food materials anyway. Based on extensive phylogenetic considerations Cimino and Ghiselin (1999) argue that a common evolutionary pathway in chemically defended nudibranchs may start with a capacity to deal with food toxins, which develops into defensive sequestration of the toxins and in some species is followed by emancipation from a food source by biosynthesis.

Historical perspectives are also beginning to illuminate the ecological events that may facilitate diversity in chemical defences. For example, Termonia et al. (2002), recently argued that the evolution of dual or multiple defences may open an important evolutionary pathway leading to diversification. They argue that in phytophagous species the creation of dual defences frees animals to move between host plants and 'increases the evolvability of host affiliation and of chemical protection' (Termonia et al., 2002). In a phylogenetic analysis of *Platyphora* leaf beetles, Termonia et al. (2002) found that all species studied sequester amyrins from their hostplants and convert them to defensive saponins. A 'P.A. clade' within this group additionally sequester pyrrolizidine alkaloids from their host-plants, giving them a dual defence against enemies. The possession of a dual defence appears to have enabled continued diversifiction in the members of the clade; some moving to new host plants to gain pyrrolizidine alkaloids, others returning to the ancestral state of 'amyrin-only' sequestration. A similar story has been found in other groups studied despite fundamental differences in chemistry (see data and references in Termonia et al., 2001), such that Termonia et al. (2002) suggest 'that such transitions among single-defence metabolisms via dual chemical strategy is a general pattern in the evolution of chemical defence and host-plant affiliation'.

### 5.5.2 Theoretical approaches to the evolution of defences

Since secondary defences lead more or less directly to predator avoidance, the principles of their evolution are at least superficially straightforward. Indeed, as R. A. Fisher (1930) noted, certain types of defence 'increase the chance of life of the individuals in which they develop'. Thus, if secondary defences cause predators to break off attacks before serious damage is done to a prey, they may evolve simply because defended individuals have better chances of surviving attacks than undefended individuals. Such defences are sufficient to enhance survival without additional ecological conditions being necessary. Fisher's examples of such 'sufficient' defences that could aid individual survival included stings, disagreeable secretions and odours, and tough flexible bodies. There is growing evidence that chemical defence in aquatic (Lindquist and Hay, 1996) and terrestrial prey (Wiklund and Järvi, 1982; Sillén-Tullberg, 1985*a*; Wiklund and Sillén-Tullberg, 1985) can indeed secure high levels of post-attack survival. Furthermore this argument would apply to any effective and appropriately induced morphological as well as chemical secondary defence, so long as the cost/benefit ratio were favourable. Clearly palatability (in the sense of gustatory stimulation) is very important: the bitterness of some toxins (such as iridoid glycosides) could help to make the defence sufficient to raise the chances that an individual survives attack.

However, there are at least two reasons why explaining the evolution of chemical defences is more complex than this. First, if defences are costly it is not clear how they can evolve if without them most prey are already well protected by cryspsis. This problem has been formally modelled and we come to it later in the chapter.

A second reason that explains the evolution of defences is not straightforward, and is, as Fisher noted, that some kinds of defence actually do little to protect the individual animal under attack: they are aversive to predators, but are insufficiently effective to save a prey's life. Fisher saw that 'nauseuous flavours' or 'distastefulness' presented a special problem because individuals may suffer mortal injury before a predator experienced an aversive, nauseous, or emetic reaction from which it can learn. Fisher wrote that:

> since any individual tasted would seem almost bound to perish, it is difficult to perceive how individual increments of the distasteful quality beyond the average level of the species, could confer any individual advantage (Fisher, 1930: 159).

However, he saw that when prey (specifically insect larvae) exist in family groups the phenomenon that we now know as kin selected altruism could explain the evolution of such 'nauseous' defences.

> with gregarious larvae the effect will certainly be to give the increased protection especially to one particular group of larvae, probably brothers and sisters of the larvae attacked. The selective potency of the avoidance of brothers will of course be only half as great as if the individual itself were protected: against this is to be set the fact that it applies to the whole of a possibly numerous brood. (Fisher, 1930: 159)

Fisher's insight, that evolution would favour the death of one individual that raised the survival chances of many siblings, is obviously of considerable historic importance to evolutionary biology. However, the importance of kin selection in the initial evolution of defence is more controversial. One problem with Fisher's explanation is that the current state of defence in any species may not reflect its ancestral state. As defended prey become common and are increasingly avoided by predators, optimal investment in defences may decrease (Leimar et al., 1986), in which case some of the species Fisher had in mind may actually have been highly protected at the outset and subsequently reduced investment in secondary defences. Some light is likely to be shed on this problem by the recent growth of phylogenetic studies in this area.

A second problem with Fisher's argument is that aggregation may be unnecessarily risky for edible prey whose main line of protection is crypsis. When grouped together, prey may be more conspicuous and quickly eaten by a hungry predator. Thus, unless such tight aggregations are well hidden (such as being positioned under a leaf) or predators satiate very quickly, aggregation is not readily expected in prey species prior to the evolution of the secondary defence itself. Aggregation may therefore follow the evolution of a defence, rather than precede it. When defences evolve, aggregation may have a number of benefits including dilution of predatory risk (see Box 5.1).

## Box 1 When and why should defended prey aggregate?

First, aggregation and egg clustering in terrestrial arthropods may suit some purpose other than enhancing prey defences (a variety of hypotheses exist here, including: enhanced exploitation of a host plant, decreased energy costs to ovipositing females, prevention of egg dessication, enhanced thermoregulatory efficiency, see discussions in e.g. Fisher, 1930; Clark and Faeth, 1997, 1998; Denno and Benrey, 1997; Hunter, 2000); here defences facilitate aggregation (Leimar et al., 1986), but aggregation may not in itself necessarily enhance protection.

Alternatively, aggregations of defended prey may bring forward the point at which predators cease attacks, or satiate, on a defended form (because of a variety of different proximate defensive mechanisms) (Leimar et al., 1986; Mallet and Joron, 1999; Vulinec 1990). Aggregations may therefore reduce mortality in a defended species regardless of the relatedness of the individuals in an aggregation. In this context, the questions of when and how large to form an aggregation have been modelled by Sillén-Tullberg and Leimar (1988). Developing the work of Turner and Pitcher (1986), Sillén-Tullberg and Leimar's model describes the effect of aggregation on mortality. It assumes that prey are not especially mobile and are sufficiently small that predators can eat several or more members of a group in quick succession. Sillén-Tullberg and Leimar modelled the probability that a group of defended animals would be detected by a predator as

$$R(n) = A(1 - e^{-n/n'})$$

where $R(n)$ is the rate of detection of group size $n$, $n'$ a threshold group size at which $R$ begins to level off, and $A$ the asymptotic level toward which probability of detection moves as $n$ increases (see also: Riipi et al., 2001). The proportion of the group that is attacked before the aversive experience prevents the predator attacking further is given by

$$P(n) = [1 - (1 - h)^n]/nh$$

where $h$ is the probability on each attack that the predator becomes inhibited from further attacks.

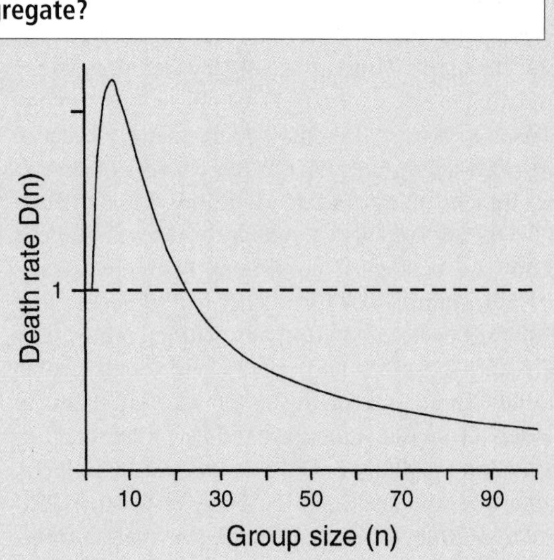


**Figure 5.1** The curve shows the rate of death from predation $D(n)$ as a function of the size of an aggregation. $D(n)$ decreases below the death rate for a single, solitary individual at $n = 23$. Redrawn from Sillén-Tullberg and Leimar (1988).

Sillén-Tullberg and Leimar showed that in this model death rate (the product of $R(n) \cdot P(n)$) increases with group size, up to some critical point, after which it falls continuously toward zero. Thus there is a range of group sizes between two and some threshold of group viability, within which individuals in groups suffer higher death rates than solitary individuals (Figure 5.1). Beyond this viability threshold (where a group of particular size and solitary individuals are equally fit) mortality reduces in a non-linear fashion. Once gregariousness is favoured, ever larger group sizes further enhance survival. The model therefore predicts a bimodal distribution of prey; solitary or in large groups, depending on the values of parameters, especially $h$ and $n'$. Sillén-Tullberg and Leimar suggest that this prediction is consistent with strongly bimodal distributions found between species of Pierine, Nymphaline and other Lepidoptera. The Sillén-Tullberg and Leimar model is important because it formally predicts and quantifies the advantages of aggregation, crucially without the need for any relatedness between individuals.

However, the role of kin selection in the evolution of defence should not be rejected for a number of reasons. First, it is possible to envisage plausible instances in which some level of kin association exists within undefended, edible populations. Tight spatial proximity is in fact neither necessary nor sufficient for Fisher's model to work. Fisher himself wrote:

The institution of well defined feeding territories among many birds in the breeding season makes it possible to extend the effect produced on gregarious larvae to other cases in which, while not gregarious in the sense of swarming on a plant are yet distributed within an area that usually falls within the feeding territory of a single pair. (Fisher, 1930: 159)

The Fisherian kin selection model therefore works if: (i) regardless of the physical distance between prey, individuals distribute themselves so that nearest neighbours are likely to be family members; and (ii) part or all of a family grouping must encompasses a locality containing only a few predators which can quickly learn to avoid the defended forms.

A second important reason that Fisher's explanation may in fact pertain is that chemicals that provide protection from predation may have other primary uses that favour their initial evolution: sexual communication mechanisms (see, e.g. Weller et al., 1999), subjugation of prey, antiparasite/parasitoid effects, antimicrobial properties, wetting or waterproofing agents, intraspecific conflict, etc. (see Whitman et al., 1990). Similarly, certain morphological traits such as spines or tough skins may have primary functions other than predator deterrence (such as interspecific aggression, protection from abrasion, etc.) but may also aid in survival from predation.

If such traits arose primarily for these or other reasons, and provided some defence against predators (i.e. they were 'proto-secondary defences'), then two things may follow. (1) There may be further selection for refinement of the phenotype as a defensive trait by individual selection if refinements increased the probability that a prey survived attacks. (2) Aggregation as an anti-predator device may become possible and profitable where it was dangerous before. If aggregation were favoured then kin selection could facilitate the refinement of the trait into a more efficient and effective secondary defence. This argument would also apply to nascent secondary defences that result from food items that provide some protection by residing in a prey's gut (Bowers, 1992). The defences may be sufficiently aversive to predators to make aggregation beneficial as a means of diluting predator threats. Aggregation may then enable the evolution of refinements of a defensive system by kin selection, so that the complex adaptation necessary for effective sequestration of defensive secondary metabolites can evolve with or without individual selection.

### 5.5.3 Formal modelling of the evolution of defences

If secondary defences commonly preceded the evolution of prey aggregation and conspicuousness then we have to be able to explain how defences initially evolved in relatively cryptic prey. It is easy to see how this would follow when defences incur no costs and enhance individual survival. However, it is a less straightforward question when defences incur costs. A surprisingly small number of theoretical treatments consider this problem explicitly in relation to constituitive animal anti-predator defences. In the most important published model, Leimar et al. (1986) constructed the components of an individual predator's psychology and behaviour (varied learning rates and sensory generalization in order to calculate attack probabilities) and the properties of individual prey (effectiveness of unprofitability in terms of individual survival and effects on learning rates, costs of a defence, degree of conspicuousness) and prey populations (size and degree of clustering of prey as a proxy for kin selection). Leimar et al. combine these components into a model that calculates the fitness of prey morphs in which parameters such as conspicuousness and effectiveness of defences vary continuously. For a specified set of parameters they deduced the evolutionarily stable strategy (ESS) values for pertinent variables such as unprofitability and conspicuousness. Leimar et al. demonstrated the intuitive finding that for costly defences to evolve there must be some level of individual and/or kin mediated advantage.

The model predicts that there can be a single monotypic ESS for defence, provided that the degree of exposure of prey to predators is large enough, that is, that crypsis is not extremely effective. This, again, confirms the intuition that costly toxins should not be favoured in prey that are already very well protected by crypsis. An important point in this model is that small changes in parameter values (e.g. conspicuousness, kin grouping) cause similarly small, gradual changes in optimal levels of defence (or unprofitability). Hence, Leimar et al. conclude that a major transition from edible to defended status may be preceded by some major change in conditions, removing the effectiveness of crypsis. Given that this is one of the few models of the evolution of constituitive anti-predator defences in animals it is worth listing a set of further questions that in our opinion should be addressed in future theoretical treatments.

First, the prediction of a single optimal level of defence in the ESS model devised by Leimar et al. (1986) fits poorly with the observation of high levels of variation found within populations. Thus one important direction for future work is the extension of Leimar et al.'s (1986) model to include intraspecific variation in defense which may occur because of feeding constraints, physiological constraints (such as body size and age; Eggenberger et al., 1992; Eggenberger and Rowell-Rahier, 1993), recency of usage of a secretion (Blum, 1981; Holloway et al., 1991), passage of substances through the gut, or genetic constraints (such as linkage) that make variation in defensive levels between individuals essential. Similarly, it is important to know how and to what extent the prediction of one single optimal level of defence changes when defences are not strictly continuous. Thus, Speed and Ruxton (unpublished) found that if there is a sharp discontinuity between two states of secondary defence/no defence within a population, there may be a single ESS frequency of the defended form that falls between zero and one (see also models and discussions of defences in plants by Till-Bottraud and Gouyon, 1992; Augner and Bernays, 1998). What is needed now is a more thorough exploration of ESS solutions when the strict assumption of continuous defence is relaxed.

A second and related aspect of secondary defences that warrant serious empirical and theoretical investigation is the potential for apostatic selection of defensive compounds within and between species. Pasteels and Gregoire (1984) found that sawfly predators (*Tenthredo olivacea*) preferred to attack defended *Chrysomelid* species that were familiar to those that were also defended but unfamiliar; predators may here be following a 'better the devil you know' rule, which if common would generate frequency dependent (pro-apostatic) selection for diversity of chemical defence within and between defended species. If general, this result could help to explain the remarkable diversity of types and concentrations of defensive constituents found within and between many species, but, clearly more empirical and theoretical work is necessary.

A third way in which models of secondary defences may be developed is by giving greater consideration to the benefits that defences provide. In any given habitat, secondary defences may evolve in response to an existing predatory pressure, but once evolved they may enable the newly defended prey to shift the set of microhabitats over which it moves, enabling it to make better use of its environment, especially with respect to foraging, thermoregulation and mating. Hence, when modelling the evolution of defences, the fitness and indeed the life-history consequences of enhanced habitat exploitation following raised levels of individual survival must be considered. Secondary defences may allow prey to inhabit microhabitats in which they are conspicuous, to become larger and to compete more effectively for resources (see discussions in Pasteels et al., 1983; Lindquist, 2002). As Schmidt (1990: 389–390) writes 'An effective sting allows bees and wasps to exploit, while in full view of many birds, such rich sources of food as floral nectar and pollen; allows wasps to actively and conspicuously search vegetation and the ground for prey and allows ants to patrol plants and open soil in search of food.' We return to the possibility that secondary defences evolve in order to widen niche in the context of warning signals in Chapter 7.

## 5.6 Summary and conclusion

The premise of this chapter is that if the form of a defence is an important determinant of the class and form of a warning display, then it is important to (1) consider the range of possible secondary defences that prey can take, and (2) to outline the factors that determine the form of secondary defences taken by particular prey. Secondary defences are made up of traits that fall into three broad (and not mutually exclusive) categories: behavioural, morphological, and chemical. For each class there is a large and growing literature examining the costs and benefits that accrue from each. However, to better understand how the form of a secondary defence is determined, we also need to consider, the range of predators that prey meet, and the historical and ecological constraints that limit the evolution of body form for any prey species. Nonetheless, cost/benefit analyses may be key to our understanding of variation in defensive forms.

CHAPTER 6

# Signalling to predators

## 6.1 Introduction

We have already considered some mechanisms where characteristics of prey act to misinform predators or suppress the flow of information to predators, such as cryptic colouration. Here we consider the idea that under some other circumstances prey may send information to predators that is of mutual benefit. In particular, we have in mind situations where a prey individual honestly informs the predator that it has a particularly low chance of successfully attacking that particular individual, and so it should avoid any costs involved in launching an attack. The origin of this idea is variously credited to Zahavi (1977), Baker and Parker (1979), or Woodland et al. (1980). We first of all consider whether such signals are theoretically viable before critically evaluating the available empirical evidence for these signals.

## 6.2 Signalling that an approaching predator has been detected

Signals that inform predators that they have been detected have been suggested for several prey species. The basic idea is that a predator has a reduced chance of successfully capturing an individual if that prey individual becomes aware of the approaching predator before the final attack is launched. Hence, the predator may benefit from being informed that it has been detected, since it may now be optimal for the predator to cancel its attack. This cancellation could be beneficial to the predator if the attack involves an investment in time or energy, or if there is some other potential cost to the predator, such as risk of injury. If this signal does cause the predator to cancel its attack, then the signaller also benefits in saving the cost of evading the attack and/or the risk (however small) of being captured in that attack. Since both parties appear to benefit, the evolution of such a signal seems uncontroversial. However, we might ask what stops prey individuals from cheating, and signalling that they have detected the predator when they have not. If signalling was energetically inexpensive, this would seem an advantageous tactic, reducing the predation risk of cheats compared to non-cheats. As cheats prosper, so the signal would have little value to the predator, which should be selected to ignore it. Hence, the signal can break down because of cheating by prey. Bergstrom and Lachman (2001) developed a formal model to explore the conditions required for the signal to be maintained in the face of such a danger from cheating. We describe this model in detail below.

A game between predator and prey is played in a series of rounds. For each round, the predator is either in the vicinity of a single prey individual and hence able to attack (hereafter, 'present') or not. The predator is present with fixed probability $\alpha$. However, in a given round, the prey does not know with certainty whether the predator is present or not. Rather, it detects sensory stimuli (e.g. sounds, movements, etc.) that are caused both by the predator (if it is present) and by other components of its environment. These stimuli are characterized by a random variable $x \in [0,1]$. If $x$ takes a high value (range $B$ in Fig. 6.1), then the prey is certain that the predator is present; if $x$ takes a low value (range $A$), then it is certain that the predator is not present. There is, however, an intermediate range of values of the stimulus where the prey is uncertain. Specifically, $x$ for a given round is drawn from different probability distributions according to whether the predator is present or

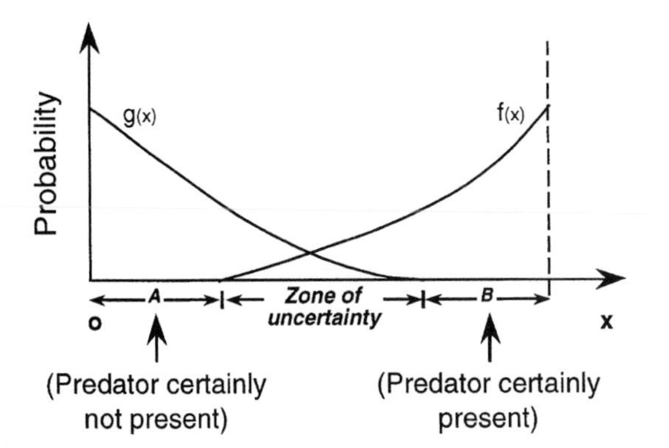


**Figure 6.1** The situation modelled by Bergstrom and Lachman (2001).

not ($f(x)$ and $g(x)$, respectively). There is some overlap between the two distributions, so there is a range of $x$ values for which the value of $x$ does not unambiguously inform the prey whether the predator is present or not (the 'zone of uncertainty' in Fig. 6.1).

The model assumes that signalling costs the prey. Specifically, if a prey individual signals in a given round then its probability of surviving that round is multiplied by a factor $(1 - c)$ for some $0 \leq c \leq 1$. This cost has to be interpreted with care. The cost is paid regardless of any effect that the signal may have on the predator to which it is directed (hereafter, the primary predator). Hence this mortality cost is not associated with the primary predator. Bergstrom and Lachman suggest that signalling attracts a different type of predator (hereafter, a secondary predator), attacks by which are responsible for the mortality cost of signalling. This interpretation justifies a situation where the cost is incurred whether or not the primary predator is present. If the primary predator had to be present for a cost to be imposed, then signalling would break down; prey could signal indiscriminately without heed to whether they thought a predator was present or not, without incurring prohibitive costs (since signalling would be cost-free when the predator was not present).

All attacks by the primary predator require the predator to make an investment ($d$). However, attacks differ in their probability of success ($t$). Bergstrom and Lachman argue that, for signalling to be stable in their model, the probability of success must be negatively correlated with the sensory stimulus $x$. This is a key result of the model, and formalises the idea that the prey can gain information about whether a predator is likely to be present from the stimuli it detects ($x$) and reduce its risk from an attack ($t$) by being forewarned of the danger. Both $x$ and $t$ are generated from a bivariant distribution $f(x,t)$ for each round when the predator is present.

These rules translate into the pay-off matrix shown in Table 6.1. The predator is unaware of the values of $x$ (the strength of stimulus received by the prey, and hence the prey's knowledge of the predator's presence) and $t$ (the probability that an attack will succeed). The only information that the predator can use to decide whether to attack or not is whether the prey signals or not. The prey is aware of $x$ but not $t$. Hence, the prey's decision to signal or not will be based on $x$ and knowledge of the frequency with which predators occur (defined by fraction of rounds in which the predator is present ($\alpha$)). A successful attack brings the predator a reward of one, not attacking brings no reward (but saves the investment of $d$).

The prey gains nothing from signalling if the predator is not present, but pays a price in raised mortality because of the signal attracting secondary predators (described by the value of $c$). Part of the prey's strategy should therefore be not to signal when it knows that the predator is not present, this occurs for the range $A$ of $x$ values in Fig. 6.1. If it is worth the prey signalling for one value of $x$, then it is worth it signalling for all higher values of $x$ (where it is more likely that the predator is present). Hence, the prey's strategy can be defined by a critical value $x_c$. If $x$ in the current round is below $x_c$ then the prey does not signal, otherwise it signals. We can see that costs of signalling ($c$) must be greater than zero for the signalling equilibrium to be stable. If there is no cost to signalling, then the prey could always signal, in which case the predator should ignore the signal as it contains no useful information about the prey's estimate of the likelihood of the predator being present. However, $c$ should not be so high that prey never benefit from signalling.

The predator's strategy depends on the values of two related components: its probability of attacking

**Table 6.1** The pay-off matrix for the model of Bergstrom and Lachman (2001)

| **Payoffs per round** | | | | | |
|---|---|---|---|---|---|
| **Predator present** | | | | **Predator not present** | |
| **Prey signals** | | **Prey does not signal** | | **Prey signals** | **Prey does not signal** |
| ***Attack*** | ***No attack*** | ***Attack*** | ***No attack*** | | |
| **Prey** | | | | | |
| (1-c)(1-t) | 1-c | 1-t | 1 | 1-c | 1 |
| **Predator** | | | | | |
| t-d | 0 | t-d | 0 | 0 | 0 |

when the prey signals and its probability of attacking when there is no signal. Since the predator has no way to distinguish one non-signaller from another, or one signaller from another, then it should operate an all or nothing policy with respect to attacks on signallers and non-signallers (i.e. either attack all or no individuals of each type). If all values of probability of an attack in a specific round being successful ($t$) are greater than the investment made in each attack ($d$), then the predator should always attack regardless of any signalling. Similarly, if all $t$ are less than $d$, then it should never attack. We are interested in cases where it should attack one group but not the other. Recall that a key requirement for stable signalling in the model is that $t$ is negatively correlated with $x$. Now, high values of $x$ induce the prey signal. Hence, the average value of $t$ from rounds where the prey signal (call this average $t+$) is less than the average value from rounds where no signal is given ($t-$). The predator should attack when it is most beneficial to it, so if it attacks one group but not the other, then it should attack non-signallers.

Bergstrom and Lachman demonstrate that there is a set of necessary conditions that are sufficient to produce a stable system in which (i) only prey that are reasonably sure that they have detected a predator signal and (ii) the predator responds to that signal by not attacking. These conditions are that

**(1)** there is a cost to signalling
**(2)** this cost to signalling is not so high that signalling is never profitable
**(3)** prey have some means (even if imperfect) of gauging the likelihood of a predator being present, and hence their risk of attack; that is, those prey that are most 'concerned' about predation are actually those that are most at risk
**(4)** prey that strongly suspect the presence of the predator are more difficult to capture than those which have lower levels of suspicion
**(5)** the costs to the predator of attacking are not prohibitively high.

Thus, it seems that signalling of detection of the predator is theoretically possible, although under quite a specialised set of circumstances. This is particularly true as there is a tension between satisfying conditions (3) and (4) above. Condition (4) requires that the more sure a prey individual is that a predator is present the less vulnerable to capture it is if the predator attacks. Condition (3) requires that prey that are more 'concerned' about their predation risk are actually those most at risk from predation. This 'concern' is the product of the prey's estimations of the likelihoods that a predator is present and that and attack will be successful (conditional on a predator being present). These conditions act in opposition to each other. As certainty of the predator's presence increases, the prey's concern increases so condition (3) requires that the probability of capture by the predator must increase. However, as certainty of the predator's presence increases, condition (4) requires that the probability of capture (conditional on the predator being present) must decrease. Satisfying these two conditions simultaneously is not impossible, but requires that the values assigned to system parameters lie in a relatively narrow region of parameter space.

Two aspects of this model that may make it challenging to apply to real systems are the description of time as a sequence of rounds and the definition that $t$ (the probability that an attack is successful) and $d$ (the cost to the predator of attacking) are considered to be in equivalent units. These restrictions to the model's applicability were removed by Getty (2002). As can be seen in Table 6.1, in the original model if the predator is present but does not attack then it receives a pay-off of zero. This can be considered as the opportunity cost of pursuit. If we move from a series of rounds of a game to a continuous time setting, then Getty demonstrates that the fundamental conclusions of the original paper are retained except that this opportunity cost now changes. The predator should pass up opportunities to attack signallers and resume searching for non-signallers whenever the opportunity cost of pursuing signallers exceeds the expected benefit from attacking a signaller. Getty's modifications therefore make the model more useful in the sense that the opportunity cost of attacking a signaller is now costed as the expected value of the alternative activity (finding and attacking a non-signaller). The result is that parameters such as the expected time to find a non-signaller become important with the further effect that population density of prey and search ability of the predator become explicit parameters of the model. Getty's work also provides a re-interpretation of the model in terms of commonly considered foraging concepts like diet choice, and concepts from signal-detection theory such as the receiver operating characteristic (ROC) curve.

## 6.3 Signalling that the prey individual is intrinsically difficult to catch

One benefit of signalling may arise simply by informing the predator that it has been detected. An alternative (yet not necessarily mutually exclusive) role of signalling may be to indicate to the would-be predator that the signaller is difficult to catch. A key theoretical work on honest signalling by prey is that of Vega-Redondo and Hasson (1993), and looks at the maintenance of signals of 'quality' rather than 'alertness'. Rather than communicating that it has detected a predator, the prey signals that it is a particularly high-quality individual that would be difficult to capture in an attack, and so the predator would be better served by passing up the chance of attacking it. Their model is structured as follows.

In each round of the game the predator confronts a group of $N$ prey individuals drawn randomly from the entire population. The prey population is made up of two prey types, differentiated by their qualities $q_1$ and $q_2$. We assume that $q_1$ is greater than $q_2$, and that this denotes that $q_1$ individuals are intrinsically more difficult for predators to catch than $q_2$ individuals. Prey can signal to the predators, this signal ($s$) must be selected from the continuous range $[0, s_{max}]$. In the example used by Vega-Redondo and Hasson, $s$ represents the distance between predator and prey at which the prey flees. The constant $s_{max}$ is then the maximum distance at which approaching predators can be detected by the prey, and small values of $s$ are intrinsically more risky for the prey than higher values of $s$, and so are more likely to indicate a high-quality individual. It is on the basis of these signals that the predator chooses which prey to attack. The predator attacks only one individual from each group. We assume that if two individuals signal alike then (even if they are of intrinsically different quality) they are both equally likely to be the individual attacked. The pay-offs from a round of the model are zero to any prey item in the group of $N$ that is caught by the predator and one to all those that are not caught (either because they were not the one selected for attack or were selected but the attack failed). The predator gets a reward of one if it attacks successfully and zero if it is unsuccessful. The final part of the model is the function $h(s_j,q_i)$ which is the probability that a targeted individual of quality $q_i$ that signals $s_j$ and is targeted by the predator, escapes the attack.

Vega-Redondo and Hasson's key finding is that there is an evolutionarily stable signalling equilibrium where the two prey types issue different signals, and the predator responds differentially to those signals. The $q_1$ individuals issue signals $s_1$, and the $q_2$ individuals issue signals $s_2$, where $s_1 \neq s_2$. Indeed, in the case where $s$ reflects the distance to which predators are allowed to approach: $s_1 < s_2$ (i.e. high-quality individuals allow the predator to approach more closely). At this signalling equilibrium, if the group contains individuals of both signal types, then only those of the $s_2$ type have a non-zero probability of being the one targeted. This probability is the same for all such individuals, and is the inverse of the number of $s_2$ signallers. Only if the group is entirely composed of $s_1$ signallers, will one of them be targeted, each individual having a $1/N$ probability of being the one targeted.

The equilibrium exists and is evolutionarily stable providing $h(s,q)$ satisfies the following conditions:

**1.** An attack on a higher quality individual is less likely to be successful than an attack on a poor quality individual if both signal identically (i.e. if both allow the predator to approach equally close). That is, the signal ($s$) is related to a quality of prey that affects their vulnerability to predators ($q$).
**2.** If $s$ increases then the relative decrease in the probability of an attack being successful is greater for the higher-quality individual than the lower-quality one. Thus, the cost of a more intensive signal (a smaller $s$) is greater for a low-quality individual than for a high quality individual.
**3.** If $s$ becomes small enough then the predator's attack is always successful.
**4.** $h(s,q)$ converges to zero as $s$ increases, and does so faster than its derivative.

These conditions would have to be modified (but in a very intuitive way) if the signal is one where increasing signal value is associated with increasing quality, for example, stotting height or stotting frequency (see Section 6.5.1).

What keeps the signal honest is the risk that all individuals in a group will be vulnerable to attack, including a low quality individual that pretends (by its signal) to be a high-quality individual. If by chance, such an individual ends up in a group of high-quality individuals then it has a $1/N$ chance of being the one selected for attack. If this occurs, then this individual pays the price of its deception through a high probability of predator success, because it has allowed the predator to approach so closely. This means that both a large herd size $N$ and a relatively low frequency of high-quality individuals strengthens the incentive for low-quality individuals to cheat. However, providing the four conditions above are met, the signalling equilibrium will remain stable, although the signals will be driven to higher $s$ values by such circumstances. If both signals are greater than $s_{max}$, then the signalling equilibrium will break down and both types will do their best to flee as soon as they detect the predator. Hence, there is a requirement that $s_{max}$ is sufficiently big to incorporate the two signals. Ecologically, we can see that if $s_{max}$ is very small then signalling breaks down because the optimal strategy for all prey types is to flee as soon as the predator is detected. However, there may also be a restriction that $s_{max}$ cannot be too big, if the prey are too good at spotting the predator, then the analysis becomes more problematic and is not explored in the original paper.

Notice that at equilibrium, the low quality individuals do not necessarily flee as soon as they detect the predator. A mutant that did this would always be the one targeted by the predator in preference to the other low-quality individuals that allow the predator to close a little more before bolting.

There is much that can be done to develop this model. It seems likely that predators can only discriminate signals if they are sufficiently far apart, and this minimum separation distance is likely to change with signal intensities. Prey individuals need not associate at random. It may be possible for cheating to take root more effectively if cheats band together or preferentially associate with low quality individuals, to avoid or reduce the risk of being the targeted individual. There may be costs to both predator and prey individuals involved in an unsuccessful attack. Prey populations are likely to be composed of more than two quality types, and predators may meet prey singly rather than in groups. All of these elaborations are worth exploring, but even the original model warrants further investigation. The original paper demonstrates that a signalling equilibrium can exist, it does not explore the nature of that equilibrium and how it is affected by plausible parameter values and specific functional forms.

## 6.4 Summary of theoretical work

There is a quantitative theoretical basis for considering that both signals that predators have been detected and signals that individual prey are hard to catch or subdue are at least theoretically possible. Notice, for all the modelling discussed in this section that we would only expect signals to be honest 'on average'. Johnstone and Grafen (1993) argue that we should always expect some low level of cheating. Hence, evidence that cheating sometimes occurs is not evidence that the signal carries no useful information to the predator 'on average'. Indeed, the fact that palatable (Batesian) mimics can persist strongly suggests that they do not completely undermine the effectiveness of the signal generated by the unpalatable model (see Chapter 10).

## 6.5 Empirical evidence from predators

Since it does appear that honest signalling between prey and predators is theoretically possible, we now turn to the empirical evidence that such signalling actually occurs. Since the key function of these signals is to alter predator behaviour, we start by considering studies where a natural predator was observed. We are interested not only in evidence that such signals occur but also in considering whether they occur in the circumstances predicted by the theory introduced at the start of this chapter. (A thoughtful consideration of the particular requirements for effective empirical study in this area is provided by Caro (1995)).

### 6.5.1 Stotting by gazelle

The behaviour most commonly cited as a signal to predators of individual prey's escape ability is stotting (jumping with all four legs held stiff, straight, and simultaneously off the ground). This has been observed in several mammalian species but has been investigated most fully in the Thomson's gazelle (*Gazella thomsoni*). Caro (1986*a*,*b*) identified 11 non-exclusive potential functions of stotting. Two of these involved signalling to the predator: either that it has been detected or that the prey individual would be difficult to catch. Caro suggests that cheetah need to approach within 20 m of gazelle in order to have a chance of a successful attack. Stotting during a chase occurred when the gazelle was generally further than 60 m from the cheetah, and occurred mostly towards the end of unsuccessful chases. These observations are compatible with stotting being a signal of prey athleticism to the predator. The outcomes of 31 cheetah hunts are reproduced in Table 6.2. These data may suggest that stotting is related to both failed chases and the cheetah giving up a hunt without giving chase (although sample sizes are too small to be certain). Again, this interpretation is consistent with stotting being a signal of fleetness. Surprisingly, Caro initially favoured the 'predator detection' hypothesis over the 'high quality' hypothesis (although both could operate). He realized that if stotting only functions to indicate to the cheetah that the gazelle has detected it, then explaining its occurrence during chases is challenging. He suggested that 'the act of fleeing itself may be insufficiently unambiguous to inform the predator that it has been detected'. This seems unlikely to us (although gazelles often make quick dashes to gain respite from biting insects; Tim Caro, personal communication). Crucially, Caro notes that the 'high quality' hypothesis 'is damaged primarily because

**Table 6.2** Outcome of 31 cheetah hunts reported by Fitzgibbon

| | Hunt abandoned | Chase occurs but unsucesssful | Chase sucesssful |
|---|---|---|---|
| ***Gazelle stotted*** | 5 | 2 | 0 |
| ***Gazelle did not stott*** | 12 | 7 | 5 |

so few fleeing members of a group (less than 20%) stott and that stotters do not get captured any less than non-stotters' (see Table 6.2). The fraction of individuals stotting does not seem relevant to us. The results shown in Table 6.2 do not have the sample sizes to persuade us that it is safe to assume that stotting during a chase has no influence on the eventual outcome of that chase. However, it is true that they provide no evidence in support of such an association.

Fitzgibbon and Fanshawe (1988) demonstrated that the Thomson's gazelles that African wild dogs (*Lycaon pictus*) selected to chase from a group stotted at lower rates than those that dogs did not select to chase. However, cause and effect are unclear, it could be that gazelles react to being chased by stotting less, rather than dog's selecting those that stott less. Fitzgibbon and Fanshawe argue against this alternative explanation as follows:

> In the eight cases when gazelles were observed both before they were selected and while being chased, four increased their stotting rate when selected and four decreased it. This implies that the gazelles are as likely to decrease their stotting on being hunted as they are to increase it, and that the dogs are selecting on the basis of stotting rate. In addition, wild dogs were seen to change the focus of a hunt from one gazelle to another on five occasions, and on four of these the gazelle preferred was stotting at a lower rate.

Hasson (1991) points out that Fitzgibbon and Fanshawe's data also shows that the one occasion where the dogs switched to a faster stotting individual is also the case where the two gazelles' stotting rates were most similar. Fitzgibbon and Fanshawe also report that those gazelle that were selected but which outran the wild dogs were more likely to stott and stotted for a longer duration than those that were successfully captured. Although this is suggestive of stotting being a signal of ability to escape, there is an alternative explanation: animals that escape may not be more able to stott but simply have more opportunity to do so, as they are not as closely pursued. Together, these studies of stotting behaviour come closer to convincing evidence of signalling of ability to evade attack rather than awareness of the predator's presence. However, that stotting is a signal (of any kind) to predators has not been demonstrated to the exclusion of all other candidate explanations, although it certainly remains highly plausible.

### 6.5.2 Upright stance by hares

Holley (1993) reported on observations of naturally occurring interactions between brown hares (*Lepus europaeus*) and foxes (*Vulpes vulpes*) on English farmland. He observed 32 occasions in which feeding adult hares were approached by a fox from a distance. On 31 of these occasions, as the fox closed to 20–50 m of the hare, the hare stood bipedally, ears erect, directly facing the fox, turning its body so as to remain facing the fox as it moved, continuing until the fox moved away. This behaviour was not seen consistently in other contexts during 5000 h of observation of hares. Hence, Holley suggested that this behaviour was a signal to the fox that it had been detected. In support of this interpretation, Holley pointed out that none of the 31 'signalling' hares were attacked. He further argued that such signalling is plausible because adult hares are sufficiently fleet that foxes are only successful in catching fit hares when catching them unawares, allowing capture before the hare is able to accelerate to its maximum speed. On five occasions, a fox appeared near hares as a result of using cover to approach undetected. In these five cases, the hare did not show this 'signalling' behaviour, but either moved away from the fox or adopted a crouched stance that Holley suggested allows fast acceleration if required. This is interpreted as the hares not signalling in circumstances where signalling of detection may not deter the fox, because it has already managed to get near to the hare. Holley also reports that in 28 encounters between hares and domestic dogs, this 'signalling' behaviour was not shown. He interpreted this by suggesting the domestic dogs are coursers, willing to pursue hares over long distances in order to exhaust them, are thus not reliant on surprise like foxes, and so are less likely than foxes to be concerned about being detected by a hare. Hence, such signalling would not be expected to deter dogs from attacking.

Holley considers alternative explanations for this bipedal behaviour. He argued that the bipedal behaviour is unlikely to serve solely to improve the hare's view of the fox, because it is not used against dogs. Further, he suggested that in low vegetation the fox is visible to the hare at much greater distances than the distance at which this behaviour is induced (never more than 50 m). He also argued that bipedal behaviour is unlikely to serve solely as a signal to conspecifics because it was not given to dogs, because the hare turns its body so as to constantly face the fox, and because on 13 of the 31 occasions when it was used the signaller was solitary. We consider this study as strongly persuasive that the hare's behaviour serves at least in part as a signal to the fox that it has been detected. The turning of the body to face the moving fox is hard to reconcile solely with signalling to conspecifics and is strongly suggestive of signalling to the fox. A follow-up study, using videotape, that allowed quantification of this aspect of behaviour would be useful. Such a study could also search for evidence that hares detect foxes before they have closed to 50 m, perhaps detection can be identified by head turning or increased vigilance. However, given that the original study observed 32 encounters between foxes and hares in 5000 h of observations, such a study may be very challenging.

### 6.5.3 Push-up displays by lizards

Leal and Rodriguez-Robles (1997) reported that a lizard (*Anolis cristatellus*) performed conspicuous 'push-up' displays, in which the body is moved up and down in a vertical plane by flexion and extension of the legs, in response to a snake predator. Snakes have been demonstrated to stop an approach to a lizard in response to this signal (Leal and Rodriguez-Robles, 1995). This earlier study also demonstrated that lizards that signal are attacked significantly less often than those that do not. These lines of evidence are certainly suggestive that the push-up display acts, at least in part, as a signal to predatory snakes. Leal (1999) provided evidence that the intensity of the 'push-up' signal given in response to predators by this lizard species is correlated with individual physiological condition as measured by endurance capacity, whereas the intensity of this signal given in intraspecific non-predatory contexts was not similarly correlated. Leal further argued that since struggles between a captured lizard and the snake can last several tens of minutes, and can result in the escape of the lizard, the endurance capacity will correlate strongly with how easily an individual can be subdued. This suggests that the push-up signal may be an honest and reliable signal of escape ability. Although these studies do not definitively demonstrate that the push-up signal is costly, the authors argue persuasively that such signals are costly both energetically and in terms of increased conspicuousness attracting secondary predators.

### 6.5.4 Singing by skylarks

Cresswell (1994) reported on extensive and carefully recorded naturally occurring predation events by a raptoral bird (the merlin, *Falco columbarius*) attacking a songbird (the skylark, *Alauda arvensis*). The merlin clearly selected a skylark for pursuit before any song by the skylark was heard. If singing by the skylark was heard (by Cresswell), then it started very soon after the pursuit began. For pursuits where the merlin gave up without capture, merlins chased non-singing or poorly singing skylarks for longer periods compared to skylarks that sang well: these chases often exceeded 5 min in duration over many kilometres so that the costs of not signalling or responding to the signal were high for the skylark and merlin, respectively. Merlins were more likely to catch a non-singing skylark than a singing one. The study also showed that skylarks that did not sing on attack (and therefore probably that could not sing) were more likely to attempt to hide from the merlin rather than outrun it. The author himself admitted that one drawback to this study is that he does not demonstrate that singing whilst being pursued is costly, although other studies have shown that singing in skylarks in the absence of pursuit is energetically demanding. However, in all other respects this study comes very close to a convincing demonstration of pursuit deterrence signalling.

### 6.5.5 Predator inspection behaviour by fish

In an ingenious set of laboratory experiments, Godin and Davis (1995*a*) demonstrated that predatory cichlids (*Aequidens pulcher*) were less likely to attack and kill guppies (*Poecilia reticulata*) that approached and inspected them than those that did not inspect. On this basis, the authors speculated that an inspecting prey individual may signal to the predator that it has been detected and so an attack is unlikely to succeed. Whilst this is plausible, alternative explanations are possible (see below). Godin and Davis also suggested that inspection may be a way to signal an individual guppy's high quality (i.e. high capacity to evade an attack) to the predator. They make this argument on the strength of previous findings that larger, better armoured and high condition prey are more likely to inspect (Godin and Davis, 1995*a*). Godin and Davis's original experiments were subject to further scrutiny in two subsequent papers (Godin and Davis, 1995*b*; Milinski and Boltshauser, 1995). From our perspective, the important outcome of this scrutiny is a small but significant caveat to the conclusions of the original paper. It is unarguable that the predators were selective in which prey they attacked in the original experiments. It is not absolutely clear what particular quality (or qualities) are used by predators to select prey for attack, however this quality is certainly strongly associated with inspection behaviour. The most parsimonious explanation is that inspection itself is taken as the signal that predators respond to: with predators being less likely to attack an individual that inspects than one that fails to inspect. However, this is not the only explanation. It could be that the signal that the predators use to discriminate individuals is something quite different, some more general measure of 'quality', such as body shape or colouration. It may be that high-quality individuals are more likely to inspect than low-quality individuals, and that predators pay no heed to inspecting but actively select to attack low-quality individuals (selected on the basis of, say, body shape or colour). While there is no evidence for a 'third variable' that functions as a signal of general quality, such variables cannot at this point be ruled out. Support for the third variable interpretation comes from the study of Milinski et al. (1997) using a live predator (pike, *Esox lucius*) and dead fish (sticklebacks, *Gasterosteus aculeatus*) whose position could be manipulated by the controller. Dead fish whose behaviour mimicked inspection did not experience reduced likelihood of attack in this study. If we accept that predators' behaviour toward dead fish is indicative of behaviour towards live ones, then this study suggests that pike at least do not use inspection behaviour as a basis for prey selection.

The upshot of all this endeavour is that we are still unsure if predator inspection behaviour by fish is or is not a pursuit deterrent signal, although it does remain a very plausible possibility.

### 6.5.6 Calling by antelope

Tilson and Norton (1981) suggested that alarm duetting (almost simultaneous rendering of an alarm call) between a mated pair of klipspringer (*Oreotragus oreotragus*: an African antelope) was a predator deterrence signal. The evidence for this was the observation of 13 occasions when a jackal (*Canis mesomelas*) 'lost interest in' stalking the antelope after alarm duetting started. However, antelope climbed a rocky outcrop before beginning the duetting, and jackals were never observed to attempt to follow them onto outcrops. This suggests the alternative interpretation that alarm calling is directed towards conspecifics, with this being initiated once a pair has reached the safe haven of higher ground, and the jackal giving up because of the antelope reaching the rocky outcrop rather than in response to vocalisations. An experimental approach using a recording of duetting in concert with dummy or tethered antelope might be fruitful, as it would be possible to explore the extent to which jackal behaviour is affected by vocalisation in isolation from any effect of movement of antelope to rocky habitat. At present there is no substantial evidence for predator–prey signalling in this system.

### 6.5.7 Fin-flicking behaviour by fish

Brown et al. (1999) suggested that the fin-flicking behaviour exhibited by a characin fish (*Hemigrammus erythrozonus*) in response to an alarm pheromone emitted when a conspecific's skin was broken functioned as a predator deterrent. In laboratory experiments, they found that a predatory fish was slower to attack fish that fin-flicked. However, the signal appears to be a little different from the 'pursuit deterrent' signals considered in this section, in that the predator still always attacked the signalling prey. It may function more like the startle signals discussed in Chapter 13. As such, it would be interesting to explore whether the delayed attack observed in Brown et al.'s experiments is subject to a habituation effect, when the predator encounters the signal on repeated presentations. Such habituation would be more easily accommodated by explanation based on a startle effect than one based on communicating information to the predator. The information contained in the putative signal is also unclear. It may be that the fin flicking actually functions as a warning that indicates to the predator that a predation event has recently taken place. Recent predation increases the likelihood that another predator (which may be a danger to the focal predator) is nearby. Such information may induce the more circumspect attack strategy exhibited by Brown et al.'s predators. Again, this study falls well short of providing convincing evidence of predator–prey signalling because alternative explanations cannot be ruled out.

## 6.6 Studies where predator behaviour is not reported

A number of studies that report on prey behaviour but not that of predators are regularly cited as evidence of pursuit-deterrent signalling. We feel that since the key objective of the signalling studied in this section is modification of predator behaviour, that such studies are very unlikely to be able to produce unambiguous evidence of such signalling (see Caro, 1995 for further discussion on this). However, this does not make studies that do not report predator behaviour useless. Predation in the wild is generally unpredictable in space and time, and predation events in laboratory settings raise ethical issues. Hence, well-designed studies that do not involve observation of predation events have a useful role in identifying promising study systems and eliminating some alternative functions of observed signals.

### 6.6.1 Tail-flicking by rails

One of the earliest suggestions of quantitative evidence for a pursuit deterrent signalling comes from observation of tail-flicking behaviour by eastern swamphens (*Porphrio porphyrio*: a large bird of the rail family) when approached by a human (Woodland et al., 1980). Flicking rates were highest in the birds nearest the approaching human; birds' tails were orientated towards the human and flicking rate increased with decreasing separation between human and bird. Woodland et al. interpreted these observations as evidence that tail flicking was a signal directed towards predators. Craig (1982) challenged this interpretation, suggesting that tail flicking was more likely to be a submissive signal directed at dominant conspecifics. He suggested that the reason that birds nearest the approaching predator signal most was that these birds were likely to be low quality birds at the periphery of groups. These peripheral subdominants have increased need to signal submission to dominants when the human 'predator' drives birds closer together. The closer the human gets, the more signalling is required as the birds are driven closer together when they bolt for nearby water. Craig also suggested that the apparent orientation of the tail towards the human was simply a necessary consequence of the bird walking away or preparing to do so.

Craig's alternative interpretation certainly complicates the situation, but further work on this system may be worthwhile since there are other observations in Woodland et al. (1980) that the alternative explanation does not easily accommodate. For example, when birds on reed beds were approached by boat, 'birds would remain virtually on one spot, standing erect, looking back by rotating their body so that the rump flash was directed towards the intruder. Again the flicking rate increased as the intruder approached'. This observation seems easier to reconcile with tail-flicking being a signal directed to potential predators rather than one directed to conspecifics.

Alverez (1993) observed that moorhens (*Gallinula chloropus*: another member of the rail family) tail-flicked at a higher rate when facing into a group of individuals (so that the tail-flick signal was less visible to other group members), than when facing away from the group. This falls well short on demonstrating that tail-flicking is a signal directed at predators, but is certainly consistent with such a hypothesis. However, it is also consistent with signalling to conspecifics. Alverez suggested that tail-flicking is associated with higher vigilance rates, and birds may be more vigilant for conspecific aggression when facing into the group. Ryan et al. (1996) also claimed that tail-flicking by the Australian Dusky Moorhen (*Gallinula tenebrosa*: another bird of the rail family) was directed at predators, but their only 'evidence' for this is that tail-flicking increased with scanning rate. As our discussion above should make clear, this does not imply that tail-flicking in this species is a signal to predators.

In conclusion, tail-flicking in rails may be a signal directed towards predators, but conclusive evidence is lacking, particularly as an alternative interpretation based on signalling to conspecifics has been hypothesised. One puzzle about tail-flicking is what keeps the signal honest, since no obvious time or energy cost is apparent, birds can feed and signal at the same time. Alverez (1993) suggested that the cost is an added risk of predation from secondary predators, as envisaged by the theory of Bergstrom and Lachman (2001). This hypothesis could be testing with flicking and non-flicking model birds.

### 6.6.2 Tail-signalling by lizards

Dial (1986) is often cited as empirical evidence supportive of prey signalling to potential predators. However, this paper's evidence amounts to demonstrating that of two lizard species (*Cophosaurus texanus* and *Holbrookia propinqua*), the one living at lower populations densities (*C. texanus*) produced a tail display more readily in response to predatory threats. We do not consider this conclusive proof that such signals are not aimed at conspecifics, and so the conclusion that these results provide evidence of a predator-directed signal are premature.

Hasson et al. (1989) also suggested that tail-wagging in a lizard species (*Callisauus dranconoides*) was a pursuit-deterrent signal. Their evidence is as follows. Wagging occurred more in individuals near to a refuge. Wagging increased with ground temperature, which was postulated to correlate with escape ability. Wagging was less common in a situation where the experimenters hypothesized that the lizard should consider itself to be in extreme danger. Lizards wagged more when moving than when stationary, which the experimenters suggest is because their risk of being detected by a predator is higher when moving. This seems contrary to the increased wagging near refuges and at higher temperatures, both of which were considered indicative of reduced predation risk. This study certainly constitutes a stronger argument than that of Dial for tail-wagging in lizards being a predator-directed signal.

Cooper (2000) raised further concerns about the interpretation of both the studies discussed in this section, but he does consider predator deterrence to be a likely function of at least some tail wagging behaviours in lizards. Very similar evidence to that presented by Hasson et al. is presented for another species (*Leicephalus carinatus*) in a study by Cooper (2001), which was careful to consider and eliminate plausible alternative functions of the behaviour. Hence, we agree with Cooper that pursuit-deterrence is currently the most plausible explanation for some tail-wagging in lizards, but further work on eliminating other potential explanations is required.

### 6.6.3 Calling by Diana monkeys

Zuberbuhler et al. (1997) suggested a perception advertisement function to so-called 'long distance' calls given by Diana monkeys (*Cercopithecus diana diana*). The evidence for this is that calls are given only in response to predators whose hunting success depends on unprepared prey, such as leopards and crown hawk eagles, but not in response to pursuit hunters, such as chimpanzees and humans, which can pursue the caller into the canopy. Calling was regularly combined with approaching the predator. This evidence falls a long way short of demonstrating the hypothesised function of signalling to predators. In particular, it does not eliminate alternative explanations based on communication with conspecifics, perhaps warning them of approaching danger. Humans and chimpanzees are reported to use acoustic cues when searching for prey, which may explain the lack of calling in response to these predators. The authors themselves acknowledge that evidence for the behaviour of predators in response to these calls would greatly strengthen their case.

### 6.6.4 Snorting in African bovids

Caro (1994) explored the behaviour of several types of African bovids as he approached them on foot. He suggested that snorting signals awareness of a predator. In the topi (*Damaliscus korrigum*), hartebeest (*Alceaphus buselaphus*) and wildebeest (*Connochaetes taurinus*), individuals nearest the approaching human in a group were more likely to snort and snort rate increased as the human closed. In topis, he found that snorting declined as he walked away from them. These observations are consistent with snorting being a predator-directed signal, but do not demand such a conclusion. Another display (which he calls leaping) was considered as a signal of condition because it was more associated with animals that appeared to Caro to be in better condition based on body shape. The same interpretation was given to two other behaviours (prancing and stotting), on similar evidence, and because animals were considered to orientate themselves so that they were flank-on to the intruder when displaying. Again, these results certainly are consistent with these behaviours being predator-directed signals of condition, but observation of predators would be required to elimate alternative explanations based on signalling to conspecifics.

### 6.6.5 Tail-flagging by deer

Another very commonly cited signal is tail-flagging in white-tailed deer (*Odocoileus virginianus*). Caro et al. (1995) provide a synthesis of the available data on this. They conclude that tail-flagging may indicate quality, on the grounds that flagging animals ran faster than those that did not. However, when no studies have used natural predators and studied their response to this behaviour, and Caro et al. suggest that alternative explanations based on crypsis exist, this remains only conjecture.

### 6.6.6 Barking by deer

Reby et al. (1999) suggest that barking in roe deer (*Capreolus capreolus*) is directed at predators. The evidence for this is that in a population of free-ranging roe deer, when the experimenters came across a deer, it was more likely to bark when alone than when in a group. However, it could simply be that the bark is directed at conspecifics, and there was less need to signal when in a group because either the predator is less of a threat to grouped prey or because if one group member can see the predator then it is likely that all group members can. Further, this study does not consider that the animals in groups may be intrinsically different in sex, age, or internal state than those found alone, and this introduces the strong possibility of confounding factors in the study.

## 6.7 Conclusion

Signalling by prey of both of individual quality and awareness of a predator are both theoretically possible, although both lines of theory could be usefully extended. It is also important to note that any given signal can simultaneously have a variety of meanings ('I can run fast' is almost inevitably intertwined with 'I've seen you') and that the most appropriate interpretation may be context dependent. There are also some studies that provide reasonably persuasive evidence that such signalling does actually occur in nature, although many studies that purport to demonstrate such signalling are far from conclusive. As yet, there has been no definitive study, that (in particular) demonstrates that signalling is costly to the signaller and identifies the information that is transferred to the predator in the signal. However, suitable study systems (e.g. hare/fox and skylark/merlin) have been identified, which could provide the basis for a new generation of studies explicitly aimed at testing predictions of the theory outlined at the start of this chapter. Both these systems would be difficult to experimentally manipulate and impossible to transplant into the laboratory. Such considerations may suggest that the push-up displays of lizards in Section 6.5.3 and the inspecting fish of Section 6.5.5 may be worthy of further consideration.

## CHAPTER 7

# The form and function of warning displays

When a skunk with its characteristic black and white stripes stomps or raises its tail end right up in the air, the meaning of this display, to any but the most oblivious of predators, is clear: avoid me or pay the costs of a very unpleasant and unprofitable experience. Describing an encounter with a skunk, A.R.Wallace noted that:

> Owing to its remarkable power of offence the skunk is rarely attacked by other animals, and its black and white fur, and the bushy white tail carried erect when disturbed, form the danger signals by which it is easily distinguished in the twilight or moonlight from unprotected animals. Its consciousness that it need only be seen to be avoided gives it that slowness of motion and fearless aspects which are, as we shall see, characteristic of most creatures so protected. (Wallace, 1889: 233)

It seems obvious that since well-defended prey and their potential predators benefit by avoiding each other, signals will evolve that ensure mutual avoidance. In addition, the ubiquity of warning displays makes their existence seem too ordinary to merit much attention. In truth, though, warning displays have been the focus of intense and justifiable attention of biologists since the time of Wallace (1867, 1889) and Darwin (1887). Besides the question of why a defended organism should signal its unprofitability, there are two particularly important questions about warning displays that have received considerable attention. First, what factors affect the design of warning displays? Second, how do warning displays evolve? In this chapter we consider the first question, in the next chapter we consider the second. We consider both questions in some detail; describing and evaluating the broad theoretical and empirical basis of studies of warning displays. However, before considering theoretical and empirical work, we first, briefly consider the concept of warning displays and the history of the idea.

## 7.1 Characteristics of aposematic warning displays

Given that attacks are costly to the prey in terms of survivorship, injury, or simply the time taken to see-off a threat, it remains important that they are prevented, even by prey with highly effective secondary defences. Defended prey can, of course, achieve this by crypsis but an alternative is to provide predators with some indicator that defences are present. Identification of a primary function of such warning signals arose from discussion between Darwin and Wallace. In 1867 Darwin wrote to Wallace seeking an explanation as to why 'some caterpillars are so beautifully and artistically coloured' (Darwin, 1887). Conspicuous, colourful larvae were a problem for Darwin since the colouration of immature animals could not be explained by his theory of sexual selection. Wallace (1867, 1889) realized that in order to avoid unnecessary attacks, prey with secondary defences might need traits that enhance the capacity of predators to distinguish noxious from edible prey. In his text *Darwinism* he wrote that:

> the animals in question are possessors of some deadly weapons, as stings or poison fangs, or they are uneatable, and are thus so disagreeable to the usual enemies of their kind that they are never attacked when their peculiar powers or properties are known. It is, therefore, important that they should not be mistaken for defenceless or eatable species of the same class or order since they might suffer injury, or even death, before their enemies discovered the

danger or uselessness of the attack. They require some signal or danger flag which shall serve as a warning to would-be enemies not to attack them, and they have usually obtained this in the form of conspicuous or brilliant colouration, very distinct from the protective tints of the defenceless animals allied to them ( Wallace, 1889: 232 )

Furthermore, Wallace (1867, 1889) predicted that birds and other predators would reject conspicuous prey whilst accepting cryptic prey. Darwin was famously pleased by Wallace's explanation, writing 'I never heard anything more ingenious than your suggestion, and I hope that you may be able to prove it true.' (see Darwin 1887). Subsequently, Poulton (1887) reported that the hypothesis was 'proved true' at least in the sense that investigators were able to find a correlation between bright colouration and the unacceptability of prey to selected predators. In fact, Poulton did more than anyone to formalize and develop Wallace's insight as evidenced in his classic text 'The Colours of Animals' (Poulton, 1890), adding the suggestion that warning displays not only improved discrimination in educated predators, but also enhance learning and retention of memories about the unprofitability of defended animals (p. 160). Together with his friend Arthur Sidgwick, Poulton defined a nomenclature for the range of anti-predator defences he had described (Poulton 1890: 336). Of these, only the terms 'cryptic' and 'aposematic' remain in use. 'Aposematic' referred to 'an appearance that warns off enemies because it denotes something unpleasant or dangerous'.

Poulton (1890: 188) noted a number of traits present in some or all prey animals he considered to be aposematic, specifically these were: contrast of a prey's colours against a background, possession of two or more contrasting colours within a prey's pattern (p. 169–70), conspicuous behaviours (p. 188), including an absence of hiding and escape behaviours (p. 169–70) and sluggishness, (p. 175). Furthermore, Poulton did not believe that bright colouration itself was always necessary in aposematism; thus he wrote that being 'freely exposed rather than conspicuous' (p. 170) could act as a warning display. One way that this could happen would be by gregarious aggregation, an idea that Poulton attributed to Müller (1877). It is easy to see how each of these traits could facilitate easy and accurate recognition by educated predators as Wallace hypothesized.

In addition to naming the phenomenon of aposematism, Poulton made a number of seminal points about the phenomenon that show a sophisticated understanding of evolutionary and ecological processes. Thus, Poulton noted that aposematism is not a necessarily fixed state: the same prey could over a short duration be both aposematic and cryptic, by moving over a range of natural backgrounds. In addition, Poulton pointed out that aposematic displays may direct predators to the least vital parts of their bodies (p. 204–205) and that 'unpleasant qualities may be concentrated in these conspicuous parts', such as the defensive tussocks of caterpillars, 'with very unpleasant results for the enemy' (p. 196 and see Chapter 5). Poulton further proposed that sexual displays may serve as prototype aposematic cues, being developed into characteristic warning displays by natural selection (p. 191). Though he recognized that certain odours were associated with visually conspicuous displays, Poulton (p. 171–172) also considered that the pungent odours of insect prey could be explained as a noxious trait in itself, rather than as part of a warning signal.

Although Poulton wrote about aposematic displays as primarily visual phenomena, empirical research, particularly over the last 20 years (as well as the observations of generations of entomologists) has provided good grounds for extending this view; we now know that many aposematic displays probably function by simultaneously stimulating more than one sensory modality. In addition to visual components, warning displays often consist of combinations of sounds, smells, tastes and possibly even prey textures. A good example of a multi-modal warning display is the North American porcupine *Erethizon dorsatum*. This animal relies on the possession of numerous quills in its tail as a highly effective mechanical defence. If attacked, a porcupine slaps its tail against the aggressor, imbedding the quills in its flesh; once dampened by the flesh, the quills expand and embed themselves further into the victim. This defence is advertised aurally (by tooth clacking), visually (when threatened the animal arches up, erecting its quills, revealing black/white areas on its tail) and with odour (when quills are erected a warning odour is also released, Li et al., 1997). Multi-modal signals that warn of chemical defences are also well known; many chemically defended insects utilize pyrazine odours in combination with visual displays (Woolfson and Rothschild, 1990) and some accompany sights and smells with hissing or other sounds. Pyrazines are associated with warning displays in plants as well as animals. Thus Woolfson and Rothschild (1990), note that pyrazines are present at high levels in chemically defended plants in which 'the odour may become obnoxious, arresting, and highly distinctive'. It may be the case that the components of a display work synergistically so that they are only really effective in warning off predators if all components are present (Marples et al., 1994).

### 7.1.1 Aposematism does not require complete avoidance by predators

Wallace's prediction, that conspicuous defended larvae would be avoided by predators was quickly investigated so that by 1887, Poulton was able to summarize a large number of investigations by Weir, Butler, Vernay, Weismann, and Poulton himself, demonstrating the unacceptability of aposematic insects to predators (Poulton, 1887). Reviewing this material in his text 'Animal Colouration', Frank Beddard, a Zoologist at Guy's Hospital, complained that the development of aposematism theory was hindered by the frustrating inconsistency of the animals in their diet choice (Beddard, 1892). Some predators in these and other studies would avoid a prey on one day, only to sample it the next. In truth these were not, on the whole, carefully controlled investigations in the modern sense; rather, insect prey were presented to predators (lizards, birds, and some mammals such as marmosets and capuchins) with little control over experimental conditions.

Evidence for variation in predator diet choice led to a remarkably persistent strain of scepticism that lasted at least until the late 1930s. Some claimed that if there was evidence that predators ate aposematic prey, then their chemical defences *could not* afford protection (e.g. translation of Heikertinger, 1919; Schuler and Roper, 1992). Famously, McAtee (1932*a*), proposed the 'principle of proportionate predation', in which predators were viewed as unselective, simply removing prey in the proportion that they encountered them. McAtee (1932*a*) who apparently doubted the soundness of major components of the Darwinian enterprise (McAtee, 1932*b*), came to this conclusion after finding numerous aposematic prey in the guts of birds. In an intemperate exchange with Hugh Cott in *Nature* he wrote that supporters of the view that predators selected prey were 'only bolstering a single pillar of a structure that . . . is adjudged by many biologists as doomed to collapse' (McAtee, 1932*b*). However, Cott (1932, 1940) essentially put the controversy to bed with a comprehensive review of the evidence showing that predators were indeed selective in their diet choices. Emphasizing the earlier arguments of Poulton et al., Cott pointed out that predators do indeed both eat *and* avoid aposematic prey, but do so in a selective manner, according to their levels of hunger and the presence of alternative prey (to quote Cott 1940: 276, 'If animals are indiscriminate feeders . . . then aposematic appearances can have no more biological meaning than the marks on the moon'). Complete avoidance is not therefore necessary for aposematism to work; aposematism merely has to provide lower mortality than crypsis for the warning signal to be beneficial.

### 7.1.2 Conspicuous animals are not necessarily aposematic

Having recognized conspicuous displays as warnings to predators, naturalists for more than a century followed Wallace's suggestion to examine the correspondence in nature between conspicuous phenotypes and effective secondary defences. Warning displays are now well known in terrestrial arthropods (see review in Cott, 1940; Edmunds, 1974; De Cock and Matthysen, 1999; Mallet and Joron, 1999) and vertebrates including mammals (e.g. Lariviere and Messier, 1996), and snakes (e.g. Smith, 1975; Brodie, 1993; Brodie and Janzen, 1995). Warning displays may also be present in marine environments, especially in nudibranchs (see discussions in Edmunds, 1991; Guilford and Cuthill, 1991; Tullrot and Sundberg, 1991), and also in larvae and embryos of meroplanktonic animals (those who spend only some part of their life cycle in the plankton; Lindquist, 2002). Potential examples are also found in birds (e.g. Wallace, 1889; Baker and Parker, 1979; Dumbacher and Fleischer, 2001) and plants (Hamilton and Brown, 2001; Lev-Yadun, 2001).

However, it should be stressed that it is not always easy to discern whether or which if any components of an animal's external traits are aposematic. An informative example is the sea butterfly or pteropod *Clione antarctica*. This animal is known to be chemically defended (Bryan et al., 1995) and it is relatively large, colourful and conspicuous. Furthermore, the population of Antarctic sea butterflies blooms during a 5–6-month period when sufficient light penetrates such that visual inspection by predators is possible. Is the colourful appearance of the animal part of an aposematic display? The answer is; apparently not. The bright orange colouration of the digestive gland reflects a wavelength of 600 nm, beyond the likely limits of visual sensitivity of predatory fish (Bryan et al., 1995). Hence, though it looks superficially like aposematic colouration, the colour patterning of this animal probably has nothing to do with predator deterrence.

Even when it is well established that traits have evolved to deter predation through warning displays, it may be a mistake to assume that the phenotypes in question are purely aposematic in function. The black/yellow stripes of bees and wasps are widely recognized as ritualized warning signals, but in fact also function to facilitate social interactions in some wasp species (Parrish and Fowler, 1983). The external appearance of any prey animal may therefore perform several functions beyond predator–prey signalling (see e.g. Stamp and Wilkens, 1993); especially intraspecific sexual communication (Mallet and Singer, 1987), social facilitation (Parrish and Fowler, 1983), thermoregulation (de Jong and Brakefield, 1998), UV protection (Pasteels et al., 1989) and parasite defence (Wilson et al., 2001; Wilson et al., 2002). Though crucially important, this point has been repeatedly overlooked by many theorists, a fact that may limit the generality of several mathematical models of aposematism.

## 7.2 Design of aposematic displays I: why conspicuousness?

Conspicuousness, in whatever sensory modality, describes a set of stimuli that attracts a predator's attention, thereby facilitating detection. Wallace recognized that defended animals would benefit by providing distinctive 'flags of danger' that would prevent predators mistaking them for edible species that are often cryptic. One way that they can do this, Wallace argued, was to be bright and conspicuous. However, this hypothesis on its own does not necessarily explain why conspicuousness, rather than some other trait, is so often heightened in aposematic animals. Sherratt (2002*a*) and Sherratt and Beatty (2003) recently suggested that there may be something special about conspicuousness that makes it a common component of warning displays: conspicuousness incurs the cost of increased attention from predators and this may be too expensive for prey that are not protected by secondary defences (see discussion in Speed and Ruxton, 2002).

Conspicuousness may therefore be important because it confers some degree of signal reliability on an aposematic display. More generally, Sherratt and Beatty argued that aposematic signals should: (1) enable predators to distinguish prey with secondary defences from those without them and (2) they should do so in a manner that is rendered reliable on the grounds that it imposes costs that can only be paid by individuals with good secondary defences. Conspicuousness itself is not always necessary to meet these criteria. For example, it is possible to imagine that some examples of prey aggregation do not increase the conspicuousness of an individual but do impose excessive risks for edible prey that lack secondary defences. Similarly, slow flight and sluggishness may or may not enhance conspicuousness, but they may be traits that cannot be safely adopted by many prey that lack noxious characteristics. Defended prey may even be willing to pay an energetic cost to achieve slow, conspicuous flight (Srygley, 2004).

The reliability of an aposematic signal as an indicator of unprofitability depends to a large part on the effectiveness of filtering by predators. Since sampling unfamiliar prey may be costly, the optimal behaviour of one predator depends on the extent to which others have already acted as filters, removing edible conspicuous prey while rejecting conspicuous defended individuals before causing them serious damage. In an analytical model Sherratt (2002*a*) showed that the ESS for predators was often mixed, with some always attacking, others attacking cautiously and others not at all. The end result is a world in which conspicuousness is reserved for prey that have effective secondary defences.[3]

Consistent with a view that conspicuousness can only be sustained by well-defended prey, Endler and Mappes (2004) found that when predators varied in their levels of susceptibility to a prey's defences, then conspicuousness is not necessarily favoured. Thus, if a chemical defence is very effective against some predators, but ineffective against others, prey may be much less likely to evolve an aposematic signal. Endler and Mappes (2004) provide an excellent review of a large body of literature that indicates high levels of variation in the willingness of predators to take chemically defended and aposematic prey. One good example is the monarch butterfly (*Danaus plexippus*), for several decades the paradigm species in the study of aposematism. This colourful, aposematic butterfly contains cardiac glycosides that are toxic to many bird species. However some birds, such as the black-headed grosbeak (*Pheucticus melanocephalus*), are relatively insensitive to the toxicity of the butterflies and do use the monarch as a food source (Brower, 1988). Similarly chemical defences of caterpillars that are effective against birds may be ineffective against parasitoids (Gentry and Dyer, 2002). When defences vary in their actions against multiple predators, conspicuousness may be useful as a signal to some predators, but invite attack from other predators and parasites. Endler and Mappes modelled a large range of scenarios, evaluating the likelihood that conspicuousness would evolve. Though complex in detail, broad conclusions were that when the proportion of predators that will learn to avoid defended prey is low (less than 0.5), then conspicuousness is not favoured. Conversely, when such predators are common (greater than 0.5) and the conspicuous signal will be reliably used by a large number of predators, then there is a raised tendency for prey to become conspicuous.

However, the presence of some degree of signal reliability does not necessarily mean that warning displays are handicap signals. In many cases warning signals incorporate no components of the defence being advertised. For instance, the bright colouration of a butterfly's wing is not conferred by the same chemicals that make it toxic. When displays and defences are decoupled in this way, the form of an individual's display does not necessarily indicate anything about its underlying unprofitability, in which case many aposematic displays can easily be faked by defenceless cheats. For this reason many warning displays can not function as handicap signals (see Guilford and Dawkins, 1993). Empirical support for this view has recently been reported by Gamberale-Stille and Guilford (2003), who tested the hypothesis that if the conspicuousness of a warning display acts as a handicap signal, predators should use conspicuousness rather than colour as a cue for unprofitability. Young chicks (*Gallus gallus*) in fact attended to colour as a means of learning about unprofitability rather than conspicuousness.

Batesian mimics are perhaps the best-known corrupters of aposematic signalling (though see also Chapter 12). These are members of undefended species that share a warning display with defended species (see Chapter 10). One evolutionary response of defended prey to Batesian cheats might be to enhance conspicuousness (Sherratt, unpublished) to an extent that makes anything other than almost perfect mimicry unprofitable for undefended prey. Hence, heightened conspicuousness may not

[3] The idea that predators may act as filters has a long heritage. Thus, Beddard (Beddard, F. E. 1892. *Animal Colouration: an account of the principle facts and theories relating to the colours and markings of animals*. London: Swan Sonnenschein and Co., p. 173), discussed the idea that brilliant colouration could have been the normal colouration for caterpillars but 'the advent of bird-life proved a disastrous event for these animals and compelled them to undergo various modifications, except in the case of those which combined brilliant colouration with uneatableness'.

merely be a way of discriminating defended from cryptic prey, but also a means of escaping parasitic mimics.

### 7.2.1 The opportunity costs of crypsis

Crypsis may bring benefits of low detection rates, but it is likely also to impose opportunity costs on some prey species, limiting their capacity to travel over a range of heterogeneous microhabitats in order to fully exploit available opportunities. This may be true even for prey with cryptic phenotypes that are optimized to provide maximum benefits across heterogeneous habitats (Merilaita et al., 1999). However, prey may be much better able to exploit environmental opportunities if they possess secondary defences (Sherratt, 2002*a*; Speed and Ruxton, unpublished). A related explanation for the prevalence of conspicuousness in aposematic displays is therefore that optimal conspicuousness, in the sense of optimal apparency to predators, may be higher for prey with secondary defences than for those that lack them. Conspicuousness by exposure (or apparency) may be considered a prototype warning display. Poulton himself (1890: 170) noted that 'being freely exposed' is tantamount to a warning, writing that a freely exposed caterpillar's colours 'although sober, do not harmonise with those of the food plant' and may thus convey that the prey has an 'unpleasant quality'. If prey are already conspicuous then the addition of 'aposematic' traits may incur relatively small additional costs of conspicuousness, but add significant benefits from enhanced recognition, memorability etc.

In addition, prey with high levels of optimal apparency to predators may acquire traits that have several functions, one of which is aposematic with only a small additional cost of conspicuousness. One example is the metallic elytra of the chemically defended leaf beetles that may be both aposematic and provide a means of UV protection (Pasteels et al., 1984). Once a prey is freed from the constraints of crypsis it may be free to further heighten its level of conspicuousness for a variety of reasons unrelated to predation. An interesting prediction that follows is that the optimal conspicuousness of defended prey may be closely related to the effectiveness and strength of its defences. This prediction matches the recent findings of a positive correlation between strength of defence and degree of conspicuousness in the dendrobatid frogs (Summers and Clough, 2001).

We note that for other species, the opportunity costs of crypsis might be small, in which case optimal conspicuousness may be low, even for prey with effective secondary defences. None the less, such prey may evolve aposematic signals with some conspicuous elements, especially if these are cryptic at a distance but ensure accurate recognition from close up . The possibility that the same phenotype is cryptic or aposematic according to the distance of deployment has been discussed many times in the literature (see, e.g. discussions in Harvey et al., 1982; Marshall, 2000), but has not, to our knowledge, been explained as an outcome of optimization of the apparency of prey to predators.

### 7.2.2 Forms of secondary defence and the need for conspicuous components of warning displays

Another explanation for the conspicuousness of all or part of a warning display is that it functions to draw a predator's attention to the presence of a 'visible' secondary defence (Poulton, 1890). Some defensive traits may be manifest externally and be evaluated by predators without attacks taking place. Spines, claws and inducible morphological defences may be evaluated from a distance by predators and aposematic 'amplifying' traits may help to draw a predator's attention to them and aid in their evaluations. In such cases, the warning display contains *both* a manifestation of the secondary defence itself and some 'directing' or amplifying traits that draw attention and discourage attacks. Hence the defence is to some extent self-advertising. Components of warning displays that include the defence as part of the advertisement may be reliable, provide accurate information as to the quality being advertised and be very hard to fake. If these criteria are met they may therefore serve as 'reliable indices of quality' (see discussion in Maynard Smith and Harper, 2003).

Self-advertisement of external, morphological defences has recently been proposed by

**Figure 7.1** *Hyalophora cecropia* caterpillar, sharp spines can be visually detected presumably before many predators (and parasitoids) attack the caterpillar. The colours of the tubercles from which the spines protrude may provide distruptive crypsis from a distance and, alternatively, may draw a predator's attention to the spines at close proximity. Photo courtesy of Jim Kalisch, Department of Entomology, University of Nebraska-Lincoln. (See also, Plate1).

Lev-Yadun (2001) as an explanation of the use of conspicuous markings on and around thorns in higher plants. This function of conspicuousness may be widespread. In a number of insect larvae, for example, sharp spines may be accompanied by bright colouration. In the cecropia moth caterpillar (*Hyalophora cecropia*: Fig. 7.1: see also, Plate 1), such colouration may, as described above, provide disruptive crypsis at a distance, but draw a predator's attention to the sharp spines from up close (see Deml and Dettner 2004).

In some cases in which secondary defences are advertised there may not even be a need for an additional warning display. Pike, for example, prefer to feed on crucian carp (*Carassius carassius*) that have an uninduced shallow-bodied morphology compared to induced forms with deeper morphologies, since induced carp require longer handling times (see Brönmark and Miner, 1992; Nilsson et al., 1995; Brönmark et al., 1999). What we do not currently know is whether or not the design of the induced morphology in these fish is affected by factors that enhance evaluation of their handling times by predators.

Self-advertising secondary defences are most obviously morphological defences with a clear visible manifestation. However, in a number of instances chemical defences too have some component of self-advertisement. Thus in some cases animals make samples of their internally stored chemical defences external by reflex-bleeding or some other form of secretion. Defensive secretion can serve purely as a defence (as shots fired at a predator, e.g. the explosive sprays of the bombardier beetle; Eisner and Aneshansley, 1999), and/or as a signal of defensive potential, (as warning shots fired in the air). One example of a 'warning shot' may be reflex-bleeding in the seven-spot ladybird

(*Coccinella septempunctata*, Marples et al., 1989; Holloway et al., 1991). Here the sight, smell, and taste of the defensive secretion may serve as a self-advertisement, warning the predator that the rest of the prey is toxic. Even when chemicals are stored internally, conspicuous aposematic signals may direct a predators' attention toward internally stored defences that are localised in the least vulnerable parts of a prey animal, and again some degree of conspicuousness may be necessary for this to be effective (e.g. Brower and Glazier, 1975; Penney, 2002). Brower (1984) saw that the taste of a prey could serve an aposematic function (and see Lindquist, 2002). In a related discussion, Turner (1984) argued that if a toxin did not also confer distastefulness, then this trait might evolve in order to enhance individual survival from attacks. It may also be that distinctiveness as well as aversiveness of a taste or smell may evolve as an aposematic signal.

We conclude that unprofitability may be frequently signalled by conspicuousness rather than some other trait for at least two major reasons; first because it confers reliability, especially for prey that gain from being freely exposed in their environments; and second because it directs predator attention to the possession of some aversive component. In the next section we consider that conspicuousness may in addition generate special responses in predators that heighten avoidance and hence heighten aposematic protection.

## 7.3 Design of aposematic displays II: the psychological properties of predators

If attacks on unprofitable prey are costly for both predators and prey then we should expect a signalling system to evolve that matches the form of a warning display with the psychological characteristics of the predator such that prey avoidance is enhanced. At its most fundamental this means that if, for instance, relevant predators see in colour but not in the UV, then prey will evolve warning displays that function only in colour. Thus, the general form of warning displays must lie within the operational boundaries of the perceptual systems of relevant predators (see Endler, 1978, 1988; Endler and Mappes, 2004). In addition, we should expect some kind of match between signaller and receiver that extends beyond the purely perceptual. There are at least four major components of predator psychology that may affect the way that predators and aposematic prey interact. These are: (1) the capacity to show unlearnt wariness of prey items; (2) a capacity to learn to avoid defended prey; (3) memory retention of learnt avoidance, and (4) recognition processes. It is easy to see that a prey's survivorship is maximized if it can present a predator with a warning display that: (i) enhances unlearnt wariness, (ii) accelerates avoidance learning that occurs if wariness fades, (iii) reduces any tendency for predators to forget, and (iv) maximizes accurate recognition, so that attacks are not made in error.

Avoidance of prey obviously brings benefits to predators if it prevents attacks on unprofitable food items. However, avoidance also incurs costs if it leads predators to forego profitable opportunities, such as attacks on nutritious prey items of similar appearance. It seems likely that unlearnt persistent avoidance of a prey will be optimal when the prey species is highly dangerous and sufficiently abundant that the risk of death from an attack outweighs opportunity costs of failing to attack. However, when the threat from a defended prey is relatively small and the costs of avoidance high, it might be optimal not to manifest unlearnt avoidance responses. When predators do not show unlearnt wariness of aposematic prey, we might expect that learning and memory retention are both heightened so that predators quickly protect themselves when prey do turn out to be harmful.

It is possible to envisage optimisation of the forms of predator behaviours and the warning displays of prey as separate processes. However, it seems likely that at some stages there have been co-evolutionary interactions in which, for example, a change in predator behaviour elicited changes in prey phenotypes, stimulating further evolution in predators (Turner, 1984; Sherratt, 2002*a*). Whether predators and prey simply converge toward some mutually beneficial state, or whether there are long periods of unidirectional evolution by predators or prey is a complex question, highly dependent on the conditions of individual species. Whatever the precise mechanisms of co-evolution, there are good theoretical explanations to explain why the design of warning displays may enhance the willingness of predators to avoid defended prey. For most of the rest of the chapter we therefore consider the evidence that warning signals do indeed enhance wariness, learning, memory retention and recognition in a manner consistent with enhancement of predator avoidance and to greater or lesser extents, with predator–prey co-evolution. There is a large body of relevant experimental evidence (e.g. Lindström 2001), and hence the following sections are quite detailed. Readers who do not require a detailed account may wish to move to our summary of these studies (Section 7.3.5) and their implications.

Finally, having reviewed the predator psychology literature, we return to the question of conspicuousness, asking the question what came first, conspicuousness or special predator responses to conspicuousness?

### 7.3.1 Unlearnt wariness

*Unlearnt avoidance of colour patterns*

For a long time it was believed that predators did not have instinctive aversions to aposematic displays (e.g. Lloyd Morgan, 1896; Cott, 1940). A series of studies have now, however, established that naive predators do indeed show unlearnt aversions to aposematic prey, or more generally feeding biases away from certain prey forms. Nonetheless, these unlearnt avoidances can be plastic and affected by subtle environmental cues; they also range from very intense distress reactions through to mild temporary avoidance and cautious handling of prey.

Unlearnt preferences for different coloured foods were known in birds since the 1950s (see Schuler and Roper, 1992) and have been investigated in research directed at poultry production as well as aposematism (e.g. references in Jones, 1986; Mastrota and Mench, 1994, 1995). In a number of cases there is good evidence for very strong unlearnt avoidance reactions, particularly to venomous, aposematic snakes (Smith, 1975, 1977), showing localized geographical specialization (Smith, 1979). Sometimes reactions to snakes can be very strong indeed. Caldwell and Rubinoff (1983), for example, demonstrated that naive herons and egrets had strong distress reactions to presentations of the yellow-bellied sea snake (*Pelamis platurus*); the animals 'flew erratically around the aviary, trying to escape by scratching at the gate with feet and bill . . . [subsequently] . . . all birds remained immobile except for feeble attempts to get out of the cage.'

Demonstrations of strong, unlearnt avoidance reactions to highly dangerous prey are to be expected, if as we suggested above, the strength of avoidance is optimised in relation to the threat posed by prey. However, there are no experiments that test the durability of these very strong responses to repeated exposure. On grounds of animal welfare this seems a good thing indeed; however, it does mean that we do not have a test of the general prediction that very strong unlearnt avoidances are also likely to be very durable.

When there are graded avoidance responses it is possible to examine how the components of aposematic displays contribute to unlearnt wariness in predators; therefore most of the empirical work on avoidance signals and unlearnt responses over the last 35 years has been focused on bird-insect systems. Coppinger (1970) showed that a range of naive, hand-reared birds (blue jays, *Cyanocitta cristata*, red-winged blackbirds, *Agelaius phoeniceus*, and grackles, *Quiscalus quiscula*) showed a marked, unlearnt avoidance of novel foods (initially mealworms) and of red/black but palatable lepidoptera (*Anartia amathea*) compared to similarly sized brown and white lepidoptera (*Anartia jatrophae*). Since the birds had been fed on brown food pellets, Coppinger attributed his results to the relative novelty of the colours in *A. amathea* rather than the colours themselves; however, some authors (e.g. Curio, 1976) suggested that since unlearnt avoidance of the black and red pattern cannot be ruled out, Coppinger's experiment may in fact be an early and important example of unlearnt avoidance of specific colours. To investigate the matter further, Schuler and Hesse (1985) presented domestic chicks (*Gallus gallus*) on their first feeding with a choice of mealworms painted either with black/yellow stripes or with a uniform olive green. While chicks pecked at these two prey forms equally, they ate significantly fewer black/yellow striped mealworms, and hence showed an unlearnt 'proceed with caution' reaction to the aposematically patterned mealworms.

In order to determine whether the effect was caused by particular colours and/or their particular patterns along the mealworm, Roper and Cook (1989) developed this experiment by presenting chicks with four types of mealworm, again on their first feeding: green (control, mixed from yellow and black paints) and one of : (1) black/yellow stripes, (2) plain black, (3) plain yellow, and (4) 'bicoloured': half yellow/half black prey. Furthermore, Roper and Cook (1989) considered the ingestive process in terms of pecks and 'pick-ups', in which the prey was lifted in the bill. Measured relative to the green control prey, chicks were more averse to black than to black/yellow striped prey, significantly reducing contact by all behavioural measures with black, but reducing only the number of pick-ups with yellow/black striped prey. In contrast, by some measures of ingestion, chicks preferred yellow mealworms and bicoloured mealworms to green. These results strongly suggest that chicks have quite specialised avoidance mechanisms that are not always based simply on colour, but rather on colour and form of an aposematic display.

Though unlearnt, Roper and Cook (1989) established that these preferences were in fact susceptible to environmental modification. Chicks that were reared in cages with black and yellow walls and floors (rather than the standard grey walls and white floors) changed their preferences: reversing the aversion to black/yellow striped prey (and also reducing the aversion to black prey, increasing preferences for yellow prey and for yellow/black bicoloured prey). Hence, while the animals showed unlearnt preferences against some aposematic colours and patterns (black, stripes) but not others, these 'avoidance instincts' were not fixed and could easily be modified by the rearing environment.

*Further effects of visual stimuli on unlearnt avoidance reactions*

An interesting question concerning unlearnt avoidance is whether predators avoid particular colour patterns, or whether the degree of contrast with the background contributes to the unlearnt avoidance. To examine this question Roper and Cook repeated parts of their first experiment using different colours. Relative to olive green prey, they found strong unlearnt aversions (to plain black prey), less intense aversions to other patterns (plain red), and some mild preferences (for red/yellow striped prey, red/yellow and red/black bicoloured). Since all prey were presented against white backgrounds, Roper and Cook ruled out the possibility that differences in contrast against the background were likely to cause the preferences shown by the chicks. Later Roper (1990) presented red and brown mealworm prey on red, brown and white backgrounds; brown mealworms were preferred in all conditions (and see Sillén-Tullberg, 1985*b*). Thus, based on a visual signal alone, there is evidence that colour per se rather than degree of contrast determines unlearnt biases in feeding preferences in chicks.

Another important visual component determining the efficacy of warning displays may be their size. A series of investigations on phobic reactions in *Gallus* chicks by Gamberale and Tullberg (Sillén-Tullberg, 1990; 1996*a*; Gamberale and Tullberg, 1998) found good evidence that wariness in naive birds increases with prey size and with the size of prey aggregations (Gamberale and Tullberg, 1998). An important finding was that the effect of aggregation was apparently colour-specific; aggregation-induced phobias were not generated by non-aposematic insects. Indeed when the prey were edible mealworms, the attraction of aggregations increased with their size. Why should predators be particularly wary of aggregated prey? One explanation is that aggregation is itself a component of many aposematic displays, as Müller (1877) first suggested. If, as we argued in Chapter 5, prey without secondary defences are unlikely to form conspicuous aggregations, then aggregation itself may be a reliable indicator of unprofitability. In relation to aggregations, the optimal level of wariness for naive predators may often therefore be quite high.

In summary, these experiments allow us to conclude that birds do show quite specific, unlearnt wariness to visual stimuli that are associated with aposematism. However, as we now explain, warning displays often function in more complex ways to provoke unlearnt avoidance in predators.

*Multi-modal displays enhance unlearnt biases and wariness*

There is considerable evidence that signals which work in more than one sensory modality may be more effective than monomodal signals (see review in Rowe, 1999). Warning displays may consist of one or all of visual, auditory, olfactory, gustatory, and behavioural components. Pyrazine is a common olfactant in insect displays: it is volatile and pungent (Woolfson and Rothschild, 1990). Some multi-modal warning displays have quantitative effects on predator behaviour and are more effective at deterring predators than the sum of their parts (Marples et al., 1994). Hence, the common and understandable assumption of naturalists (e.g. Poulton, 1890) that an aposematic signal is primarily a conspicuous visual display (a 'warning colouration' in Poulton's phrase) may lead to a misinterpretation of the structure of the signal altogether.

One instructive case study in multi-modality is the question of contrast in the generation of unlearnt phobias. Roper and Cook (1989) and Roper (1990) found that specific prey colours provoked unlearnt avoidance reactions, but contrast against a background had no such effect. However, Lindström et al. (2001*c*) performed a similar experiment presenting chicks with matching or visually contrasting food items (green and purple food crumbs on either green or purple backgrounds) in the presence/absence of a pyrazine olfactant. The presence of pyrazine enhanced unlearnt bias against the contrasting food items compared to the cryptic items (Fig. 7.2), even in groups of birds that were familiar with the contrasting food items before contact with pyrazine. Although the experimental systems differ, the contrast in results is instructive. If we ask birds to show unlearnt responses to warning displays and only present one component of the signal, we may considerably underestimate the existence and intensity of their avoidance behaviours (see also Kelly and Marples, 2004).

A number of other experiments show that unlearnt wariness in chicks can be provoked by particular kinds of novelty within an aposematic signal. For example, the presence of a novel pyrazine odour biased food choices away from aposematic colours, even when the foods themselves were familiar, having been eaten on up to 48 occasions before testing (Rowe and Guilford, 1999). Furthermore, olfactants can also interact with the novelty of colours. Using naive *Gallus* chicks, Marples and Roper (1996) showed that the presence of pyrazine and other odours interacted with the novelty of (blue or green) colours of food or drink to greatly heighten neophobia: attack latencies were considerably lengthened by the presence of odours only if the colour of the food and drink presented was novel. Furthermore, novel odours themselves can heighten neophobia for novel colours even when the odours are not associated with aposematic displays (Jetz et al., 2001). In an extension of this work Rowe and Guilford (2001) report experiments in which the buzzing sound of a bumblebee biased foraging preferences of chicks against novel coloured food items and also against familiar yellow food (relative to green). As Rowe and Guilford (2001) point out, unlearnt biases may extend beyond sight, sound and smell, to incorporate taste (and even tactile stimuli). Hence, the role

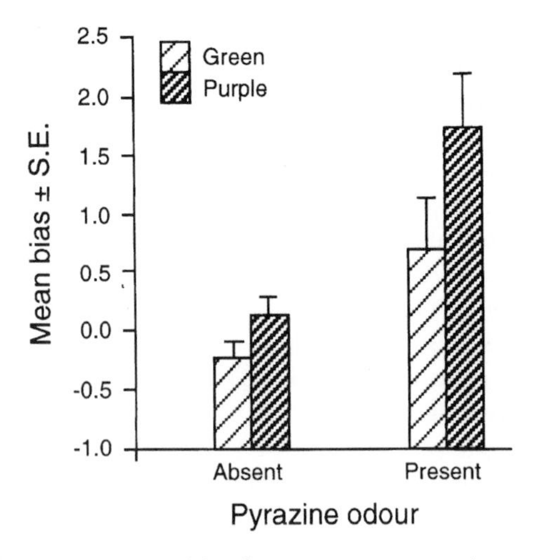


**Figure 7.2** Mean attack bias for or against green or purple conspicuous crumbs in the presence and absence of pyrazine odour. Dark shaded bars represent purple backgrounds (green conspicuous crumbs, purple cryptic crumbs). Light shaded bars represent green backgrounds (purple conspicuous crumbs, green cryptic crumbs). (Attack bias is the number of cryptic crumbs attacked minus the number of conspicuous crumbs attacked.) Redrawn from Lindström et al. (2001*c*).

of multi-modality extends clearly beyond olfaction and visual stimuli in a manner still to be evaluated in detail.

*The presence of familiar foods enhances avoidance of novel items*

A final and important point about wariness of novel prey items is that predator aversions to novel foods are heightened by the presence of familiar, edible food items. Notably, Marples et al. (1998) familiarized 19 wild blackbirds with a coloured pastry bait (green, blue, or red) and then offered them a choice between the familiar bait and a novel coloured bait. About one-third of the birds showed no fear of the unfamiliar coloured bait and recruited it quickly into their diets. However, others took much longer: one-third of the birds took more than 20 trials to recruit the novel bait into their diets, with one stubborn bird needing 125 presentations. This is an example of dietary conservatism; a prolonged reluctance by a predator to handle and then to incorporate a new food item into its diet (Marples et al., 1998; Marples and Kelly, 2001; Thomas et al., 2003; Kelly and Marples, 2004). In a related experiment Lindström et al. (2001*b*) presented great tits with unpalatable and novel artificial prey alongside edible, cryptic artificial prey. Here the aposematic signals were black and white symbols carefully designed to be either of high or low levels of conspicuousness. When the cryptic prey were familiar, the birds showed a strong feeding bias away from the conspicuous prey and toward the cryptic prey, especially if the novel prey were highly conspicuous.

Birds are presumably reluctant to incorporate new prey into their diets because they already have access to a profitable and familiar set of prey and because they also have a learnt (or indeed, unlearnt) tendency to treat novel conspicuous items as if they could be harmful. The birds presumably avoid novel prey until something changes, for instance that hunger passes some threshold. In addition, birds may also begin to attack if persistence of a novel prey for some minimum time is an indicator that the prey is not likely to be transient in the predator's habitat. Finally, it may be that some birds avoid the novel prey until conspecifics appear to profit from taking them. Boyden (1976), for example, reported a case of such 'eavesdropping' in reptiles when he noted that 'other lizards, if nearby, would also avoid the butterfly on this cast if it was left alone by the first lizard'. Whatever the precise reasons, the remarkable duration of dietary conservatism in at least some birds may imply that avoidance of novelty (with learnt and unlearnt components) does make an important contribution to the evolution of warning signals (Thomas et al., 2003; Thomas et al., 2004).

*What can we learn about wariness?*

These experiments suggest that at least some species of birds have specially evolved wariness responses which are evoked by fairly specific kinds and combinations of stimuli. Birds are wary of novel prey that have particular colours or colour patterns especially if they are accompanied by odours and sounds. In bird–snake systems wariness can be manifested as extreme distress reactions that lead to high levels of avoidance, whereas in bird–insect systems avoidance may wane much more quickly. Furthermore, predators become risk-averse and avoid novel prey items if there is an abundance of edible, familiar food present.

However, it is important to note a number of limitations to our knowledge of the wariness reactions of birds, especially because so many experiments use domesticated chicks (*Gallus gallus domesticus*). These animals have been bred for diminished fear of new food items (Marples and Kelly, 2001) and comparisons with other species indicate that levels of wariness in non-domesticated birds may be much higher (see Schuler and Roper, 1992; Kelly and Marples, 2004). In addition, there may be effects on wariness of age (e.g. *Parus major*, Schuler and Hesse, 1985; Wrazidlo, 1986; Lindström et al., 1999*a*), mode of presentation of the food (Schuler and Roper, 1992; Roper and Marples, 1997*a*; Gamberale-Stille and Tullberg, 2001), and prior experience with other novel food items (Shettleworth, 1972; Schlenoff, 1984; Jones, 1986; Roper, 1993; Marples and Kelly, 2001). A more detailed review of factors that limit our understanding of wariness is provided in Speed (2000). These limitations mean that we should be cautious in assuming that particular demonstrations will always apply when naive predators encounter defended prey. However, we should, nonetheless, conclude that there is persuasive evidence that some bird species do show heightened unlearnt wariness of aposematic, especially multi-modal warning displays.

### 7.3.2 Aposematism and predator learning

*Enhancements to learning*

When unlearnt avoidances have waned and predators begin to attack aposematic prey, learning takes over as a primary determinant of avoidance behaviours. It has been repeatedly demonstrated that warning displays can accelerate learnt avoidance and thereby reduce the number of prey martyred to predator education (e.g. Gittleman and Harvey, 1980; Roper and Wistow, 1986; Riipi et al., 2001). The mechanisms by which warning displays accelerate learning in predators are complex and have been very well reviewed by Turner (1984) and Guilford (1990*a*) amongst others. Here we consider how warning displays can accelerate learning by two major mechanisms. First, the conspicuous appearance of a prey may change the predator's rate of attack and thus, by a behavioural mechanism, facilitate avoidance learning. Second, the particular features of warning displays may enhance the attention that predators give to their prey so that at each sampling event predators commit more to memory than they would in the absence of a warning display. The crucial distinction between these two mechanisms is that the second 'attentional' mechanism works regardless of the rate at which prey are encountered by predators: it therefore provides a density-independent enhancement to avoidance learning. These attentional enhancements to learning may to some extent be specialized, adaptive responses of predators, enabling them to learn about the environment, while minimizing exposure to unprofitable experiences. In contrast, the behavioural mechanism relies only on the general feature of perceptual systems, that conspicuous stimuli are by definition more easily detected, and as such it may not represent any special adaptation in predators. We consider both behavioural and cognitive mechanisms by which aposematic prey enhance learning in turn.

*Behavioural enhancements to learning*

By definition, conspicuousness of a prey enhances its chances of detection by a predator and, other things being equal, this will increase rates of attack on aposematic prey by predators that have lost their initial wariness (Matthews, 1977; Guilford, 1990*a*). Predators probably find attacks on some prey particularly aversive if they attack them in quick succession, rather than at long intervals. One good model system is the threshold for toxicosis demonstrated in blue jays being fed on monarch butterflies that contain cardenolides (Brower et al., 1968; Mallet, 2001*a*). Taken too slowly, the toxins may have little aversive effect on a bird. Indeed, as Brower et al. (1968) noted, eaten at a low rate the butterflies could be used as a food source; but taken too quickly they cause an emetic reaction which facilitates rapid avoidance learning by a predator. Clearly conspicuousness, in the sense of a contrasting signal, could raise detection rates in this manner. However, we note that distinctiveness itself might raise initial detection rates, regardless of the conspicuousness of a prey item against its background. Gregarious aggregation too, even without other conspicuous displays, would shorten the gap between encounters and aid in avoidance learning (see discussions in Fisher, 1930; Guilford, 1990*a*) in which case an aposematic display within individual prey may not be necessary (Müller, 1877; Poulton, 1890) or may be used to affect some other component of predator psychology such as recognition or memorability.

Raised attack rates on easily detected prey could have several related effects on learning including: bringing a predator's toxin levels to the point where an aversive reaction and learning become inevitable; using fewer individual prey to obtain this result and/or increasing the aversiveness of the experience, thereby generating stronger and more durable aversions (Gittleman and Harvey, 1980; Gittleman et al., 1980).

Systematic experiments with chicks have shown higher levels of attack on conspicuous than cryptic prey items both in terms of total number attacked within a trial (Gittleman and Harvey, 1980; Gittleman et al., 1980) and in terms of the rate at which a fixed number of prey are taken (e.g. Roper and Wistow, 1986; Riipi et al., 2001). In both cases subsequently higher rates of learnt avoidance were recorded, but it is not clear that enhanced learning necessarily follows simply from enhanced encounter rate. An alternative explanation is that the birds learnt faster because contrasting food items presented a more salient stimulus that accelerated learning independent of attack rates. Nonetheless, it does seem likely that raised attack rates would aid avoidance learning especially in systems in which toxins are slow-acting and in which one individual prey is insufficient to cause an aversive reaction on its own.

*Cognitive effects*

Here we group together varied qualities of aposematic displays that have in common the capacity to increase the rate of acquisition of long-term memories for prey avoidance by predators, independent of the number of attacks that a predator makes over a time interval. When a predator learns about an unpleasant aposematic prey it forms a long-term memory of the warning signal (or conditioned stimulus, CS) and of the prey's nastiness (or unconditioned stimulus, US). This is known as excitatory Pavlovian learning (see reviews in Davey, 1989; Pearce and Bouton, 2001). Aposematic displays can accelerate Pavlovian learning because of their absolute qualities (presence of particular components in a signal) or because of qualities dependent on context at presentation (novelty, distinctiveness, contrast against a background). This latter category has received the most empirical attention and we therefore consider it in most detail here.

*Novelty*

The concept of novelty is apparently simple—some set of stimuli that a predator has not met before—but is in truth quite subtle. For instance, the whole display may be novel, but its components familiar; or the whole display may be familiar, but to some extent forgotten. We therefore propose a simple operational definition here: novelty describes *a set of stimuli that are treated at least as very unfamiliar by a predator*.

The hypothesis that novelty can have cognitive effects that enhance avoidance learning and thereby facilitate the survival of novel aposematic prey should be testable independently of particular colours or degrees of contrast in visual signals. Such a test was reported by Roper (1993) who familiarized chicks with either red or blue fluids and subsequently trained them to avoid a familiar or an unfamiliar distasteful fluid. The rate of avoidance learning was highest when the training colour was novel, regardless of which colour was used (see also Shettleworth, 1972).

In a seminal discussion, Turner (1984) suggested that when predators ingest several different food items and subsequently become ill they need some reliable way to identify the cause of the aversive reaction in order to avoid it in future. Clearly novelty is one label on a food item to which predators should pay attention. It may be that predators are prepared by evolution to blame aversive reactions on any novel aposematic prey item recently eaten. There is an abundance of evidence from the psychology literature that taste, smells, and visual stimuli can act in this way; even if there are delays of 24 h between experience of the novel stimuli and illness. Indeed Turner's hypothesis explicitly drew on the discovery of taste avoidance learning (Garcia and Koelling, 1966) in which rats associated the novel taste of saccharin with sickness induced by X-irradiation (see discussion in Davey, 1989).

*Distinctiveness*

A second property of warning displays that may accelerate learning is their distinctiveness relative to cryptic individuals. Aposematic signals may enhance the rate of learnt discriminations between defended (aposematic) and edible (cryptic) prey. Here it is primarily important that signals are distinct rather than necessarily conspicuous; however, it may be that conspicuousness is one very good way of being distinct from edible, cryptic animals (Wallace, 1867, 1889). Laboratory studies with pigeons and chicks as predators have indeed shown that distinctive components of signalling stimuli can facilitate learnt discriminations between rewarded and unrewarded (or unprofitable) events. Kraemer (1984) reported an experiment in which pigeons learnt which of two visual signals predicted food and which predicted no food. Signals were small illuminated dots projected on a screen, and the birds learnt the discrimination most readily when the signals were most distinct (i.e. different colours, numbers of dots).

In the context of warning displays Gagliardo and Guilford (1993) found that discrimination learning by young chickens (*Gallus gallus domesticus*) between a yellow distasteful food and a green edible food was substantially accelerated if more than one unpalatable food item is present. Hence prey aggregation may enhance discrimination learning. Similarly with great tits (*Parus major*) as predators Riipi et al. (2001) demonstrated a strong interaction between distinctiveness of prey and their group size on discrimination learning. The birds could not learn to distinguish edible from distasteful prey when their signal was similar, however, when their appearance was dissimilar, they not only learnt the discrimination but did so most quickly when prey were grouped into aggregations (Fig. 7.3, see also Mappes, 1997*a*). Even if there are some effects of encounter rate in these experiments, they do provide compelling evidence that aggregation enhances the speed of learnt discrimination.

As well as variation within a modality, visual discriminations between edible and unprofitable, toxic prey may be enhanced by the presence of a warning odour: Roper and Marples (1997*b*) found that both vanilla odour and almond odour could facilitate a learnt discrimination between two visually identical fluids, one of which was distasteful. Similarly, Guilford et al. (1987) showed that pyrazine could be used as a cue in discrimination learning. Extending the scope of multimodal studies, Rowe (2002) recently demonstrated that a sound could enhance learnt discriminations by chicks between rewarding and unrewarding prey items. This experiment suggests that we have still much to learn about multi-modality and warning displays.

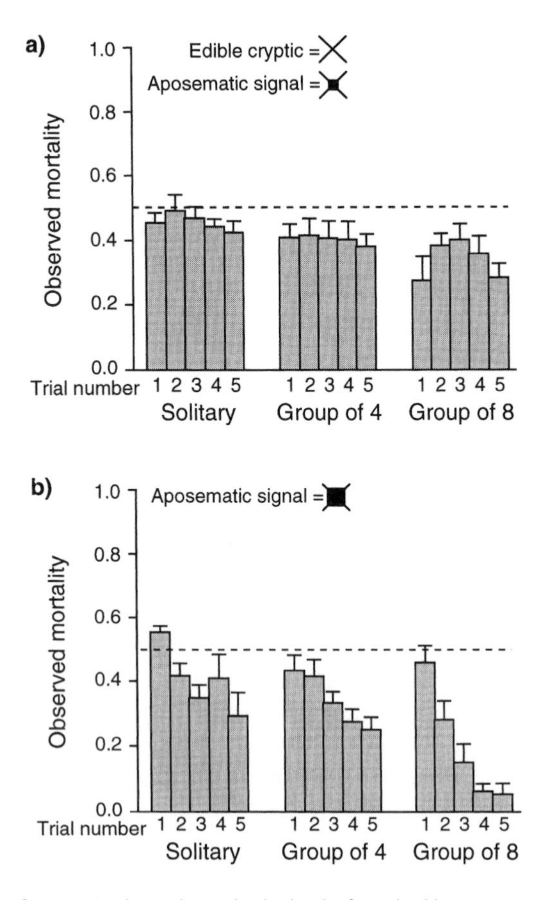


**Figure 7.3** Observed mortality (with SE), of unpalatable aposematic prey as the proportion of unpalatable items from all prey that were eaten. The dashed line represents equal mortality for palatable cryptic and unpalatable aposematic prey. Over sets of five trials, birds were presented with solitary edible cryptic prey and with distasteful aposematic prey in groups of varying size. In (a), the aposematic signal is of low distinctiveness; in (b), it is of high distinctiveness. Redrawn from Riipi et al. (2001).

*Contrast against a background*

There is some good evidence that visual contrast may accelerate simple (i.e. non-discriminative) avoidance learning. Thus, Roper and Redston (1987) presented chicks with conspicuous or cryptic beads that were either distasteful or not. Chicks were allowed a single peck in training, and multiple pecks for a defined period in subsequent testing. Birds presented with conspicuous, distasteful beads pecked least frequently at retesting, regardless of whether the background in question was red or white (a similar result is noted in Roper, 1990). While this appears to be definitive evidence that contrast per se enhances avoidance learning it should be noted that to make it distasteful, the bead was coated with methyl anthranilate. Later Marples and Roper (1997) noted that since methyl anthranilate has a detectable odour for humans, it might affect learning in chicks.

In fact Marples and Roper (1997) demonstrated that chicks can detect the odour of methyl anthranilate, that they can use it as a discriminative stimulus and, in the first demonstration of its kind, that this odour itself had aversive qualities for chicks. Though Roper and Redston's (1987) experiment (and Roper, 1990) is usually considered a demonstration of single trial learning from a monomodal signal, we can not exclude the possibility that the presence of the olfactant affected learning. The strong odour of methyl anthranilate may have interacted with detectability, biasing learning perhaps by accelerating it in the conspicuous condition. We therefore conclude that contrary to common belief, a definitive demonstration that visual contrast alone can accelerate simple avoidance learning, irrespective of the rate at which prey are attacked, remains elusive.

*Absolute qualities of warning displays*

It may (or may not) be that particular colours, odours, sounds, tastes—or combinations of these in whole signals work to accelerate avoidance learning irrespective of other contextually determined qualities (i.e. novelty, distinctiveness, conspicuousness, etc.). However, the evidence that such effects are widespread is not presently compelling. Thus, although Roper and Marples (1997*b*) found that almond enhanced the rate of avoidance learning for a familiar coloured fluid, but vanilla did not, the odours were novel in both cases and we simply do not know whether similar effects pertain when chicks have met the odours before.

Roper and Cook (1989) demonstrated that there was no effect of prey colour/colour pattern on the strength of one-trial avoidance learning. Hence, while colour generated unlearnt biased preferences for and against particular colours, there was no apparent effect of particular colours/patterns themselves on avoidance learning. This presents the surprising conclusion that visual stimuli which generate unlearnt wariness do not enhance learning. Roper and Cook postulated the ingenious idea that since animal learning takes place when events are surprising (see, e.g. Davey, 1989; Pearce and Bouton, 2001) animals that are appropriately prepared by evolution will not be surprised, and therefore will not learn rapidly, when certain aposematic prey are indeed aversive. These results hint at the intriguing possibility that prey animals may in some circumstances face trade-offs in signal design; some components enhance learning but diminish unlearnt wariness and vice versa.

More success in demonstrating that physical qualities of aposematic displays can affect avoidance learning has been demonstrated with visual aggregation of prey. Taking perception of the entire group as a warning stimulus in itself (Müller, 1877; Poulton, 1890) the sheer size of a perceived aggregation (not only in visual, but also olfactory and auditory terms) is an absolute quality of a warning signal (regardless of its colour, distinctiveness, contrast, type of odour, etc.) that may affect learning rates. In addition to enhancing learnt discrimination, aggregations may indeed enhance learning rates. Notably, Gamberale-Stille (2000) demonstrated a greater rate of learning by *Gallus gallus* chicks with aggregated than solitary prey (though see Gamberale and Tullberg, 1996*a*). The coincidence of a warning stimulus and an aversive stimulus in a predator's attention may thereby help to accelerate or strengthen avoidance learning by predators (see Edmunds, 1974; Gagliardo and Guilford, 1993). Alternatively, it may be that birds generalize from an aversive experience with one individual to the remaining individuals within an aggregation. Birds may give up attacks early because there is a high chance that if one individual is aversive, the rest of the aggregation may also be aversive (e.g. Alatalo and Mappes, 1996; Riipi et al., 2001; Sherratt and Beatty, 2003). Whichever is the case, it does seem that size of a prey aggregation can enhance learning, even when it is not a learnt discrimination.

### 7.3.3 Memorability

Warning displays may include components that prevent or delay forgetting (Speed, 2000). In behavioural terms forgetting of long-term memories is manifest in the reversal of learnt behaviours as time passes (Spear, 1978; Bouton,

1994; Kraemer and Golding, 1997) or as physical contexts change (Bouton, 1993). In a number of cases, forgetting seems to happen not because memories themselves are lost, but because they become harder to retrieve (or activate) as time passes. However, it remains possible that either animals really do lose memories over time or that animals remember events very well, but for subtle reasons of behavioural optimization (not evident to experimental psychologists), choose to behave as if they have not retrieved certain memories.

Good memorability features of warning displays are those that provoke low rates of forgetting, and conversely poor memory features generate high rates of forgetting. It is thus possible to consider two warning stimuli which might generate similar levels of learnt aversion, but different forgetting rates (Speed, 2000). The signal with good memorability features would be forgotten more slowly, and thus provide the prey with greater protection than one with poor memorability features. It is easy to see that good memory retention is generally beneficial for defended prey, but for predators it may be optimal to reverse learnt avoidance if sufficient time has elapsed and/or a minimum number of prey have been avoided. Though we might expect memory to decay in relation to the ecological importance of an event, there are surprisingly few data relevant to aposematism (see review in Speed, 2000). For instance, we know of no published data that properly test whether forgetting rates are directly related to the intensity of a reinforcer. None the less memorability may be heightened by at least three mechanisms.

First, memorability for experience with a prey may be affected by the specific qualities of its warning signal (conditioned stimulus) and the strength of its reinforcing property (unconditioned stimulus). Thus, Roper and Redston (1987) showed that aversive conditioning to an unpalatable object is accelerated and forgetting is decelerated if the conditioned stimulus is conspicuous rather than cryptic (Fig. 7.4; note again though that the presence of the malodourous methyl anthranilate at learning means that some degree of multi-modality may be involved). Similarly, Roper and Marples (1997*b*) found that relative to a vanilla odour, almond odour accelerated avoidance learning and seemed to decelerate subsequent forgetting. Unfortunately, there are few similar studies and none that determines whether individual signal components interact in their effects on memorability.

Second, warning displays may generate particularly unforgettable discriminations between defended aposematic and edible cryptic prey. Hence, distinctive components of warning displays may be good memorability features whose distinctness provides protection against 'discrimination forgetting'. One good example is reported by Kraemer (1984), but more studies are undoubtedly necessary.

Third, warning displays may jog predator memories (see examples in Speed, 2000). Memory-jogging may consist of one or all of: temporarily halting forgetting; reversing the effects of previous forgetting, and slowing down any subsequent forgetting. If seeing and avoiding conspicuous aposematic prey leads to memory-jogging it would therefore oppose the detrimental effects of predator forgetting and increase avoidance of defended prey. In many habitats seasonality means that relatively long-lived predators will not meet many otherwise

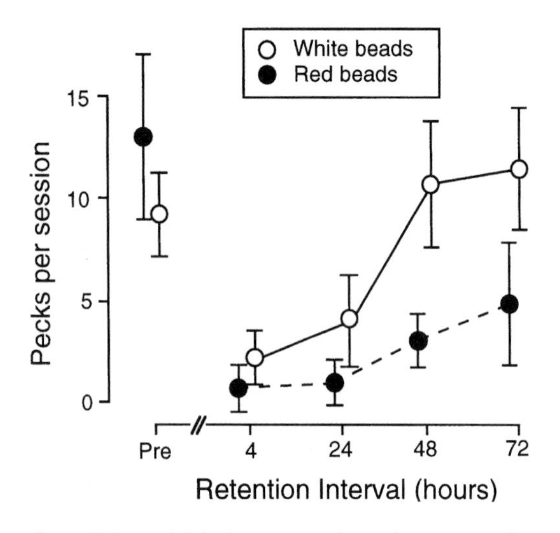


**Figure 7.4** Attack behaviour measured as pecks per session for chicks that (a) meet red or white beads for the first time (Pre) and (b) meet them again at one of the retention intervals after a single experience with a distasteful bead. All beads presented on a white background. Redrawn from Roper and Redston (1987).

familiar prey for some months, and here memory-jogging may be particularly important.

Compared to cryptic phenotypes aposematic displays may be particularly good as memory-jogging devices for two reasons. First, if as is usually the case, they are more conspicuous than related cryptic displays they will be detected and hence 'jog' memories more frequently. Relatively little is known about the controls of memory-jogging, so that it may be that memories are reinstated more easily if reminders are spaced in time at some optimal, presumably fairly frequent duration. Second, aposematic displays may have particular components (or sets of components) that are intrinsically good at memory-jogging. Thus, Rothschild (1984) suggested multi-modality in the display of defended hymenoptera may function as a particularly effective *aide memoire*. However, there are very few studies that directly test specific hypotheses about memory-jogging and aposematism.

In sum, the question of aposematic signal design and memorability is important and promising but in need of systematic experimental investigation.

### 7.3.4 Recognition

*Recognition and warning displays*

Here we consider the fourth and final component of predator psychology that may be affected by warning displays; how predators use information gained from their learning experiences to recognize and identify aposematic prey. Some of the features that enhance learning, may also aid recognition after learning is complete; distinctiveness is one such feature. Wallace's original explanation of aposematic displays was essentially that they facilitate recognition and discrimination from cryptic, edible prey (and see Fisher, 1930). In an extension of this hypothesis, Guilford (1986) proposed the distance-detection hypothesis; that conspicuous and distinctive warning displays may reduce recognition errors in experienced predators; and furthermore, that recognition accuracy would be enhanced by the fact that since conspicuous prey will be detected from greater distances than cryptic prey they can be examined for longer before an attack (Guilford, 1990*a*). Although intuitively appealing, direct experimental evidence is not extensive; thus Guilford (1986) demonstrated that chicks would make fewer recognition errors if they had longer to observe defended prey before an attack. However, with great tits as experimental predators he failed to show an improvement in error rate with distance (Guilford, 1990*a*). However, Gamberale-Stille (2000) recently explored the effects of visual contrast on prey recognition, demonstrating that experienced chicks would make fewer attacks on the defended, aposematic heteropteran bug *Tropidothorax leucopterus* when it contrasted with its background than when it matched its background. Furthermore, error rates increased in the presence of competitive conspecifics when decisions presumably have to be made by the birds more quickly.

A second, important way that recognition systems may affect signal design is by biased generalization. In *unbiased* generalization predators would show their highest levels of learnt avoidance in relation to the warning displays used by a prey. However, in biased generalization, they may show greater post-learning avoidance to exaggerated versions of the original display. Thus a mutant prey that was, for instance, more conspicuous than the rest of its aposematic conspecifics might gain greater protection than them. This could happen simply because predators generally show greater, supernormal responses to exaggerated prey. However, biased generalization could also be generated by peak-shift-like effects (Hanson, 1959), which result from discrimination learning. During discrimination learning, an animal learns that two stimuli, which vary for instance along a single stimulus dimension (such as wavelength of light), signal motivationally opposed outcomes that require different responses. However, when learning is complete the peaks of these responses lie not with the original training stimuli, but can be shifted in opposite directions along the stimulus dimension, away from the original positions of the training (e.g. Hanson, 1959; Honig, 1993). Shifting the peak of avoidance of defended prey would favour exaggeration of the distinctiveness and conspicuousness of a warning signal. There is a growing body of evidence in birds that is consistent with an expectation that predators should show peak-shift effects. In discrimination experiments learnt responses may be shifted away from initial, training stimuli to more conspicuous or larger forms (see e.g. Gamberale and Tullberg, 1996*b*; Gamberale-Stille and Tullberg, 1999; Lindström et al., 1999*b*) when learning is complete. Recognition systems can therefore have a profound effect on the form of warning signals, selecting for distinctiveness and exaggerating them through peak-shift like processes.

### 7.3.5 Summary

The complexity and subtlety of animal cognition is reflected in the complexity of the experiments reviewed in the preceding paragraphs. Nonetheless, we can make four broad points about the ways in which the design of warning signals may affect the behaviour of predators.

First, unlearnt wariness is likely to be an evolved (perhaps co-evolved) phenotype that prepares naive predators for unprofitability in prey. Birds seem to have unlearnt aversions to colours and colour patterns that are typically aposematic. However, an accurate evaluation of the form, intensity and durability of this behaviour may require further, quite extensive testing with multi-modal stimuli.

Second, aposematism may enhance learning by accelerating the frequency of predator–prey encounters, or by causing cognitive changes that cause higher learning rates regardless of the rate of encounter. This second category of enhancement to learning is complex (including traits such as novelty, distinctiveness, conspicuousness, and magnitude of a signal). It may in part represent an adaptive response in predators, such that they give more attention to reliable signals of unprofitability.

Third, there may be important effects of warning displays on memory, slowing down forgetting, perhaps reinstating memories of the aversiveness of prey by memory jogging. Currently, relevant data are lacking.

Finally, distinctiveness and conspicuousness of aposematic signals may enhance their recognition by predators, and experiments are beginning to demonstrate that this is indeed the case. After discriminations have been learnt, predators may heighten their avoidance response if characteristics of a warning signal are exaggerated. Consistent with this, biased generalization has been repeatedly shown in laboratory birds.

To end this chapter, we return to the earlier question, *why conspicuousness?* And re-examine it in the light of abundant evidence that warning signals do indeed heighten wariness and accelerate learning in predators.

## 7.4 Co-evolution: which came first, conspicuousness or special psychological responses to conspicuousness?

Although quite a lot is known about the way that warning signals work with contemporary predators there have been few, if any, analyses of the extent to which behaviours like those outlined in the preceding sections are causes or consequences of the evolution of warning displays (though see discussion in Sherratt, 2002*a*; Speed, 2003). Thus: did pre-existing biases in predators drive the evolution of signal design, or did the forms that prey presented to predators drive the evolution of their psychological systems? These questions are complex, not least because the first species with aposematic displays were in aquatic environments (Tullberg et al., 2000) and aposematic animals have a wide range of predators with different sensory qualities (Bowdish and Bultman, 1993; Kauppinen and Mappes, 2003), yet nearly all of our relevant data come from birds. Nonetheless, it should be possible to construct some framework for a preliminary evaluation of 'special' effects of warning signals and coevolution and we attempt this below. Since little is presently known about memorability, we limit our considerations to effects on learning and unlearnt wariness.

*Argument 1: Evolved prey phenotypes drove the evolution of predator psychology.* In this view predators had no pre-adaptive biases before meeting prey with warning signals for the first time. Instead the reliability of conspicuousness as a badge of unprofitability (Section 7.2) ensured that conspicuousness became a common component of aposematic displays. The association between conspicuousness and unprofitability then drove the evolution of predator psychology toward a number of special avoidance responses such as unlearnt wariness of unfamiliar conspicuous prey and acclerated avoidance learning about their unprofitability. In turn, changes in the cognitive structures of predators may affect the design of warning signals, for instance, selecting for similarity in signalling phenotypes.

*Argument 2: The special effects of aposematism on predator learning are based on pre-existing 'universal' properties in perceptual and cognitive systems.* Learning may be enhanced by aposematic stimuli through (1) acceleration of discrimination and (2) because certain stimuli themselves happen to enhance avoidance learning regardless of discrimination. Since all perceptual systems need to be able to generalize and discriminate, aposematic displays may simply originate because they happen to have properties (such as visual contrast against a background) that predators utilize in discrimination (cf. Wallace, 1867). They may also have been strongly influenced by inbuilt supernormal or peak-shift responses found commonly within predator psychological systems (see Section 7.3.4 and Leimar et al., 1986). Similarly it may be that absolute properties of, for example, brightness, size, strong odours, and noises of particular frequencies and intensities all happen to attract attention and thus facilitate information processing and learning; this would explain why multimodality is so common in animal signals (see review in Rowe and Guilford, 2001). It can therefore be argued that the design of aposematic displays utilize general properties of cognitive systems which were present at any 'initial origin' of very early forms of aposematism (see also Forkman and Enquist, 2000).

*Argument 3: The special effects of aposematism on predator learning are based on pre-existing but* ***not*** *universal properties in perceptual and cognitive systems.* This argument is essentially as 2, above, but here it is not the universal processes of psychological systems (discrimination, generalization etc.) that drive the initial form of warning signals, but instead particular psychological biases that predators possessed in particular habitats where forms of aposematism initially evolved. Such sensory biases could be completely non-adaptive by-products of psychological functioning, or could be adaptive biases generated for intraspecific interactions, or because of the adaptation to the general sensory ecology in which particular predators lived.

Which of these scenarios is most likely? At the time of writing, there has been relatively little discussion about alternative scenarios for co-evolutionary history of aposematic displays in the literature (though see related discussion in Jansson and Enquist, 2003). It seems likely that there is a strong element of signal reliability in the evolution of many aposematic phenotypes, hence Argument 1 seems likely to be important. However, universal functions of perceptual systems (Argument 2) and the influence of the particular evolutionary history of predators prior to the evolution of aposematic prey (Argument 3) cannot be discounted *a priori* either. Though it seems a trite conclusion, the complex matter of predator–prey co-evolution in the design of warning displays is just that; highly complex. As a constructive suggestion, we propose that one useful response to this complexity might be the generation of theoretical models that test the sensitivity of coevolutionary dynamics to systematic variation in the influences of signal reliability and the perceptual psychology of predators. In addition, we expect that comparative, phylogenetic work may be of importance (Härlin and Härlin, 2003). We suggest that co-evolutionary questions will be a major focus of work in aposematism over the next decade.

## 7.5 Conclusion: designing a warning display

In this chapter we have introduced warning displays as sets of signals that have the overall function of minimizing contact between predators and unprofitable prey. However, it is now well known that the means by which this end is met are multiple and highly complex. Nonetheless, some general aspects of signal design have been identified. The correspondence between conspicuousness and unprofitability can, for example, be explained by a signal reliability argument that postulates that since conspicuousness is often presumed to be costly for prey without secondary defences, it can

**Table 7.1** Experiments performed to investigate whether characteristics of aposematic stimuli affect particular psychological characteristics of predators

| **Aposematic Trait** | **Effect on Predators** | | | | |
|---|---|---|---|---|---|
| | ***A*** *Heighten wariness* | ***B*** *Accelerate learning* | ***C*** *Enhance recognition* | ***D*** *Enhance memorability* | ***Example references*** |
| Particular colours/ patterns | Yes | Yes | | | A&B (Roper & Cook, 1989)<br>A (Rowe & Guilford, 1996)<br>A (Sillén-Tullberg, 1985b)<br>A (Roper, 1990)<br>A (Lindström *et al*, 1999a) |
| Contrast of colour vs background | Yes | Yes | Yes | Yes | A (Lindström *et al.*, 2001c)<br>A&B (Roper & Cook, 1989)<br>B (Gittleman *et al.*, 1980)<br>B (Roper & Wistow, 1986)<br>B&D (Roper & Redston, 1987)<br>B (Riipi *et al.*, 2001)<br>C (Gamberale-Stille, 2001) |
| Odour | Yes | Yes | | Yes | A (Marples & Roper, 1996)<br>A (Lindström *et al*, 2001b)<br>A (Jetz *et al.*, 2001)<br>B&D (Roper & Marples, 1997b) |
| Sound | Yes | Yes | | Yes | A (Rowe & Guilford, 1999)<br>B (Rowe, 2002) |
| Novelty as characteristic of a stimulus | Yes | Yes | | | A (Marples & Roper, 1996)<br>A (Lindström *et al.*, 2001b)<br>A&B (Lindström *et al.*, 2001c)<br>A (Rowe & Guilford, 1996)<br>A (Jetz *et al.*, 2001)<br>B (Roper, 1993)<br>B (Mappes & Alatalo, 1997) |
| Size of visual stimulus | Yes | Yes | | | A (Gamberalle & Tullberg, 1996a)<br>A (Gamberalle & Tullberg, 1998)<br>A&B (Gamberalle-Stille, 2000)<br>B (Gagliardo & Guildford, 1993)<br>B (Gamberale-Stille & Tullberg, 1999)<br>B (Riipi *et al.*, 2001)<br>B (Mappes & Alatalo, 1997) |
| Biased generalisation | | | Yes | | C (Gamberalle & Tullberg, 1996b)<br>C (Gamberale-Stille & Tullberg, 1999)<br>C (Lindström *et al.*, 1999b) |

be used as a reliable discriminator. Conspicuousness, in the sense of exposure to predator detection, may be optimal for well-defended prey because crypsis imposes too many opportunity costs. In addition, many defences are to greater or lesser extents self-advertising, in which case conspicuousness may be important in drawing a predator's attention to them in order to minimize the damage caused by an attack.

Given that the conspicuous nature of many warning displays can be explained, we need further insights into the minds of predators in order to specify *which* particular forms of conspicuousness may heighten avoidance responses. Thus, do we know enough about predator psychology to design a hypothetical warning display that will maximise 'predator psychology' avoidance effects? Despite two decades of intensive study, the answer is apparently—no. Table 7.1 summarises the areas that have been subject to systematic experimentation. As can be seen from this table and the preceding discussion, we know quite a lot about wariness and learning, but rather less about recognition and almost nothing at all about memorability. There is, however, no reason to think that these relatively neglected components of signal design are less important than the others. Hence, if we want to know how to design an optimal warning signal from a psychological perspective, much more emphasis is needed on studies of signal recognition and memorability.

CHAPTER 8

# The initial evolution of warning displays

We now turn our attention away from explanations for the forms of warning signals and instead consider the evolutionary problems posed by the need to explain how aposematic displays initially evolved. There are two obvious approaches here: first, a mechanistic set of explanations, based on theoretical models and empirical simulations, and second, an historical, phylogenetic approach. Since theoretical models have been the focus of much more emphasis than the phylogenetic analyses, we consider modelling first and at some length.

## 8.1 The initial evolution of aposematism: the problem

Many theorists and some empiricists take as a starting point the assumption that at its genesis, aposematism evolved through conspicuous mutation emerging in a defended and essentially cryptic species (e.g. Sillén-Tullberg and Bryant, 1983; Engen et al., 1986; Leimar et al., 1986; Yachi and Higashi, 1998; Speed, 2001*a*). In this scenario, the evolution of warning displays seems improbable for two related reasons (Mallet and Singer, 1987; Mallet and Joron, 1999): first, new aposematic mutants that emerge in a population of defended, cryptic individuals lack protection from crypsis. They are, by definition, more conspicuous than their cryptic conspecifics and in a world of ignorant predators, are therefore exposed to higher risks of predation. Second, new mutants are, also by definition, very rare and can easily become extinct by attacks from ignorant predators. This latter problem—of rarity—can be further subdivided into at least two categories: first, the problem of the 'lone mutant' (Speed, 2001*a*), in which a single aposematic individual must survive long enough to reproduce; second, the problem of the 'advantaged minority', in which a more numerous, but still quite rare form must somehow incur sufficiently low costs of predator education to afford them higher per capita survival rates than much more common cryptic prey.

When aposematic displays are common, it is easy to see that they represent evolutionary stable phenotypes; predators rapidly learn to avoid them by killing a small proportion of the population, have frequent memory-jogging reminders, and make very few of the recognition errors that likely beset defended cryptic animals. Hence, the per capita survival rate of common aposematic animals is likely to be high when they are abundant. However, when rare, aposematism looks like a decisively unstable evolutionary strategy: rare aposematic prey may be subject to very high per capita death rates, since a large proportion (of a small number) of individuals are killed during predator education *and* they lack the protection of crypsis (see Speed et al., 2000; Lindström et al., 2001*a*). These factors conspire to make rare aposemes much less fit than their cryptic counterparts.

One approach to this problem is to suggest that the chances of a rare conspicuous mutant surviving and reproducing in a large cryptic population are so poor that some other class of explanation that somehow side-steps the problems of rarity and conspicuousness must pertain. Explanations that side-step the problematic 'initial evolution' scenario are undoubtedly important but, for a variety of reasons have received less attention than deserved. We therefore consider first, how theoreticians and some empiricists have attempted to explain aposematism in the context of the 'rarity and conspicuousness' problem (a more detailed review can be found in Speed, 2003). We then turn our attention to the equally important possibility that aposematism can more easily evolve in other contexts.

## 8.2 Stochastic–deterministic scenarios

In a seminal discussion, Mallet and Singer (1987) considered the fitness consequences for prey that would arise when there is variation in frequency and conspicuousness of alternative forms. They pointed out that in an environment in which defended prey can take two (or more) morphs, there may be a critical allele frequency above which the fitness of a single morph will be higher than for other morphs. Other things being equal (such as conspicuousness and levels of defence) if one morph is more common than another then, since the mortality costs of predator education are more widely spread, individuals of the more numerous form have the lowest per capita mortality rate and the highest fitness (see curve X in Fig. 8.1).

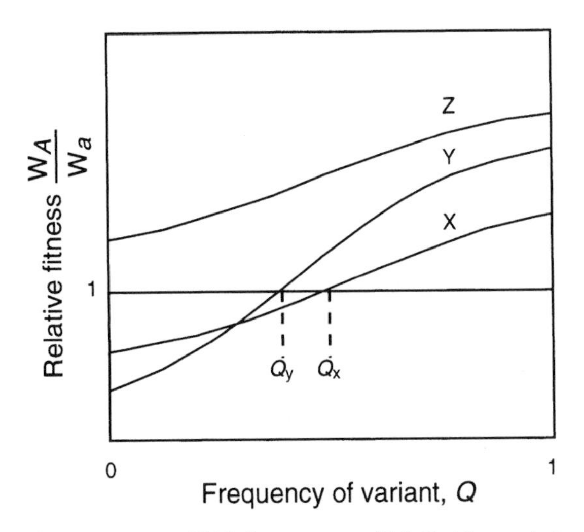


**Figure 8.1** Fitness ($W_A$) in frequency range ($Q$, 0–1) of three variant focal morphs (X, Y, and Z) relative to fitness of an alternative 'wild type' morph (Wa). X = focal morph and wild-type are equally conspicuous. Y = focal morph is conspicuous and wild type is cryptic. Z = focal morph is a Mullerian mimic. $Q_x$ and $Q_y$ represent the threshold frequencies beyond which focal morphs are superior to wild type morphs. Redrawn from Mallet and Singer (1987).

Relaxing the assumption of equality in conspicuousness, aposematic morphs could, arguably, have higher fitness than some cryptic conspecifics simply by being substantially more numerous. However, Mallet and Singer (1987: 341) suggest that aposematic superiority is most convincing if 'the number [of aposematic forms] sampled by a predator memorizing the new morph are by definition lower than the number of [cryptic forms]'. Hence, the critical frequency for aposematic advantage would be reduced if the warning signal diminishes the costs of predator education (curve Y in Fig. 8.1). Learning and memory effects can therefore make it easier for aposematism to replace crypsis, but while this lessens the problems it does not remove them: new aposematic forms still have to survive and increase to some relatively high critical level before gaining superiority over cryptic forms. To circumvent this problem, Mallet and Singer (1987) and Mallet and Joron (1999) suggested that chance events may nullify the problems of rarity and conspicuousness. In particular, the temporary absence of predators and random genetic drift processes may be sufficient to explain how new conspicuous morphs could rise toward their critical frequency (see also related models in Sundberg, 1987).

In line with these suggestions, some recent experiments demonstrate that stochastic factors may indeed facilitate the evolution of warning signals. Thomas et al. (2003) showed that dietary conservatism in birds (see Section 7.3.1) may provide a stochastic mechanism by which aposematic mutants can rise to high levels within otherwise cryptic populations. Birds are highly variable in their expression of dietary conservatism (Marples et al., 1998), which suggests that the survival values of new mutant forms may be quite heterogeneous. Should a rare form emerge in a 'safe' area of a habitat with very conservative predators then it may rise to high levels in that area. To test this hypothesis, Thomas et al. (2003) commenced their experiment by presenting a set of singly housed robins with 19 common pastry baits of a familiar colour plus one of a novel colour. After the first trial, prey were presented in the same ratio that survived at the end of the previous trial, but scaled up to 20 items. In 14 of 40 trials, the novel form rose to fixation, so that even when it was very common the birds still preferred the familiar coloured food. Furthermore, Thomas et al. (2003) showed that this was the result of active choices by the birds and not a random drift process. Similar results have been found with free-living birds (Thomas et al., 2004). Given that birds are highly variable in their dietary conservatism, these results suggest a need for spatial modelling that considers the survival of rare aposematic forms in a heterogeneous selective habitat.

An alternative stochastic mechanism that promotes the initial evolution of a rare, conspicuous morph may simply be that by chance predators do not come close enough to detect it (see models in Speed, 2001*a*). Sherratt and Beatty (2003) found evidence for this stochastic effect in a system with evolving computer-generated prey that were subject to continued predation by humans. Profitable prey (foragers gain points) and unprofitable prey (foragers lose points) all began in a similar-looking cryptic form but over generations 2–4, two novel conspicuous mutants were introduced into each of the populations (mutants were different both between species and between generations). Human predators sometimes missed the novel conspicuous mutants, enabling them to survive long enough to reproduce. In 6/10 replicates the frequency of unprofitable prey increased to more than 50 per cent of the population by generation 30, because such forms could be reliably distinguished from profitable prey. By contrast, profitable prey always retained their cryptic form.

Whatever the stochastic mechanism, it seems likely that in some populations at least, rare aposematic morphs could reach sufficiently high levels so that the deterministic processes of selection would strongly favour the aposematic prey. When an aposematic morph is at or near fixation in one locality, it may destabilise crypsis in neighbouring populations and initiate a narrow shifting cline that moves through the population, destabilising and replacing crypsis as it moves. Mallet and Joron (1999) argued persuasively (1) that evidence for the necessary levels of polymorphism and drift can be seen in contemporary *Heliconius* populations (see details in Mallet and Joron, 1999), (2) that strong density dependent selection that would favour sufficiently common aposematic forms has been demonstrated (Mallet and Barton, 1989) and (3) that cline movement is predicted based on measurements of selection pressures in natural populations (Mallet et al., 1990). This stochastic/shifting balance theory is one of the most persuasive accounts of the evolution and spread of aposematic forms in the 'rarity and conspicuousness' framework. However, it does utilize the assumption that there is 'something special' about conspicuousness (Guilford, 1990*a*); that aposematic signals already speed-up learning (and/or reduce forgetting) and so reduce the costs of predator education. As we will see in this chapter all 'rarity and conspicuousness' models assume that conspicuousness has some special effects on the behaviours of predators. Whether these responses necessarily existed at an early stage in the evolution of aposematism remains an unresolved co-evolutionary question (see discussion in Section 7.4).

## 8.3 Spatial aggregation

A second, population level approach is to consider the role that spatial and temporal aggregation of individuals has played in the evolution of aposematism. As noted in Chapter 5, Fisher originated the theory of kin selection in his explanation of the evolution of defence. Fisher deduced that the death of a few defended siblings would protect the rest of the group since predators would be unlikely to carry on attacking 'distasteful' individuals in close proximity. It is easy to see that defence alleles that are localised within kin groups have higher rates of survival, and thus reproduction, than competing alleles that do not confer defence and hence make all group members vulnerable to attack. Other authors subsequently applied this argument to the evolution of aposematism (e.g. Benson, 1971; Turner, 1975; Harvey et al., 1982), arguing that the death of one or a few individuals could generate higher levels of protection for the surviving aposematic kin than that conferred on non-aposematic individuals in the population. Enhancement of survival through kin-grouped aposematism could therefore be sufficient to offset the problems of conspicuousness and rarity. Turner (1984: 143) argued that aggregation would have a key role in the evolution of secondary defences and warning signals, suggested that the evolutionary order is 'roughly toxicity–clumping–distastefulness–blatant colour–gregariousness.'

The importance of aggregation was confirmed in a family model devised by Harvey et al. (1982) who, in line with other authors, assumed that a number of key parameters could vary: these are; the degree of prey conspicuousness, the numbers of defended prey that need to be attacked before learnt avoidance is generated and the proportions of aposematic and cryptic morphs within a population. Harvey et al. (1982) assumed that prey distribute themselves in family groups that vary in number and size within a predator's foraging range. They found that kin structures within a population could make the survival of rare aposematic morphs superior to the survival of common cryptic and defended morphs so long as certain conditions pertain. In particular, essential requirements were (1) that aposematic morphs are not too conspicuous, but (2) aposematic displays accelerate avoidance learning, so that fewer aposematic individuals die than cryptic individuals while the predator is learning avoidance, and (3) the number of prey families per predator is low, so that conspicuous mutants will in relative terms not be very rare within individual territories. In essence, aposematism will evolve if the conspicuous signal accelerates learning so much that few individuals are martyred during predator education and if the benefit of this learnt avoidance is readily conferred on the martyrs' siblings by virtue of their proximity.

These conditions place some important limitations on the generality of this family model. Most important, the degree of acceleration of avoidance learning shown by predators may be positively related to the conspicuousness of a signal. Hence, aposematic displays that are not very conspicuous may not, in fact, do much to accelerate avoidance learning (see discussions in Schuler and Roper, 1992). One way to circumvent this problem, suggested by Harvey et al. is to assume that at its initial stages, the aposematic signal is cryptic from a distance, but conspicuous from up close. This is an intriguing and important idea (discussed in Edmunds, 1974; Endler, 1988, 1991; Marshall, 2000) that has been subjected to relatively little empirical investigation (see discussion in Section 7.2).

Models such as those of Harvey et al. (1982) begin at 'Stage 2' of the evolution of aposematism, a point after 'Stage 1' in which a lone conspicuous mutant must have survived long enough to reproduce (Speed, 2001*a*). In a recent paper Brodie and Agrawal (2001) attempted to fill this gap with a model in which the aposematic phenotype is not seen in the first lone mutant but is displayed in all of her offspring. Such effects of 'maternal inheritance' are becoming better understood, and its application by Brodie and Agrawal to the 'Stage 1' problem is certainly ingenious. To know how important this scenario is however, we need to know more about the prevalence and control of maternal inheritance over some key classes of phenotype—a point well made by Brodie and Agrawal themselves. In addition, we need to evaluate whether such deterministic models are more likely than simple stochasticity (e.g. Speed, 2001*a*) as explanations for the initial evolution of the lone mutant.

It is worth mentioning here that the model of Harvey et al. relies on the spatio-temporal proximity of similar individuals (whether they are kin or not) rather than some intrinsic necessity in the model for the proximity of related individuals. This argument is subtle (Guilford, 1985), has been frequently misunderstood and has been the focus of some controversy (see e.g. Guilford, 1990*a*,*b*). In Guilford's (1985) view (and see also discussion in Turner et al. 1984), it is the aggregation of similar phenotypes that provide the primary causal conditions for the evolution of aposematism. These groups may often be family, but need not be. Thus consider a hypothetical scenario in which a defended species A with aposematic yellow/black larvae switches host plant to use the same plants as a defended cryptic species B. If a mutant arises in B that confers the yellow/black phenotype on an individual, then aposematism may well be favoured in species B, because of its proximity to similar individuals, even though they are not kin or even the same species. In a strict sense then, 'family models' of aposematism are those that imply advantage to

it was assumed that there is a variable probability that individuals can survive attacks by a predator; these models demonstrated that individual selection can aid the evolution of aposematism if bright signals are more easily learnt than cryptic phenotypes (Engen et al., 1986) and also if bright, conspicuous displays enhance reluctance to attack new prey items (Sillén-Tullberg and Bryant, 1983).

However, the models of Engen et al. and of Sillén-Tullberg and Bryant do not consider the problem that rarity poses to the explanation of the initial evolution of warning signals (see critique in Mallet and Singer, 1987: 342). These models do not explicitly consider that the *per capita* mortality risk of a rare aposematic animal may be much greater than for its much more abundant cryptic conspecific. As Mallet and Singer (1987) pointed out, if only 10 aposematic individuals are killed in the course of predator education over a season and 200 cryptic conspecifics are killed over the same time period, this looks like pretty good odds in favour of conspicuous phenotypes. However, if at the start of the season there are only 20 aposematic individuals and 1000 cryptic individuals, then the *per capita* mortality risk is much higher in aposematic morphs than in cryptic ones (i.e. 50 versus 20 per cent, respectively). Thus, since the problems of rarity and frequency dependence remain in these models only limited conclusions can be drawn.

## 8.6 Evaluations of predator psychology models

As we described in Chapter 7 it has frequently been shown that conspicuous warning displays have a range of 'special' effects (Guilford, 1990*a*) on the psychological functions of predators. Though the details are complex the effects are essentially that warning displays (whether visual, olfactory, and/or auditory) may enhance predator wariness and accelerate predator learning. Warning displays may also decelerate predator forgetting processes and may enhance the accuracy of prey recognition.

One way of explaining the evolution of aposematism is, as we have seen above, to include one or more of these effects into a theoretical model and then to see whether the evolution of aposematism can be predicted. In contrast, a few studies have more generally attempted to evaluate whether effects of conspicuousness on predator psychology can be sufficient to explain the evolution of warning displays, or whether particular environmental conditions are also necessary.

In a case of convergent thinking Servedio (2000) and Speed (2001*a*) both analysed the role of predator psychology in explanations of aposematism. These models systematically vary a range of assumptions about predator psychologies and try to identify those cases in which predator psychology is, or is not, likely to be important. Servedio (2000) used a robust, clear analytical model that investigates variation in rates of learning, forgetting, recognition errors and probability that defended prey will be released unharmed. This paper is important, both for the clarity of modelling and the ambitious description of predator psychology that it employs. In addition, Servedio made no assumptions about family grouping, and thus set predator psychology the challenge of explaining the evolution of aposematism without recourse to limiting assumptions about clustering and family structure.

Servedio (2000) found that the best scenario for the evolution of aposematism is one in which prey are very highly defended, such that complete avoidance will be learnt in a single trial and completely remembered thereafter. When these conditions pertain, the aposematic morph gains higher fitness than the highly defended cryptic morph because it suffers lower costs of mistaken attacks (cf. Guilford, 1986). Servedio also found a good case for the gradual evolution of aposematism in highly defended prey. Here, the benefit of a reduction in recognition errors that accrues to individuals of slightly increased brightness offsets the costs of small increases in conspicuousness (cf. Lindström et al., 1999*b*). In other circumstances, in which predators learn and forget gradually, new aposematic mutants could not invade cryptic populations unless they were moved to unusually high densities (by random drift and other stochastic factors—Mallet and Singer, 1987). An important conclusion drawn here is that moderately defended species are very unlikely sources of new aposematic displays; when such species do evolve aposematism it is likely to be via the evolution of mimicry.

A fundamental difference in conception of predator behaviour between Servedio's (2000) and Speed's models is that Speed excluded recognition errors but included predator wariness of new mutants (i.e. neophobia or dietary conservatism), as a tendency of predators not to attack individual prey for a defined number of encounters. The best scenario for the evolution of aposematism emerged in these models when prey were allowed to cluster into phenotypically similar ('family') groups. Here neophobia was exceptionally beneficial to rare aggregations of defended individuals because each time a predator chose not to sample an aggregation for reasons of wariness, it avoided a large number of prey. Neophobic protection is a defensive resource for prey that can be exhausted quickly; after $n$ exposures of individual aposemes in a dispersed environment $n$ prey will have been protected. However with aggregations of $x$ individuals, $n$ exposures protect $nx$ individuals. Aggregation size and the extent of neophobic rejection can thus interact to provide novel prey forms with very considerable levels of protection. Given that aggregations seem to scare predators more than dispersed formations (Gamberale and Tullberg, 1998), the protection that accrues to aggregated individuals can be very large indeed. Speed (2001*a*) also found that predator forgetting could play a crucial role in the evolution of aposematism. In line with other authors (e.g. Yachi and Higashi, 1998; Servedio, 2000), Speed found that dispersed rare aposematic prey can be superior to common cryptic defended individuals if bright, aposematic displays sufficiently retard predator forgetting rates. There have been surprisingly few empirical tests of this hypothesis (e.g. Roper and Redston, 1987) and it remains one of the most important empirical areas in this field yet to be rigorously investigated (see Section 7.3).

In summary, there are some circumstances in which predator psychology can certainly help to explain the evolution of aposematism, particularly: (1) instances in which prey are very highly defended and predators are error prone; (2) predators show neophobic wariness and prey are aggregated; and (3) warning signals generate particularly low rates of forgetting.

## 8.7 Alternatives to the *rare conspicuous mutant* scenario

Most or all of these theoretical and experimental models and simulations of aposematism take as a starting point the emergence of an aposematic mutant in a cryptic population with secondary defences. However, a number of authors (Mallet and Singer, 1987; Guilford, 1988; Mallet and Joron, 1999) argue persuasively that aposematism may arise through a variety of other mechanisms and an attractive component of several alternative theories is that there is not a strong reliance on what might be highly co-evolved components of predator psychology. These mechanisms may well provide more plausible scenarios for the evolution of aposematism than those of the more popular 'rare conspicuous mutant' scenario. In this section we describe and evaluate several alternative theories.

### 8.7.1 Sexual selection

A number of researchers, at least from Poulton (1890) onward, proposed that sexually selected traits could be used as warning signals by predators, becoming modified to serve the dual purpose of warning and sexual communication if a species acquired an effective secondary defence (more recent discussions in Mallet and Singer, 1987; Guilford, 1988). In prey without secondary defences, sexual selection may counteract natural selection and move a prey away from its original state of crypsis. Strong selection for the acquisition of secondary defences may then act with the result that the animal becomes unprofitable and conspicuous. Guilford (1988) argued that such animals should *not* be considered aposematic because their conspicuous phenotype has not evolved to deter predation. We agree however with Poulton (1890), that being freely exposed rather than cryptic can in itself serve as a warning display to predators, even if the phenotype itself has not been modified in an aposematic fashion to signal danger. However, it seems likely that in many cases further evolution will indeed optimize the external traits of a prey animal so that it functions both as a warning display that matches predator sensory baises, and also an intraspecific sexual signal.

### 8.7.2 Defences, optimal conspicuousness and apparency

Most or all of the theoretical models described above assume some state of crypsis, but give little or no attention to the possibility that crypsis may be costly, because it restricts the capacity of animals to utilise all available opportunities in the environment. If prey with effective secondary defences have high levels of crypsis, they may have to pay excessive opportunity costs that follow from restricted opportunities for thermoregulation, mate-choice, foraging etc. (see broader discussion in Abrams, 1991). This possibility has been discussed in relation to caterpillars by Stamp and Willkens (1993). For some species, possession of a secondary defence may free them from the costs of crypsis, so that the fitness benefits conferred by a secondary defence come not only via reduced predation, but also from raised nutrition, etc. Hence, as we argued in Chapter 7, possession of a secondary defence is likely to specify raised optimal levels of conspicuousness (or more generally exposure to predators). It may be that if conspicuousness and defence are continuously variable traits (e.g. Leimar et al., 1986), that they move through a series of steps, optimising conspicuousness and then defence in relation to each other, increasing both toward some ultimate optimal level defined by environmental parameters.

As Guilford (1988) argued such hypotheses might explain conspicuousness, but as with sexual selection arguments, they do not in themselves explain the evolution of adapted warning displays. However, once a prey animal has a sufficiently high optimal conspicuousness, it is easy to argue that it becomes free to optimise its external phenotype to enhance anti-predator effectiveness, for instance by modifications that produce 'signals or danger-flags' (Wallace, 1889) that enhance recognition and discrimination learning by predators. In addition, non-cryptic prey may be able to optimise their external appearances for thermal or other reasons, and these changes may be used as aposematic cues by predators. To the extent that unprotifable prey have high optimal levels of exposure to predators then scenarios of 'conspicuousness and rarity' in the formal models described above do not apply. Prey with secondary defences may already have raised levels of exposure to predators so that the costs of additional conspicuousness conferred by aposematic mutations may be rather less than many theoretical models assume.

### 8.7.3 Aposematism originated to advertise 'visible' defences

A third alternative to the scenario of the 'conspicuous mutant' is the possibility that warning displays originated (whether in water or on land) as amplifiers to 'visible' secondary defences, such as numerous sharp spines etc. As described at the start of Chapter 7, bright colourations, behaviours, sounds and smells may direct a predator's attention toward a visible, unfakeable defence (Lev-Yadun, 2001). So long as the benefits of predator deterrence are sufficiently large and the costs of reduced crypsis are small, explaining the evolution and ritualization of such warning traits is straightforward. Once warning signals are common for 'visible' secondary defences and their association with unprofitability is well known, it is easy to see that they could be taken up by prey with 'invisible' chemical or other defences as a loose form of Müllerian mimicry. The reluctance of predators to handle these mimics, combined with a greater caution (or 'go slow' handling, Guilford, 1994) means that new mimetic mutants are not overly disadvantaged when rare, and thus can spread to high density relatively easily.

### 8.7.4 Facultative, density-dependent aposematism

In a remarkable study, Sword (2002) recently demonstrated that the presence of aposematic traits in individual grasshoppers of the species *Schistocerca emarginata* is dependent on population density and the presence of chemical defence. Two genetically distinct populations of this insect feed on different host plants, one conferring gut-mediated chemical defence, the other conferring no evident chemical defences. Colour patterns were described by measuring background colour and proportion of surface blackened. Responses of colour patterns to rearing densities differed markedly between the two populations: at high densities the chemically defended individuals all had a yellow background colour, whereas the palatable individuals were much more variable (colours including brown, green and yellow, as well as heterogeneously coloured individuals). Chemically defended individuals also had much higher proportions of the body surface blackened.

Sword argues persuasively that this is an example of facultatively induced aposematism: the costs of conspicuousness are avoided when population densities and risks of detection are both low; but when populations are dense, gregariousness makes detection likely. Common aposematic phenotypes may be advantageous in defended animals that have a high chance of detection because of the benefits that come from accelerated learning, accurate recognition and high levels of memorability. Facultative aposematism may be a by-product (or alternatively an 'exaptation') of some other density dependent response in prey animals, such as pathogen resistance (Wilson et al., 2001, 2002). Nearly all thinking about aposematism has been based around constituitive traits but, as Sword (2002) argues, it may be that the constituitive nature of some or many warning displays is derived from an ancestrally facultative state.

### 8.7.5 Simultaneous evolution of defence and conspicuousness

In a fifth scenario, an insect or other prey might change host plant and newly acquire both unprofitability and conspicuousness because its crypsis no longer functions (e.g. Endler, 1991). This possibility has received little theoretical attention. However, it could work if larvae, like cuckoo chicks in a nest, imprint on and return to their natal host to oviposit with a higher than average probability (Speed and Ruxton, unpublished). In this way the traits of conspicuousness and defence can appear simultaneously, can be replicated over generations and can increase in frequency together if the survival rate of conspicuous and defended individuals exceeds that for cryptic conspecifics. Heightened frequency of return to the natal habitat may therefore serve as a mechanism by which conspicuousness and defence appear together, as has been assumed in some theoretical models (e.g. Merilaita and Kaitala, 2002).

We conclude that there are several, perhaps many ways to explain the initial evolution of aposematism without recourse to the more difficult scenario of a bright conspicuous mutant in an otherwise highly cryptic population.

## 8.8 Phylogeny and evolutionary history

In a provocative and thoughtful paper Härlin and Härlin (2003) recently argued that much of the theory of aposematism is impoverished by an absence of historical perspective. In a survey of 154 articles about aposematism recently cited in the *Science Citation Index*, only 7 per cent took an explicitly phylogenetic perspective. Phylogenetic approaches are undoubtedly important in the examination of any adaptation and Härlin and Härlin do indeed point to a weakness in the literature. It is almost certainly the case that the increasing ease within which mitochondrial and other DNA phylogenies are produced will enhance the contribution of historical perspectives to the understanding of aposematic displays. It is true also that phylogenetic thinking may persuade researchers to consider the broad multiple uses of conspicuous traits as Härlin and Härlin suggest, although it is worth noting that this has taken place in the field to greater or lesser extents anyway (see, e.g. the seminal discussion in Mallet and Singer, 1987; Guilford, 1988). It is interesting that phylogenetic approaches are relatively well advanced in the study of the ecological chemistry of secondary defences (e.g. Cimino and Ghiselin, 1999; Dobler, 2001) compared to studies of aposematic displays. Perhaps this is because aposematic displays are more nebulous and thus less amenable to cladistic analysis than molecular structures. Alternatively as Härlin and Härlin (2003) suggest there may well have been an over-emphasis on functional explanations at the expense of historical descriptions of the evolution of aposematism. Whatever the explanation, interesting and insightful phylogenetic studies are becoming more common.

The most extensive and important work on the phylogenetic history of aposematism and gregariousness has been reported in a series of papers by Sillén-Tullberg and Hunter (Sillén-Tullberg, 1988, 1993; Tullberg and Hunter, 1996). Each of these analyses strongly suggests that the evolution of secondary defence and aposematism preceded the evolution of aggregation and gregariousness in microlepidoptera. In the most recent and refined study, Tullberg and Hunter (1996) considered the presence of aposematism, mode of defence and gregariousness as discrete states for a large phylogeny. Five superfamilies were represented (Bombycoidea, Geometroidea, Papilionidea, Hesperioidea, and Noctuoidea), including a total of 825 species. Gregariousness was actually a rare trait, found in 61 species, 47 of which were aposematic. In contrast, aposematism (148 species) and repellent defence (227 species) were much more common than gregarious aggregation. Transition from a solitary to gregarious lifestyle was significantly more likely in aposematic than cryptic species and in species with repellent secondary defences than those without them. Most importantly perhaps this historical approach argues against a family grouping as a mechanism for the evolution of defences (cf. Fisher, 1930; Harvey et al., 1982).

Recent work on poison dart, dendrobatid frogs has allowed historical reconstruction of the evolution and interdependence of a number of traits. Notably, Summers and Clough (Clough and Summers, 2000; Summers and Clough, 2001) used an mtDNA based phylogeny to examine the evolution of defence and conspicuousness. Summers and Clough concluded that colouration had evolved 'in tandem with toxicity', leading to a positive correlation between level of toxicity and visual conspicuousness within animals. We have discussed mechanisms that may bring about this correlation elsewhere (see Section 7.2.1), but we note here that an historical analysis provides a remarkable insight into the evolution of aposematism that has not otherwise emerged in mathematical models. Hagman and Forsman (2004) developed Summers and Clough's approach to examine the evolutionary relationship between body size and aposematic signals, finding that conspicuousness of colouration also evolved in tandem with body size, so that larger animals have more conspicuous colour patterns. It cannot be determined from this analysis which of enhanced conspicuousness or larger size comes first. However, Hagman and Forsman argue that shifts in body size may generally precede shifts in conspicuousness because of its importance in many life-history traits (though see Nilsson and Forsman, 2003). We anticipate that aposematism and related traits will increasingly be the focus of phylogenetic studies such as these and that in the longer term, theory will be driven as much by historical insight as by functional, mechanistic modelling.

## 8.9 The evolution of aposematism: a trivial question with interesting answers?

On the face of it the problems of rarity and conspicuousness make the initial evolution of aposematism a major problem for a Darwinian evolutionary biology. However, there are now a large number of alternative explanations for aposematism, both within and beyond the 'conspicuousness and rarity' framework. Many of these explanations are plausible and may be general. This leads us to conclude that the evolution of aposematism is now a relatively trivial question, in the sense that it can be answered easily; however it is a question with a large array of fascinating answers. Uncovering each of these answers tells us as much about complex processes in evolutionary ecology as the evolution of aposematic displays.

CHAPTER 9

# The evolution and maintenance of Müllerian mimicry

## 9.1 Where Müllerian mimicry fits in

This chapter examines the phenomenon of Müllerian mimicry, in which two or more species with effective secondary defences share a similar appearance. As our book deals with both warning signals and adaptive resemblance, it was not immediately obvious where to put this material. One option was to consider Müllerian mimicry as a companion to our chapter on Batesian mimicry, in which members of a palatable species evolve to resemble an unpalatable or otherwise defended species. After all, both subjects relate to mimicry, and there is a body of opinion that the distinction between Batesian and Müllerian mimicry is often blurred. However, as will become clear, the principle mechanisms that have been proposed to explain these two mimetic phenomena are rather different. On the one hand, Müllerian mimics are assumed to be in a mutualistic relationship, sharing the mortality costs of educating naive or forgetful predators; on the other, Batesian mimics are generally believed to be parasites, degrading the aposematic protection of the unprofitable species that they resemble. Furthermore, the evolutionary dynamics of Müllerian mimicry are based on a strength-in-numbers argument, leading to some tendency toward uniformity of aposematic signaling. In contrast, as we describe in Chapter 10, the evolutionary dynamics of Batesian mimicry are based on the converse; strength through rarity and consequently diversity of form. Müllerian mimicry, then, may be best considered as an interspecific form of aposematism, in which the sharing of warning displays reduces the *per capita* mortality costs of predator education and enhances accurate discrimination between edible and unprofitable prey, as Wallace originally proposed for single aposematic species.

## 9.2 Chapter outline

The widespread observation that collections of unpalatable species share common colours and patterns demands an explanation. We will begin this chapter with a short history of Müller's proposal that such resemblances have evolved as a means of reducing the mortality costs that are incurred when naive predators learn that some species are unprofitable. We next discuss some good candidate examples of Müller's form of mimicry. Although the term 'Müllerian mimicry' often evokes images of butterflies fluttering through the canopy of Amazonian rainforest, in fact the phenomenon arises within (and between) a wide variety of taxonomic groups. One might wonder how one can confidently claim that two or more unpalatable species are indeed mimetic and not similar through shared phylogeny, so we will indicate how researchers have reached this conclusion. The central focus of this chapter however is a consideration of the experimental evidence that Müller's general theory is the correct one to explain the evolved common resemblance between unprofitable species. We then briefly examine what mathematical and computer models have to say about its properties and implications. In our final section we discuss several common questions, and try to provide answers, largely from a theoretical perspective. For instance, how might Müllerian mimicry evolve through a series of forms with intermediate resemblance? Similarly, if unpalatable species gain from adopting a common form of advertising, why do not all species with secondary defences adopt the same code, a universal warning display that unequivocally tells predators to 'back-off'. To address these questions and others we will describe some contemporary (and less-than-contemporary) theories.

## 9.3 A brief early history of Müllerian mimicry

In his classic 1862 paper, Bates reflected on the similarities exhibited by many species of neotropical butterfly and proposed that in some cases this could be explained by natural selection on palatable species to resemble unpalatable species, and thereby gain protection from predation ('Batesian mimicry', Chapter 10). However, he also noted that some common and apparently distasteful butterfly species also resembled one another, and proposed that this might be due to their shared environmental conditions (p. 508, 'I think the facts of similar variation in two already nearly allied forms do sometimes show that they have been affected in a similar way by physical conditions'). The naturalist Johannes Friedrich ('Fritz') Müller, who also worked in Brazil, considered this same puzzling phenomenon and proposed several very different solutions (including one based on sexual selection, Poulton, 1909*a*; C. Darwin letter to F. Müller, 1871; Darwin, 1887). Sixteen years after the publication of Bates' work, Müller, (1878) published an explanation for the similarity between unpalatable species that is now so widely accepted that it is often considered synonymous with the phenomenon itself. Müller's (1878) explanation was extremely simple—*defended species may evolve a similar appearance so as to share the costs of predator education.* In this same paper Müller supported his arguments with a specific quantitative example.[4]

A better known follow-up paper of Müller appeared in the journal 'Kosmos' (1879, p. 100) and was subsequently translated by R. Meldola for the Entomological Society of London. His hypothesis, later named 'Müllerian mimicry' was at first widely opposed (Poulton, 1909*a*). Bates himself was never convinced by the theory (see comments in Müller 1879; Poulton 1909*a*), although he had used similar arguments in his 1862 paper to explain particular instances of mimicry in rare heliconids (see Mallet et al., 1998; Mallet, 2001*a*), revealing an early intuitive understanding of the importance of frequency and number dependence in mimicry (see also Chapter 10).

Fritz Müller's (1879) seminal paper contains a footnote which presents not only the first formal model of mimicry, but one of the first mathematical derivations of the fitness consequences of adopting particular traits (Mallet, 2001*a*). We quote his model *verbatim* in the first part of Box 9.1, because this remains the clearest way to articulate his theory. It is worth noting that Blakiston and Alexander (1884) proposed a refinement to this model using a slightly different denominator but the end result is almost identical (Box 9.1). Müller's attempt to formally represent his arguments in quantitative terms was ahead of its time. However, it primarily serves to quantify the relative benefits of mimicry to each species by comparing the survivorship within populations of a species pair that exist either with or without mimicry. Müller's theory does not therefore directly address the evolution of shared warning displays from an initial point of rarity. Evaluating the survivorship of a rare mutant mimic of perfect resemblance, comprising $m$ individuals, is relatively straightforward. Using the same nomenclature of Box 9.1 (where $n$ is the number of each distinct form attacked in the educational process, $a_1$, $a_2$ are numbers of species 1 and 2 available), if the per capita mortality of species 1 was $n/(a_1 - m)$ then the per capita mortality of a mutant which resembled species 2 perfectly would be $n/(a_2 + m)$. Thus, a mimetic mutant of species 1 that resembled species 2 would spread from rarity so long as $a_2 > a_1 - 2 * m$ (simplifying to $a_2 > a_1$ if $m$ is small).

In this original model, Müller assumed that mimetic species had equally effective defensive qualities. However, variation in levels of secondary defence is now known to be widespread (see discussions in Chapters 5 and 11). To address this issue, Mallet (2001*a*) recently extended Müller's framework so that variation in defensive qualities could

[4] We are grateful to Jochen Jaeger, Carleton University for translating this paper.

**Plate 1** *Hyalophora cecropia* caterpillar, sharp spines can be visually detected presumably before many predators (and parasitoids) attack the caterpillar. The colours of the tubercles from which the spines protrude may provide disruptive crypsis from a distance and, alternatively, may draw a predator's attention to the spines at close proximity. Photo courtesy of Jim Kalisch, Department of Entomology, University of Nebraska-Lincoln.

A lycid beetle

An arctiid moth

**Plate 2** (a) A lycid beetle and (b) an arctiid moth. These species are both unpalatable to birds and their colour pattern similarity suggests they are Müllerian mimics. They were photographed in southern Ontario by Dr Henri Goulet, Eastern Cereal and Oilseed Research Centre, Ottawa.

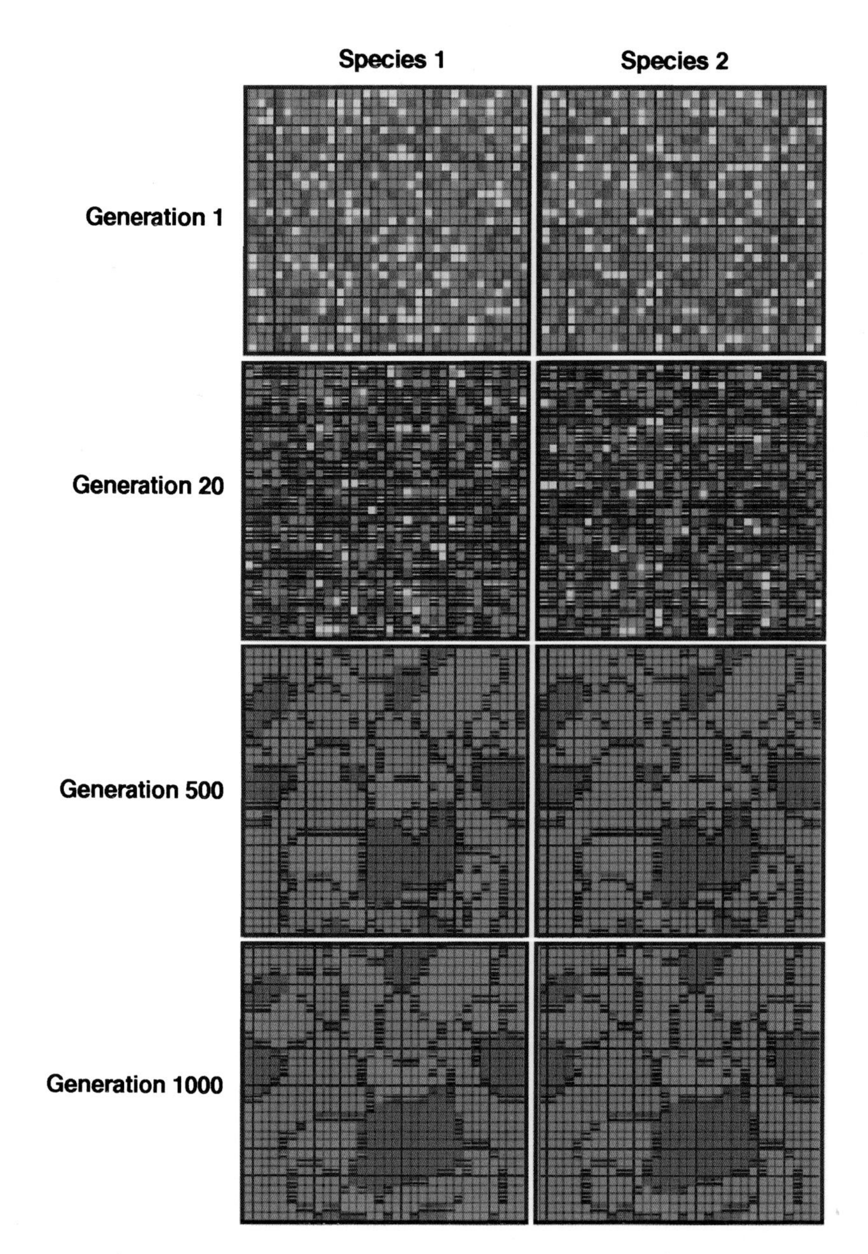


**Plate 3** The establishment of Müllerian mimicry between two unpalatable species. Here populations of the two species were distributed in a regular grid. Individuals of each species could occur in any one of six different morphs, five of which look similar between species. We assumed that the local predator community continues to forage on a given phenotype in a given cell until a combined dose of toxins is consumed. Starting with a random distribution of the most common morphs to cells, the sysem soon reduces to one in which the two species in the same cell share the same appearance, with narrow 'hybrid zones' between races of the different phenotypes.

## Box 9.1 Müller's model of mimicry

*1. Müller's theory (1879)*
Let $a_1$ and $a_2$ be the numbers of two distasteful species of butterflies in some definite district during one summer, and let $n$ be the number of individuals of a distinct species which are destroyed in the course of a summer before its distastefulness is generally known. If both species are totally dissimilar, then each loses $n$ individuals. If, however, they are undistinguishably similar, then the first loses:

$$\frac{a_1 n}{(a_1 + a_2)}$$

and the second

$$\frac{a_2 n}{(a_1 + a_2)}.$$

The absolute gain by resemblance is therefore for the first species

$$n - \frac{a_1 n}{(a_1 + a_2)} = \frac{a_2 n}{(a_1 + a_2)}$$

and in a similar manner for the second,

$$\frac{a_1 n}{(a_1 + a_2)}.$$

This absolute gain, compared with the occurrence of the species, gives for the first,

$$l_1 = \frac{a_2 n}{a_1(a_1 + a_2)}$$

and for the second species

$$l_2 = \frac{a_2 n}{a_2(a_1 + a_2)},$$

whence follows the proportion, $l_1/l_2 = a_2^2/a_1^2$ [this final expression is the 'reciprocal square rule'].

*2. Blakiston and Alexander's (1884) modification (recast using the same nomenclature)*
This absolute gain, compared with the occurrence of the species in the absence of mimicry is:

$$l_1 = \frac{a_2 n}{(a_1 - n)(a_1 + a_2)}$$

and for the second species

$$l_2 = \frac{a_1 n}{(a_2 - n)(a_1 + a_2)},$$

whence follows the proportion, $l_1/l_2 = [a_2(a_2 - n)]/[a_1(a_1 - n)]$, which is identical in form to Müller's theory if $a_1 \gg n$ and $a_2 \gg n$

## Box 9.2 Mallet's extended model

Mallet (2001a) showed that the number of individuals lost per unit time of two distasteful species could be estimated from their abundances ($a_1$ and $a_2$) and the numbers of each ($n_1$ and $n_2$) that are typically eaten by a predator when presented alone before they consume a critical dose $D$ (here $n_i = D/d_i$ where $d_i$ is the per capita dose of species $i$).

Let two distasteful species generally look very different at the outset, but allow a small number of species 1 ($= m$) to resemble species 2. Here we simply assume that by resembling an alternative species, then any resemblance to conspecifics is lost (see Fisher, 1927 for a critique of this assumption with respect to arguments propounded by Marshall, 1908). The important question is, when will this mutant have higher survivorship than the distinct species 1?

Per capita survivorship of distinct species 1
$= (a_1 - m - D/d_1)/(a_1 - m)$

Per capita survivorship of mutant
$= (m - [m (D/d_1) (D/d_2) / (m (D/d_2) + a_2 (D/d_1)\,]\,) / m$

The mutant will spread from extreme rarity when its survivorship is higher than that of its distinct conspecifics. This condition rapidly reduces to:
$a_1 d_1 < a_2 d_2$

be included. He assumed that individual prey items contain a dose of toxins and that the population of predators learn to avoid the prey entirely when they consume a combined threshold dose, rather than a fixed number of unpalatable prey. Hence, fewer highly toxic species have to be eaten before the message filters through. Under these conditions, Müller's results can be extended to cases where predators do not consume a fixed $n$ of each species, but $n_i$ for each species $i$ depending on the dose that individuals of the species contains. In Box 9.2, we recast Mallet's revised model to investigate the success of rare mutant forms. The model predicts that in cases where both distasteful species establish simultaneously then the species with the lower product of dose and abundance will evolve in appearance to more closely resemble the species with the higher product of dose and abundance. Of course, this simple view does not include a consideration of the survivorship of mutant forms that are intermediate in appearance. We leave a consideration of these important issues to Section 9.7.2.

## 9.4 Some potential examples of Müllerian mimicry

In the years following publication of his paper, Müller's theory became much more widely accepted. Poulton (1909*a*), for instance, reflected on the mimetic Lycid beetles described by Marshall (1902) 'and the heterogeneous group of varied insects which mimic their conspicuous and simple scheme of colouring. The Lycid beetles, forming the centre or "models" of the whole company, are orange-brown in front for about two-thirds of the exposed surface, black behind for the remaining third. They are undoubtedly protected by qualities which make them excessively unpalatable to the bulk of insect-eating animals' (see Fig. 9.1: see also Plate 2). As Poulton (1909*a*) pointed out, this combination included six species of Lycidae; nine beetles of five groups all specially protected by nauseous qualities, one Coprid beetle; eight stinging Hymenoptera; three or four parasitic Hymenoptera (Braconidae, a group that are themselves mimicked); five bugs; three moths (Arctiidae and Zygaenidae, known to be distasteful) and one fly. He was left to conclude that these observations 'fall under the hypothesis of Müller and not under that of Bates'. Subsequent work supports this perspective (Linsley et al., 1961), although many assemblages also appear to have Batesian mimics. Likewise and more recently, Marden and Chai (1991) cite studies (Chai, 1986; DeVries, 1987) which support the impression that (p. 16) 'Batesian mimicry appears to be rare in the Neotropics', at least in butterflies. This, of course, is just an unquantified general impression, and refers to

**Figure 9.1** (a) A lycid beetle and (b) an arctiid moth. These species are both unpalatable to birds and their colour pattern similarity suggests they are Müllerian mimics. They were photographed in southern Ontario by Dr Henri Goulet, Eastern Cereal and Oilseed Research Centre, Ottawa. (See also Plate 2).

only one area of the world. However, it suggests that Müllerian mimicry is at least comparable in prevalence to Batesian mimicry.

Below, we briefly describe some examples of similarity between defended species. Most of the examples that have been considered as Müllerian mimics are primarily thought of in this way for pragmatic reasons: they are each defended, they share common warning colours and patterns, and they occupy the same geographical areas (often the same host plant). In some cases (as in the case of the lycid beetle and arctiid moth above), the mimic species are only distantly related and the similarity can be thought of as derived. It is most unlikely that these colours and patterns are coincidentally similar, especially when close relatives of each of the mimetic species do not exhibit these characteristics. However, in other instances, the Müllerian mimics are often closely related species so that while selective pressures might act to maintain and promote similarity (particularly evident when there are co-occurring races of distasteful species), it should be noted that at least some aspects of shared appearance are likely to have arisen through common ancestry.

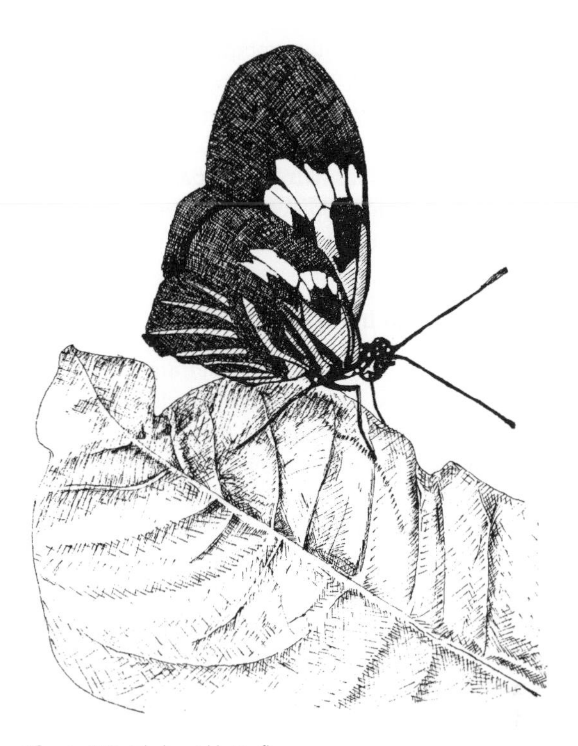

**Figure 9.2** A heliconid butterfly.

### 9.4.1 Neotropical *Heliconius* butterflies

The most celebrated example of Müllerian mimicry is that of the similarity among distasteful heliconid butterflies in South America (Fig. 9.2). The relationships among these species appear exceptionally complex: although 54 species of *Heliconius* are recognized, more than 700 names have been applied to the different phenotypic forms (Brower, 1996). Four assemblages of butterflies containing heliconiines, whose members resemble one another in some way ('mimicry rings') are typically recognized (although some authors count five or more): 'tiger', 'red', 'blue', and 'orange' (Papageorgis, 1975; Mallet and Gilbert, 1995). These rings contain up to a dozen heliconiine species, as well as mimics from other lepidopteran groups (Brower, 1996). Members of each ring tend to roost at night in similar habitats (Mallet and Gilbert, 1995) and also tend to fly in similar habitats and at the same time of year (DeVries et al., 1999), although there appears considerable overlap in the spatio-temporal distribution among members of the different mimicry rings.

Much work on heliconiids has focused on the parallel race formation (nearly 30 different colour patterns, Jiggins and McMillan, 1997) between two distantly-related members of the *Heliconius* genus: *H. melpomene* and *H. erato*. With some minor exceptions (associated with hybrid zones) the particular pattern exhibited by *H. melpomene* at any location is monomorphic and very similar to that exhibited by *H. erato*. This parallel variation in *erato* and *melpomene* across South America is also shared by several other Heliconiid species so that the exact form of phenotypes comprising a mimicry ring changes abruptly every few hundred kilometers. Clearly, some selective processes are at work to generate this close correspondence in forms, and it cannot be explained as a coincidence. We will return to Heliconid butterflies when assessing experimental evidence for the mechanism underlying Müllerian mimicry. For more information about Heliconid butterflies, readers may wish to refer to excellent reviews of aspects of this complex system that are already published. For instance, Brown (1981) reviews the biology of *Heliconius* and related genera; Turner (1971, 1981) reviews Müllerian mimicry in heliconiid butterflies, while Sheppard et al. (1985) discuss the relationship between *H. melpomene* and *H. erato* in exceptional depth.

### 9.4.2 European burnet moths

Burnet moths (family *Zygaenidae*) are day-flying brightly coloured moths with slow, drifting flight. They contain hydrocyanic acid (Jones et al., 1962) to which they are themselves resistant (Rothschild, 1971), and are rejected vehemently by wild birds (see Turner, 1971 and references therein).

Particular interest has focused on the burnet species *Zygaena ephialtes* which exists in four forms (red and yellow peucadanoid [P] and red and yellow ephialtoid [E] forms, Bovey 1941, Turner 1971) in Europe. The central and northern European populations of *Z. ephialtes* are the red-P form and resemble other red–black defended species in these locations including *Z. filipendulae*. The yellow E-form of *Z. ephialtes* occurs in southern Europe and resembles the black and white pattern of ctenuchid moths of the genus *Amata* (Arctiidae) from the *phegea* complex (Turner, 1971; Sbordoni et al., 1979). Evidence that the red-P and yellow-E forms of *Z. ephialtes* are Müllerian mimics of different models comes from several sources. As indicated above, all of these species are demonstrably aversive to birds (Jones et al., 1962; Bullini et al., 1969). More specifically, Bullini et al. (1969) (reported in Turner, 1971), conducted feeding trials using a variety of caged birds and found that those that had the opportunity to learn about the aversiveness of *A. phegea* (white–black) did not subsequently avoid attacking *Z. filipendulae* (red–black). Yet the same birds tended to avoid the yellow-E form of *Z. ephialtes*, showing some direct benefit of the shared warning pattern. Moreover, in southern areas the *phegea* complex is much more abundant than the co-occurring *filipendulae* one (Sbordini et al., 1979) and emerges earlier than *Z. ephialtes*. Since predators may learn early in the season to avoid the black and white *pheagea* moths, it may help explain why the similar-looking yellow-E morph of *Z. ephialtes* also predominates in the south.

To complete the picture, the red E-form of *Z. ephialtes* is distributed primarily in central-southern Europe and has the pattern of the *Amata* mimic but the colour of the northern forms. It is possible that this form gains selective advantage by resembling aspects of several models (Sbordoni et al., 1979; see Edmunds, 2000 for a similar theory for Batesian mimics), but this remains untested. The yellow-P form only occurs in areas inhabited by the other three forms (Bovey, 1941; Sbordini, 1979) and then only at relatively low frequency. It appears to be a non-mimetic form, and may well result as a consequence of intercrossing the different forms (Bovey et al., 1941; Turner, 1971; Sbordini et al.,1979).

### 9.4.3 Bumble bees

It is widely appreciated that female bumble bees (*Bombus* spp.) sting and that certain palatable fly species may mimic them (Plowright and Owen, 1980; see also Chapter 10). Both Plowright and Owen (1980) and R. W. Thorp (personal communication in Plowright and Owen, 1980) have proposed that the shared colour patterns of different species of bumble bee in particular geographical areas may themselves represent a case of Müllerian mimicry (see Fig. 9.3). Analyzing the morphological appearances of different species of bumble bee within England and North America (East and West Coast) they concluded that species in these areas were much more similar in pattern than their taxonomic affinities would predict. Although they describe several putative mimicry rings, it is of interest to note that seldom do all the species in one area conform to a single colour pattern (see also Section 9.4.1 on heliconid butterflies).

Several bumble bee species exhibit polymorphism. In North America, *Bombus rufocinctus* is dimorphic. One form has a red or black pile on abdominal terga 2–4, while the second has yellow or black pubescence on the fifth tergum. The first morph appears to human eyes to very closely resemble *B. ternarius* while the second morph resembles *B. vagans* (Plowright and Owen, 1980).

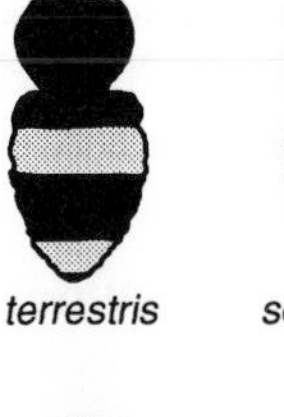
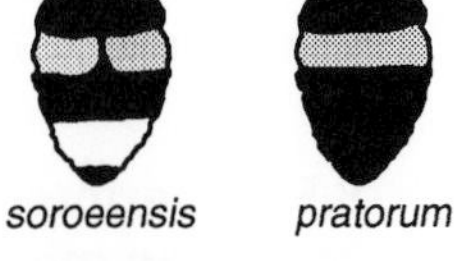

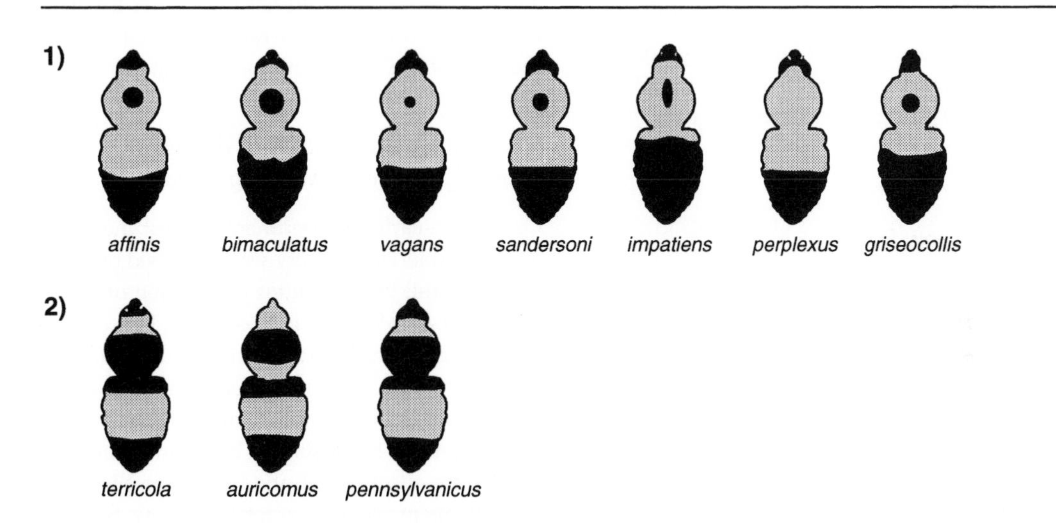


**Figure 9.3** Putative Müllerian mimicry 'rings' in bumble bees (*Bombus* spp.) in England and eastern North America. In both regions, two distinct colour patterns have been distinguished. Redrawn from Plowright and Owen (1988).

Both *B. ternarius* and *B. vagans* emerge before *B. rufocinctus* so it is likely that most potential predators will have learnt to avoid both forms before they arise. Plowright and Owen (1980) argue (with the help of a simple mathematical model) that such selection will allow a polymorphism to persist, and that it is unlikely to drift to monomorphism when the population size is large and when there is spatial heterogeneity in the distribution of *B. ternarius* and *B. vagans*.

### 9.4.4 Cotton stainer bugs (genus *Dysdercus*)

Bugs of the genus *Dysdercus* ('cotton stainers' because many species carry microorganisms which stain cotton balls as they are feeding on them) are often conspicuous, gregarious, and live on plants that are known to synthesize toxic compounds. Their close relatives (such as Lygaeid bugs) are known to be unpalatable to birds (Sillén-Tullberg et al., 1982; Sillén-Tullberg, 1985*a*,*b*). Taking these two sets of observations together, it seems highly likely that cotton stainers are unpalatable to predators (Zrzavy and Nedved, 1999). Several species of *Dysdercus* show a similarity among one another which varies from area to area (Doesburg, 1968; Edmunds, 1974), reminiscent of heliconiid butterflies. In a recent paper, Zrzavy and Nedved (1999) performed a cladistic analysis of the morphological characters of 43 species New World cotton stainer bugs and identified three putative mimicry rings ('yellow', 'median yellow–black', and 'median red–black') which also contain other bugs of different genera (Doesburg 1968; Zrzavy and Nedved, 1999). Interestingly, two highly polymorphic species (*D. obscuratus*, *D. mimus*) appear to fall into different mimicry rings in different regions, comparable to the burnet moth *Z. ephialtes*.

### 9.4.5 Poison arrow frogs

As Symula et al. (2001) report, good evidence for mimicry (whether it be Batesian or Müllerian) is found when the colour and pattern of a given species closely resembles several different model species living in different geographical regions. Bates had used similar observations and reasoning to argue that the evolution of mimicry involves the divergence of forms, rather than simply capitalizing on ancestral resemblance. An interesting example of this is the putative single species of poison arrow frog *Dendrobates imitator* which appears to resemble several different species of poison arrow frogs in Peru. By examining DNA sequence variation among members of the 'mimicry pairs' from the different regions, Symula et al. (2001) established that *D. imitator* does indeed form a monophyletic group and that it likely diverged in appearance more recently than the species it resembles. Interestingly, Symula et al. (2001) propose that *D. imitator* is more toxic (and may be able to exploit more microhabitats) than its models, suggesting that it is not a case of Batesian mimicry, and raising questions as to why it appears the 'mimic' rather than 'model'. These issues are briefly returned to in Section 9.7.1.

## 9.5 Experimental evidence for Müllerian mimicry

Here we discuss two types of evidence for Müllerian mimicry evolving via the predatory mechanism first proposed by Müller. First, we discuss the evidence that rare unpalatable forms tend to have a lower per capita survivorship when they are distinct in appearance compared to when they resemble a common unpalatable species. Second, we discuss specific evidence that naive predators typically take a few of each unpalatable form before they begin to avoid them. Finally, we ask whether these observed predatory behaviours would tend to confer a selective advantage to mutants of rare unpalatable species that resemble a more common unpalatable form.

### 9.5.1 Direct assessments of the benefits of adopting a common warning signal

One of the earliest experiments to test the advantages of Müllerian mimicry was described by Benson (1972)—a full 96 years after publication of the idea (see Fig. 9.4a)! Working in Costa Rica, Benson (1972) stained part of the forewings of a group of *Heliconius erato* butterflies, creating a non-mimetic form (red band stained black) and a mock-treated control (similar area of black stained black, in order to control for the effects of painting). The sample sizes of individuals marked were relatively low, but by re-sighting individuals at roosting sites it was possible to estimate two indices of selective (dis)advantage: mortality and beak marks. In the first year of study (1968), the non-mimetic forms had lower estimated longevity than their mock-treated controls, while in both years (1968 and 1969) the wing damage was higher in non-mimetic forms than the mock-treated controls and unaltered *H. erato* combined. Although these results are entirely consistent with Müllerian

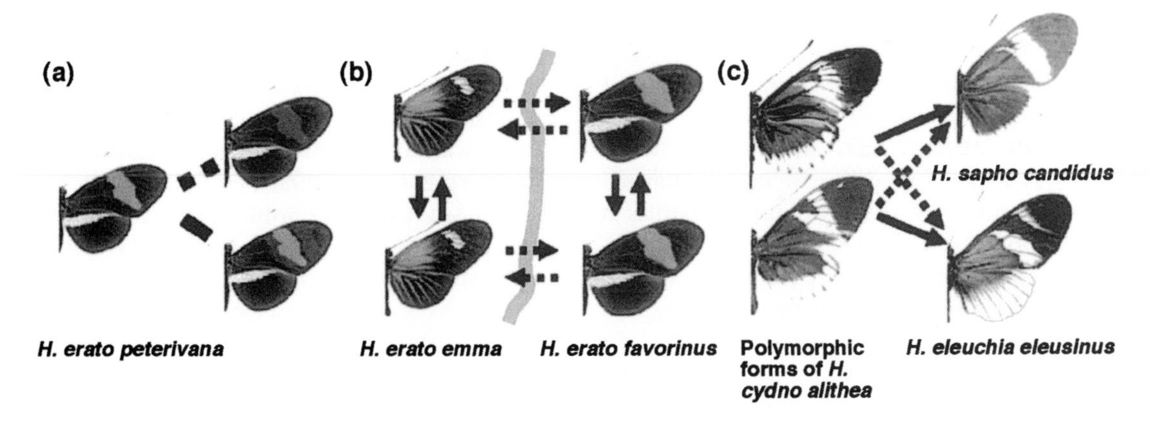


**Figure 9.4** Field tests of Müllerian mimicry. (a) In Benson's (1972) experiment the establishment success of altered and mock-treated *H. erato* butterflies were compared. (b) In Mallet and Barton's (1989) study, different forms of *H. erato* were reciprocally transferred across their hybrid zone and their establishment success compared with controls moved within each area. (c) In Kapan's (2001) experiment, white and yellow forms of *H. cydno* were separately transferred to sites dominated by white co-model (*H. sapho*) and the yellow co-model (*H. eleuchia*) and their establishment success compared. From Kapan (2001), redrawn from Nature with permission. See text for further details.

mimicry, it should be borne in mind that the disadvantage to the blackened *H. erato* may have arisen for some other reason—such as a less effective warning signal—rather than because it was rare per se (Greenwood et al., 1989).

More data consistent with the purifying (anti-apostatic) effects of Müllerian mimicry has come from modifications of the original approach taken by Benson (1972). Working in Peru, Mallet and Barton (1989) reciprocally transferred local races of *H. erato* ('postman' and 'rayed') across their narrow hybrid zone (see Fig. 9.4b), so that they could evaluate the relative success of forms that were rare (individuals transferred across the hybrid zone) with forms that were common (controls caught, marked, and released on the same side of the hybrid zone). Overall, proportionately fewer experimental butterflies were recaptured than controls. However, the residence time (mean known lifespan) of experimental and control butterflies that were recaptured was not significantly different, suggesting that selection occurred relatively soon after release. Among recaptures, a higher proportion of experimental individuals than controls of *H. erato* carried beak marks suggesting that predation (most likely by jacamars) played some role in generating differences in establishment rates. Indeed, the difference in survival rate between experimental and control releases were individually significant at two sites where jacamars were common, and not significant at the two sites where jacamars were perceived rare or absent.

In another field study, Kapan (2001) exploited the unusual polymorphism exhibited by *Heliconius cydno* in western Ecuador. Two colour morphs (yellow and white) resemble two different species: *H. eleuchia* (yellow) and *H. sapho* (white), respectively. Where *H. erato, H. eleuchia*, and *H. sapho* occur in the same locality, the yellow and white colour morphs of *H. cydno* vary in frequency. Yellow and white *H. cydno* were caught at two source sites and released in other sites which were dominated by a particular co-mimic (three sites had abundant *H. eleuchia* and the yellow *H. cydno*, on one site *H. sapho* appeared more abundant than *H. eleuchia* and both morphs of *H. cydno* were present). Treatments involved the introduction of a morph that differed from the dominant co-mimic; in the controls the forms were similar (Fig. 9.4c). If predators quickly learn to avoid morphs, then one might expect that there would be little difference in survivorship of controls and treatments when they were released at high density. To test whether the density of release affected results, different numbers of butterflies were set free (low and high densities). Overall, a lower proportion of treated (rare morph) individuals were re-sighted than control individuals, while experimental butterflies had both lower initial establishment rates and higher subsequent disappearance rates. However, release density also affected survival differences: while there was a significant difference in the estimated life-expectancy of treated and control butterflies at low density, there was no such effect at high density. This again suggests that learning was very quick (in contrast with the interpretation of experiments using artificial prey, see Section 9.5.2). Of course, we still need to understand how polymorphism in *H. cydno* can persist when one might expect that distasteful butterflies should converge on a single shared warning coloration. Kapan (2001) suggested that the solution may lie in geographically variable selection (see Joron et al. 2001, for example), or relaxed selection at higher density, but we leave consideration of polymorphism in Müllerian mimicry systems to Section 9.7.6.

Pinheiro (2003) recently conducted an interesting experiment in which he released live mimetic neotropical butterflies (participants in Müllerian mimicry rings) and non-mimetic butterflies close to predatory tropical kingbirds (*Tyrannus melancholicus*) and cliff-flycatchers (*Hirundinea ferruginea*). To evaluate the role of the likely prior experience of these predators, he conducted his experiments in three different Amazonian habitats (rainforest, city, and low growth 'canga' vegetation), where only particular species of butterflies co-occur. The highest proportion of sight-rejections by birds occurred at the forest site where cliff fly-catchers selectively attacked the cryptic and palatable species and avoided almost all the Müllerian mimics on sight. By contrast, the cliff fly-catchers in the canga did not discriminate among butterflies and attacked most mimetic and non-mimetic species. Tropical kingbirds in the city also attacked most non-mimetic species, but did tend to sight-reject some species of Müllerian mimic. Overall, Pinheiro's (2003) results not only illustrate the great selective value of carrying the right kind of morphological signal (coupled with appropriate behaviours such as slow flight), but emphasizes how the previous experiences of birds with unpalatable prey influences their subsequent discriminatory behaviour. One gets the impression that this type of fundamental work should have been done a long time ago.

Finally, in an experiment primarily aimed at investigating the benefits of aggregation in the evolution of warning signals (see Section 8.4), Alatalo and Mappes (1996) found that birds which had learned to avoid eating artificial unpalatable prey items (fat laced with chloroquinine) marked with a particular symbol, subsequently showed an early tendency to avoid identical-looking prey, this time with unpalatable almonds attached. These results serve to emphasize the potential importance of temporal differences in the occurrence of Müllerian mimics (see also Section 9.4.2), which has so far been considered largely from the perspective of Batesian mimicry (see Section 10.5.2).

### 9.5.2 Proportions of unpalatable prey consumed by naive predators in the course of education

In Müllerian mimicry, predators are assumed to take disproportionately more of the rarer of several types of distinguishable prey providing toxicities are similar ( $n / a_1 > n / a_2$ if $a_1 < a_2$). However, experiments have not always shown this to be the case. In an early experiment Greenwood et al. (1981) presented wild birds with yellow and red pastry baits (in a 1 : 9 ratio and in a 9 : 1 ratio), both of which had been made similarly unpalatable using quinine sulphate. While birds showed a frequency-independent preference for red baits they showed no tendency to consume disproportionately more of the rarer form of unpalatable prey. In a parallel experiment, Greenwood et al. (1981) allowed domestic chicks to feed on dyed unpalatable (and palatable) crumbs, and the chicks actually took disproportionately more of the more common crumbs! In a subsequent paper, Greenwood et al. (1989) presented yellow and red pastry baits (made equally distasteful with quinine hemisulphate) at different ratios (1 : 9 and 9 : 1) to garden birds, this time with separate controls involving similar pastry baits without quinine. In this experiment disproportionately more of the rare forms of distasteful baits were indeed attacked by birds, although absolutely more of the more common form were taken (in effect, the $n$ was different for the two forms).

In more recent experimental tests of Müllerian mimicry with wild birds (Speed et al., 2000), the proportion of uniformly distasteful pastry baits attacked fell markedly when their abundance increased (from circa 80 down to circa 30 per cent in one experiment, and from 100 down to circa 50 per cent in the other). Lindström et al. (2001*a*) similarly found that the attack rate of captive great tits on rare unpalatable artificial prey (a quinine-soaked piece of almond in a paper shell with square pattern) decreased on a per capita basis with increasing density available. Interestingly, the authors found that total number of unpalatable prey consumed increased with absolute density, as Greenwood et al. (1989) had also found.

The reasons why predators in the studies of Greenwood et al. (1989) and Lindström et al. (2001*a*) did not appear to take a fixed total number of prey before learning they were distasteful, are unclear. Greenwood et al. (1989) suggested that this 'variable *n*' phenomenon may have resulted from predators leaving the study area before they were fully educated to avoid the rare form, but after they were educated about the common form. Mallet (2001*b*) similarly suggested that a possible explanation for Lindström et al.'s results was that learning was not complete after the 2-day trials. However there are other explanations. First, the more common unpalatable prey might be sampled more frequently because there is more to lose if some of them turn out to be palatable. Second, there may be an element of shared learning even for distinct unpalatable prey—once birds encounter the more common conspicuous prey and find it was reliably distasteful, then they may learn more quickly to avoid the rarer conspicuous prey. Of course we do not need fixed number killed, *n*, for Müller's general mechanism of 'success in numbers' to work. What matters is that predators take a declining proportion of available unpalatable prey as their density increases, even if they take higher absolute numbers. Indeed so long as the number of individuals consumed in the course of education does not increase as fast as $a$ itself then the ratio of $n/a$ will always be monotonically decreasing. In fact, Holling (1965: 38) had anticipated Greenwood and Lindström's results more than 20 years earlier, and showed that Müllerian mimicry could nevertheless evolve in systems with variable *n*: 'He [Müller] presumed a certain fixed quantity of prey were required for the predator's education, whereas the model predicts that a variable quantity is required depending on the density of the distasteful prey. The increased number of contacts occasioned by increased density is balanced and finally nullified by a gradual increase in the efficiency of learning.'

Finally, in a recent experiment Rowe et al. (2004) attempted to measure the selection pressure exerted by avian predators on unpalatable prey when these prey exhibited different degrees of morphological similarity. Unpalatable and palatable artificial prey (a slice of almond, made unpalatable by soaking in chloroquinine phosphate) were placed under a card carrying a particular symbol, and presented to individual wild-caught great tits. A total of 20 palatable prey and 20 unpalatable prey were presented, or just 40 unpalatable prey alone. The unpalatable prey were presented either as a single form (cards marked with identical symbols), or as two forms (which were either similar or dissimilar in appearance). The authors found that while the unpalatable prey exhibited less per capita mortality when there were more of them presented (and consequently no palatable prey presented!), the extent of pattern similarity between co-mimics had no detectable influence on their survivorship. Here two patterns (however dissimilar) did not appear any more difficult to learn than just one. One reason for their intriguing result may be that the great tits were not sufficiently challenged: all the birds had to do was remember at most one palatable prey type, rather than the two unpalatable prey types. Perhaps selection pressure promoting Müllerian mimicry only becomes intense when there are many different unpalatable *and* palatable prey types, and the birds are therefore forced to apply general rules as to which prey types should be attacked, and which ones should be left. We will return to this issue in Section 9.7.5.

Given that: (1) no experiment has ever shown that predators take a equal number of distinct distasteful prey species, independent of abundance, before learning they are each distasteful, (2) the possibility that learning is still being exhibited when the experiment finishes, and (3) the fact that most of the above experiments have only investigated mortality at two or three densities of unpalatable prey (and using one or two 'species'), then there is a clear need for further work. In particular, we need to: (1) run longer experiments (or at least confirm that learning is complete), (2) elucidate the nature of the relationship between $n$ and $a$ in more detail (3) present predators with more challenging tasks involving many forms of unpalatable *and* palatable prey.

## 9.6 Models of Müllerian mimicry

As we have described in Section 3.3, the first formal model of Müllerian mimicry was described by Müller (1878) and outlined in more detail in Müller (1879). Since then there have been numerous mathematical models proposed, many of them controversial in terms of their structure and/or predictions. Appendix A outlines all of the published models that we could find that deal with Müllerian mimicry, or the relationship between Müllerian and Batesian mimicry. In some cases it is questionable whether a certain article includes a formal mathematical model at all (arithmetic arguments are frequently used)—we have cited these where appropriate, but are conscious that we may have missed many other semi-quantitative arguments.

Müller's theory depends much more on learning than does Bates' theory, so it is not surprising that there has been a lot of discussion on the precise form of learning (and forgetting) assumed, and its implications. What is perhaps more surprising is the lack of consistent predictions among these mathematical models. Almost all models agree that two distasteful species that are equally unpalatable will mutually gain from resembling one another (and that the less numerous is likely to gain more). However, when two unpalatable species differ significantly in their unpalatability then it is not entirely clear whether the less unpalatable species should be regarded as a Batesian parasite ('quasi-Batesian', Speed 1993*a*), a commensal, or a Müllerian mutualist: answers differ according to the nature of the assumptions made. This observation raises some extremely important issues, and the subject area continues to attract controversy. For this reason, we have devoted Chapter 11 to addressing it. Another thing to note when viewing Appendix A is how few formal models have considered intermediate phenotypic states—although many of the published models are of an evolutionary nature, almost all have assumed perfect Müllerian mimicry at the outset, rather than its evolution. We will present our own mathematical model with intermediate states, a formal extension of the arguments of J. R. G. Turner and P. M. Sheppard, in Section 9.7.2. Other mathematical and computer models are invoked throughout the following 'questions and controversies' section as appropriate.

## 9.7 Questions and controversies

### 9.7.1 Which is the model and which is the mimic?

This is perhaps the most commonly asked question relating to Müllerian mimicry, and one that has a long history (Müller, 1878 himself addressed it). Of course, it is quite possible that all species within a Müllerian mimicry complex can be considered 'co-mimics' (or 'co-models'), particularly if there is some form of mutual convergence in appearance among all forms. This was essentially Müller's (1878) belief (English translation): 'the question of which one of two species is the original and which one is the copy is an irrelevant question; each had an advantage from becoming similar to the other; they could have converged to each other', and probably represents the most common contemporary response. Dixey (1909) was an early proponent of this view, labeling such reciprocal advantages of mimicry in 1894 as 'diaposematism'.

However, it is also quite possible that the evolutionary dynamic is more typically one-way evolution ('advergence') than convergence. Box 9.2 makes it clear how selection on the rarer (or less unpalatable) species to resemble the more common form can occur, but not vice versa (albeit under a simplifying set of assumptions). When the possibility of advergence was first raised by Marshall (1908), using a logic very similar to that described in Box 9.2, it was almost immediately criticized by Dixey (1909). Dixey (1909) was courteous throughout, but his frustration with Marshall's (1908) sentiments boiled over in his closing paragraph (p. 583):

> However, Mr Marshall has now shot his bolt. It has failed; and the upholders of the large and comprehensive principle of Müllerian mimicry, including its corollary of diaposematism or reciprocal influence, may await with equanimity the delivery of attack from any other quarter.

Dixey (1909) effectively argued that a mutant of a common species which resembled a rarer species may not lose all of its protection from resembling the common species—on the contrary, it may effectively gain the best of both worlds. Fisher (1927, largely reprinted in 1930) followed up with a similar set of criticisms, arguing that Marshall and Dixey had both made extreme assumptions and that the truth lay somewhere in between. Today, it seems the empirical evidence and theoretical arguments have swung back to supporting advergence, this time with little debate. We have already encountered one putative advergent Müllerian species, the poison arrow frog *Dendrobates imitator* which appears to have evolved resemblance to different model species in different geographical areas (Symula et al., 2001). Similarly, Poulton (1909*b*) and Mallet (2001*a*) have argued that the North American viceroy butterfly, which is now known to be unpalatable to birds (Ritland and Brower, 1991) and has close relatives which look very distinct, adverged to resemble the monarch butterfly. Although the evidence is less clear cut, Mallet (2001*a*) also proposed that Müllerian mimicry among *H. melpomene* and *H. erato* may have arisen mostly via advergence of *H. melpomene* towards *H. erato*.

In a recent review, Mallet (2001*a*) considered the possibility that Müllerian mimicry may be characterized more by advergence than convergence, and gave a number of plausible examples. Synthesizing the arguments of a number of researchers, he proposed that when advergence occurs, the models in Müllerian mimicry rings might be expected to be (Mallet, 2001*a*: 786): (1) more unpalatable, (2) more common, (3) earlier (in seasonal species), (4) larger, (5) more conspicuous, (6) more gregarious, and should have: (7) a wider geographic distribution, (8) less 'fuzzy' colour patterns, (9) more ancient colour patterns, (10) less polymorphism, and (11) less overall divergence from an ancestral colour pattern than their mimics. Each of conditions (1)–(6) may clearly generate asymmetries in mimetic relationships (to the extent that one species may undergo selection to evolve towards the other and not vice versa), while properties (7)–(11) represent characteristics that may assist us in provisionally identifying putative models and putative mimics in Müllerian mimicry systems.

The bottom line is, we still do not know what proportion of Müllerian mimicry systems are the product of convergent or advergent evolution, but increasing recognition of the theoretical likelihood of advergence, and the identification of good candidate examples, leads us to think it is common. Mutants of species A and members of species B may both benefit if the mutant form of A looks like B, but the reverse is not necessarily true: mutants of B may not spread if they resemble species A. If advergence were widespread, this would make the evolutionary dynamics of Müllerian mimicry much more similar to Batesian mimicry (see Chapter 10) than previously believed. Of course, advergence and convergence may not always be mutually exclusive processes. Indeed, Sheppard et al. (1985) argued that while Müllerian mimicry might involve an initial stage of advergence (e.g. by a chance macromutation in either species), mutual convergence in both species might subsequently be selected for.

### 9.7.2 How can mimicry evolve through intermediate stages?

As noted in Section 9.6, the vast majority of mathematical models do not consider intermediate states, and this question has not been given the attention it deserves. Distributions of phenotypic appearances of Müllerian mimics have however been regularly depicted by P. M. Sheppard, J. R. G. Turner and co-workers when illustrating verbal arguments of the evolution of Müllerian mimicry through intermediate states (e.g. Turner 1981, 1984; Sheppard et al., 1985). In the following section we outline their phenomenological model, but in a slightly more formal quantitative way.

Consider members of two unpalatable species (species 1 and species 2) which have an appearance

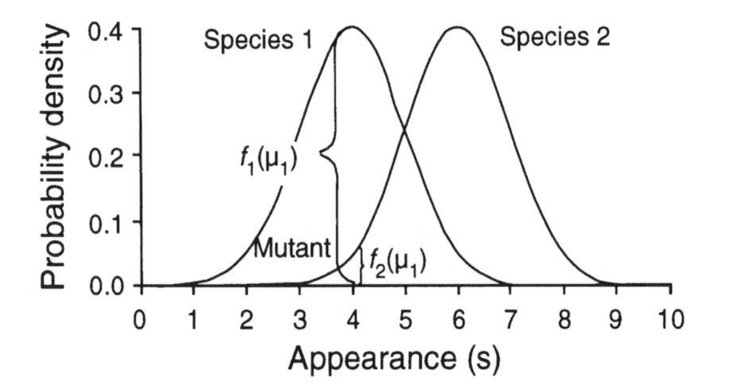


**Figure 9.5** The probability density distributions ($f_1(s)$ and $f_2(s)$) of attributes of members of two unpalatable species 1 and 2, centring around means $\mu_1$ and $\mu_2$ respectively. One measure of the distinctiveness of the two species, of use in determining how much of each species need to be sampled before their unpalatability is appreciated, is the degree of overlap in their appearance.

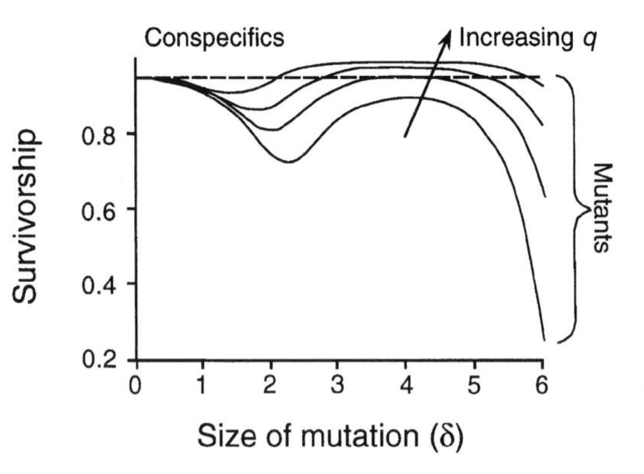


**Figure 9.6** The predicted surviroship of a rare mutant of species 1 compared with the predicted survivorship of conspecifics of species 1, when mutants differ in their signal from modal conspecifics by $\delta$. There were 100 of species 1 in total and 2 prey items of each appearance class $s$ were assumed to be attacked during the course of predator education. Here $\sigma = 1$, $\mu_1 = 3$, $\mu_2 = 7$, while $q$ was systematically increased through 0.5, 1, 2 and 5. In case (a) the two species were very distinct in appearance ($\mu_1 = 3$, $\mu_2 = 7$), while in case (b) thew two species were already similar ($\mu_1 = 6$, $\mu_2 = 7$).

('signal') $s$ that can be characterized along a single dimension. The populations of the two species have probability density functions $f_1(s)$ and $f_2(s)$ in appearance, which are normally distributed in a single phenotypic dimension around means $\mu_1$ and $\mu_2$, respectively, both with standard deviation $\sigma$ (Fig. 9.5). Let the density of species 2 be $q$ times that of species 1.

Consider now the success of a very rare mutant form of species 1 which has an appearance $\mu_1 + \delta$ ($\delta$ positive, $\mu_1 \leq \mu_2$). Müllerian mimicry is effectively a 'strength in numbers' phenomenon so, as Sheppard et al. (1985) argue, the survivorship of this mutant is likely to be higher than its model conspecifics only if there are more individuals with this specific appearance, namely if: $\{ f_1(\mu_1 + \delta) + q f_2(\mu_1 + \delta)\} > \{f_1(\mu_1) + q f_2(\mu_1)\}$. Note that such an approach allows for the possibility that a mutant can indeed sometimes benefit by resembling *both* conspecifics and heterospecifics (Dixey, 1909). Figure 9.6 compares the survivorship of the mutant and conspecifics calculated in the above manner assuming normal probability density functions for the appearances of species 1 and 2. It is readily confirmed that a rare mutant of species 1 is more likely to spread if $q > 1$ (supporting Marshall, 1908), and if this condition holds then a mutant with signal value of $\mu_1 + \delta$ will spread more rapidly the closer $\delta$ brings the mutant towards $\mu_2$. As might be anticipated, the larger the value of $q$, then the greater the 'zone' under which mutants can spread. However, a mutant of species 1 can in theory spread if $q = 1$ so long as the two species are already similar to one another in appearance—under these cases mutual convergence in both species will be selected for, although in this case the selective advantage to mutants will typically be very low.

If the mutational increment $\delta$ is itself a normally distributed parameter with a standard normal distribution, then it is easy to see how natural selection could promote the coalescence of species (typically but not exclusively, advergence of the rarer species towards the more common unpalatable species—see Mallet, 2001*a*). However, if the two unpalatable species are too far apart in appearance at the outset then almost all mutations will be at a selective disadvantage (note the dip in survivorship for Fig. 9.6) and Müllerian mimicry will not evolve.

In general, this approach predicts that more common unpalatable species will experience different selection pressures than rarer unpalatable species. It also emphasizes that intermediate phenotypes may look like no unpalatable species (in which case they are at a serious selective disadvantage), they may look like just one of the species, or they may look like both. However, we add a note of caution to Sheppard et al.'s (1985: 577) view that 'if the situation is symmetrical . . . then there will be equal selection pressure on both species to converge to a target pattern half way between the two.' By contrast, we predict that there is likely to be very little selection pressure at all to generate Müllerian mimicry in these instances: only when these two species already closely resemble one another will there be genuine convergent evolution.

### 9.7.3 Why are mimetic species variable in form between areas?

We have already seen examples of Müllerian mimics, such as *H. erato* and *H. melpomene* which, while locally monomorphic, vary considerably in appearance from area to area 'as if by touch of an enchanter's wand' (reported commentary of Bates in Müller, 1879: xxix). In recent years there has been a leaning towards investigation of Müllerian mimicry from a spatial perspective, partly with such phenomena in mind. Mallet and Barton (1989) made an early contribution by investigating cline formation in spatially distributed populations when rare forms were selected against. Sasaki et al. (2001) extended this analysis, developing a spatially explicit population model of the implications of Müllerian mimicry in heliconid butterflies. Their model was based on assuming a particular (linear) form of frequency-dependent selection (the more a wing pattern is found in a place, the more predators in that place learn to avoid them) rather than any explicit model of mimicry, and concentrated on the predicted movement of interfacial boundaries between different morph distributions. Sasaki et al. (2002) showed that where two sets of locally different forms meet, the strength of cross-species interaction determines whether the mimetic morph clines of model and mimic species coalesce, or effectively pass through each other.

Spatial approaches like the ones described above may well help researchers understand how local selection can generate regional-scale phenomena. To show this explicitly, we now briefly examine the properties of our own simple model which highlights how, despite this selection for phenotypic uniformity, regional-scale heterogeneity can still persist.

In Fig. 9.7 (see also Plate 3) we show grids of cells (32 × 32) arranged in pairs (left and right) to represent the distributions of different morphs of two distasteful species (cell *i*, *j* in each of the two grids is the same spatial location). There are a maximum of six distinct morphs of each distasteful species, and morphs 1–5 of the two species resemble one another (e.g. morph 1 of species 1 resembles morph 1 of species 2). Morph 6 (depicted in yellow) is phenotypically distinct between the two species. The morph of any given species that dominates a cell at any given time is depicted with a distinct colour (green, orange, red, blue, pink, yellow). Where no morph comprises more than 95 per cent of the individuals of that species in any given cell then the colour code for the most common morph present is employed, but patterned using horizontal lines (bars).

In each new generation of the two species, several things happen. First, a combined fixed proportion of 5 per cent of individuals move to their eight nearest neighbours each generation (edge effects are avoided by assuming a taurus). Second, the densities of species 1 and 2 in any occupied cell are brought up to their 'carrying capacities' $K_1$ and $K_2$ (= 200) by cloning at random from the available surviving parents (with a 1 per cent chance, analogous to a mutation, that offspring of any given parent have a different, randomly selected, morph). Finally, prey of both species are attacked by predators until predators learn to avoid them. The avoidance rule employed is Mallet's (2001*a*) simple dose–response algorithm. Thus, predators in a certain cell consume morphs 1–5 from both species (continually selected at random from available survivors) and only stop feeding on a particular morph when it has gone extinct from that cell, or when the predators have consumed a critically high dose of toxins (= 1) from attacking that particular morph. Individual members of each species contain a dose of 0.2. For morph 6 (different between species) each species is treated separately.

Starting with a random distribution of morphs among cells for both species (in which individual cells have only one randomly chosen morph), the system rapidly degrades by dispersion into one in which there are mixtures of morphs in almost every cell (Fig. 9.7, generation 20). However, since members of the two species in any given cell tend to survive better when they have common appearance,

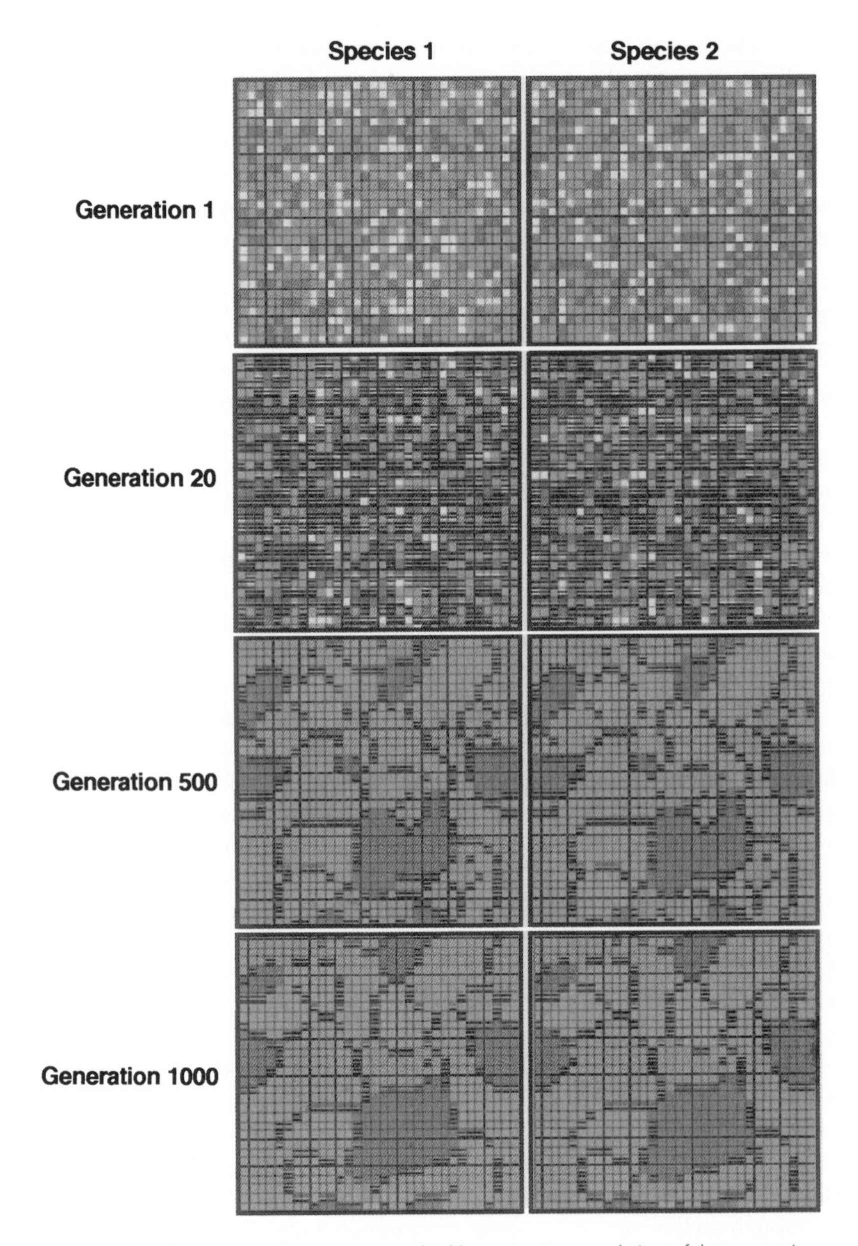


**Figure 9.7** The establishment of Müllerian mimicry between two unpalatable species. Here populations of the two species were distributed in a regular grid. Individuals of each species could occur in any one of six different morphs, five of which look similar between species. We assumed that the local predator community continues to forage on a given phenotype in a given cell until a combined dose of toxins is consumed. Starting with a random distribution of the most common morphs to cells, the system soon reduces to one in which the two species in the same cell share the same appearance, with narrow 'hybrid zones' between races of the different phenotypes. (see also, Plate 3).

those cells which happen to be dominated by species that resemble one another are highly stable (and produce more dispersers). By generation 100, very similar morphological distributions of the two species typically tend to form, which only become mixed in narrow hybrid zones (barred cells). These mimetic configurations are highly persistent and change little over time (Fig. 9.7, generation 1000).

Note that this pattern formation arises in a completely homogeneous environment and with relatively high dispersal rates of prey—all that is required to facilitate the mosaic structure is localized positive frequency dependence, coupled with chance effects. The local frequency dependence is generated by continuing selection to resemble common conspecifics and heterospecifics, while the chance effects are primarily generated by variation in starting conditions (roughly equivalent to founder effects). Of course, the random number generator in our simulation produced extreme initial variation in morph types in our model and the real world does not work like that. Several theories, not mutually exclusive, have been put forward to explain the underlying source of the geographical diversification, particularly in neotropical butterflies (see Mallet et al., 1996). For instance, K. S. Brown, P. M. Sheppard, and J. R. G. Turner have suggested a refuge explanation in which local populations diversified from one another when in isolation due to Pleistocene ice ages. H. W. Bates, W. W. Benson, and J. A. Endler have argued that such diversification might arise through local adaptation to particular environments not related to mimicry. J. Mallet and M. C. Singer place more emphasis on genetic drift as a process facilitating geographical diversification. Such a mechanism would be expected to be more powerful in small populations and in areas which are temporarily devoid of predators impacting the mimicry. Available data on genetic variation among heliconid specimens based on mtDNA (Brower, 1996, Jiggins and McMillan, 1997) does not allow one to clearly distinguish between these different theories, but future studies of the geographical variation in the genes that create colours and patterns themselves may well be more revealing (Mallet et al., 1996; Jiggins and McMillan, 1997).

Overall, it is clear that local positive frequency dependence is a factor which serves to promote local homogeneity, but it can also preserve regional heterogeneity once local differences have arisen by some other means (see also Ball, 1999). Of course, this model is too crude and general to represent any particular species system. For one thing, carrying capacities can vary from location to location and the impacts of such heterogeneity are themselves worthy of investigation. However despite the simplicity of the model, the patchwork distribution depicted is reminiscent of the distribution of the distasteful *H. erato* and its 'scrupulous' (Joron, 2003) mimic *H. melpomene*.

### 9.7.4 How can multiple Müllerian mimicry rings co-exist?

We have already considered examples of tropical butterflies (Section 9.4.1) and European bumble bees (Section 9.4.3) in which several distinct Müllerian mimicry rings appear to co-exist in one place. Given that the proposed selective benefits of Müllerian mimicry centre on reducing the burden of predator education, one might ask why do not all distasteful species evolve to have the same pattern? There are two general explanations, which are not mutually exclusive. First, the different mimicry rings may contain members that are not completely overlapping in spatio-temporal distribution, so there is little or no selection pressure for phenotypes to converge. Second, the different mimicry rings may contain forms that are so distinct, that any intermediate phenotypes are at a selective disadvantage.

As mentioned earlier, it has been proposed that the different neotropical butterfly rings may be segregated according to some subtle differences in flight height within the canopy (Papageorgis, 1975; Beccaloni, 1997). While neotropical ithomiine butterflies appear to show vertical stratification as a consequence of differences in host plant height (e.g. Beccaloni, 1997), the evidence is rather equivocal for taxonomic groups such as *Heliconius* (Mallet and Gilbert, 1995). Other candidate mechanisms that may enhance local stratification in butterfly mimicry rings include different nocturnal roosting heights (Burd, 1994; Mallet and Gilbert, 1995; Mallet and Joron, 1999) and there may be a small degree of temporal separation in flight activity (DeVries et al., 1999).

Of course, we may be looking for evidence of stratification when we don't really need to invoke it to explain multiple mimicry rings. In a seminal paper, Turner (1984) drew an analogy of an astronomical attractor, implying that the development of multiple rings was in essence an 'initial condition' phenomenon: 'In this way, like planets forming from a cloud of gas, clusters of mimetic species will arise and form what we call mimicry 'rings'. Species occupying the spaces between the rings will be pulled into them, but sooner or later these focal patterns, having absorbed the available species, will stabilize. If they differ too much from each other they will not be able to converge, for birds will never mistake one for the other.' Franks and Noble (2004) have recently made an important contribution in this area, arguing that Müllerian mimicry rings can readily evolve for precisely these reasons that Turner (1984) had in mind—intermediate mutational forms at a selective disadvantage. In their paper, Franks and Noble (2003) described an individual-based simulation model in which the appearances of unpalatable species were free to evolve by selection (imposed by predation) and found that distinct collections of unpalatable species ('mimicry rings') readily evolved. As Batesian mimicry can effectively destabilize Müllerian mimicry, Batesian mimics can chase their respective mimicry rings through cycles of colorations, increasing the chance that two mimicry rings might move within convergence range of each other. One consequence of this is that the more Batesian mimics there were in the system, the fewer the mean number of distinct rings that formed, although such predictions have yet to be tested. Their paper is important because it marks a shift in emphasis away from a simple dichotomous view of two unpalatable prey and a single predator to a more realistic community perspective involving multiple Batesian and Müllerian mimics.

We can see how Turner's arguments work by extending the model (itself based on discussions of Turner and Sheppard) outlined in Section 9.7.2. Here we now consider the evolution of the morphologies of 20 equally unpalatable species whose appearances can be represented in two separate dimensions: $x$ and $y$. We let each species in this system begin with particular mean $x$ and $y$ appearance values, which were randomly selected from an even distribution within the limits 0 and 10 (see Fig. 9.8a and c). As before, the appearance distributions of each species around their characteristic means were assumed to be normal, both with standard deviation $\sigma$. We either assumed that all the species in the system were equally abundant (Fig. 9.8a and b), or that one species had a density that was 100 times that of each of the other species (Fig. 9.8c and d). Each iteration a species was chosen at random and a mutant was generated, based on departure from the modal form. The signals of these rare mutants differed from the modal signal magnitude by $\delta_x$ and $\delta_y$ respectively, with both increments selected from a standard normal distribution (mean 0, standard deviation 1). We assumed that mutation events were exceptionally rare, so that if they occurred in any given species then they either proceeded to fixation (if the predicted per capita mortality of the mutant form was equal or higher than that of modal conspecifics) or to extinction (if their predicted per capita mortality was lower than that of modal conspecifics), before any new mutation arose. This 'invasion implies fixation' rule can be justified on the basis that as the density of mutants increase, so does their per capita survivorship.

Figure 9.8a and b shows a typical simulation for 10 000 iterations. Starting with 20 species with random phenotypes, all at equal density and with equal unpalatability, these phenotypes gradually converge by genotypic selection to reduce the per capita cost of predator education. However, in converging spaces open up which mutations have very little chance of successfully bridging. As intermediates in these instances invariably tend to have lower fitness, distinct rings of Müllerian mimics form. Note that rings formed even when all species are equally common: while an unpalatable species is unlikely to experience selection to evolve towards another single species of equal unpalatability and density, it could experience selection to evolve towards two or more of such species if these other species happen to be already similar to one another in appearance. The net effect is a gradual coalescing of phenotypic forms, but only amongst those originally started in the same general portion of phenotypic space (Fig. 9.8b). When one unpalatable species is particularly common in the system then, as expected, its own appearance did not change over time and other unpalatable species evolved towards it (Fig. 9.8c and d). Repeated simulations found that the mean number of unpalatable species

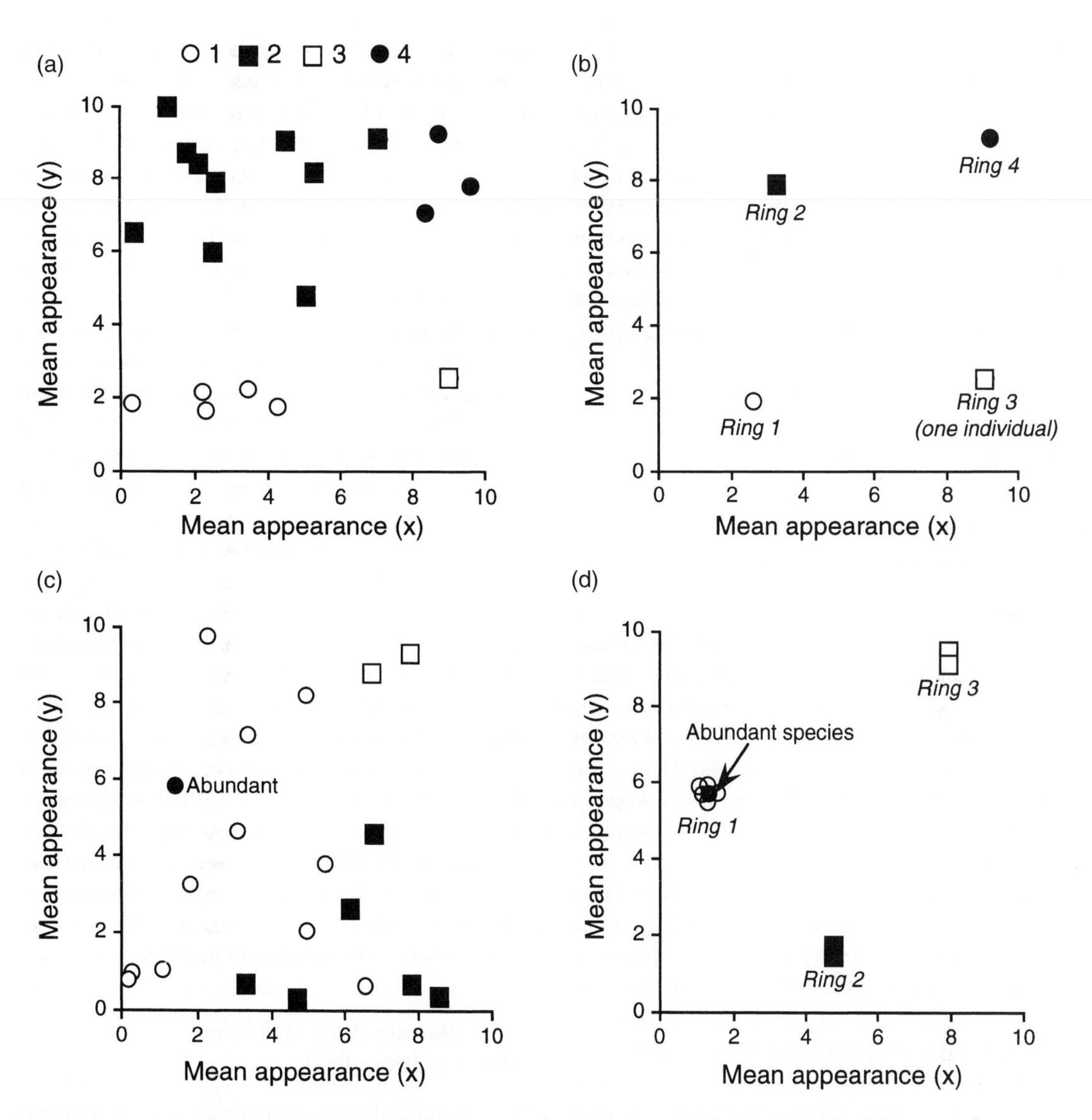


**Figure 9.8** Mimicry ring formation among 20 unpalatable species. In these evolutionary simulations we considered the success of a mutation of any randomly chosen species and allowed all members of this species to proceed to fixation if the mutation proved successful. (a) and (c) show the initial conditions (randomly selected) in two separate simulations, showing the ring to which these species were eventually drawn while (b) and (d) show the respective outcomes from these two simulations after 10000 iterations of identifying mutants. In simulation (a)–(b) all unpalatable species were equally common, while in simulation (c)–(d) on species (marked) was 100 times more abundant than each of the other species.

that evolve to resembling such a highly common species was significantly higher than the mean number of species in separate simulations that evolved close to an arbitrary species when all of the species were at equal density. However it was clear that the initial conditions (specifically, how species were originally clustered in space), in conjunction with the limited probability of large mutations, played the most important role in determining what co-evolutionary pathways were followed. As anticipated, when the standard deviation of the mutation sizes $\delta_x$ and $\delta_y$ were both set at large values then mutations were far less limited in range and a single ring consistently evolved.

### 9.7.5 What is the role of predator generalization in Müllerian mimicry?

For Müller's mechanism to generate selection for mimicry, the mortality involved during the course of predator education must be substantial. There are a number of reasons why this could be true: for instance, in some areas there may be large seasonal influxes of naive predators, or predators may be forgetful. However, as MacDougall and Dawkins (1998) recently argued, a predator's capacity to simultaneously remember the profitabilities of a wide range of prey types may be limited, and this may have a very important influence on the nature of selection experienced by prey. When there are hundreds of potential prey items, some of which are profitable and some of which are unprofitable, then unpalatable prey may experience selection 'to make life easy' on predators and thereby enhance their own survivorship. One can therefore envisage a scenario in which there is initial selection to maintain and enhance certain features (such as a colour or stripe in the right place) that happen to be shared by more unpalatable species than palatable species in a given area. Once established in collections (rings) of unpalatable species, predators might learn to use these signals as a general indicator of unprofitability, and there would consequently be even stronger selection on other unpalatable species to join the group. Note that the phenomenon of predator generalization (e.g. stripe = distasteful) does not always have to culminate in perfect mimicry—all that is required is that forms of an unpalatable species with the common signal component shared by other unpalatable prey experience higher survivorship than those which lack it. This signalling is of course open to exploitation by Batesian mimics, so one might expect Müllerian mimics evolving forms (e.g. conspicuous and/or detailed patterns) that are difficult to copy by palatable species.

As Fisher (1930: 148) observed 'to be recognized as unpalatable is equivalent to avoiding confusion with palatable species.' If predators need to adopt general rules about which prey to attack and which not, then Müllerian mimicry might be seen as little more than an extension of A.R. Wallace's theory of aposematism, in which predator confusion between edible and dangerous forms of prey is reduced by distinctive signaling. Generalization is likely to be much more necessary in those predators which face too many prey types for a predator to remember them all individually. Therefore, if generalization were important in the generation of Müllerian mimicry, one might expect to see this form of mimicry disproportionately more often in complex communities. Given the challenges of dealing with multiple prey types, laboratory and field experiments have typically manipulated only two or three prey types. Therefore, much more remains to be done to address the role of this potential key phenomenon in the generation of Müllerian mimicry.

### 9.7.6 Why are some Müllerian mimics polymorphic?

Sex-limited mimicry, sometimes seen in putative Batesian mimics, is very unusual in putative Müllerian mimics (Turner, 1978). Similarly, most unpalatable mimetic species appear monomorphic in any one region, apart from narrow hybrid zones between geographical races (Mallet and Turner, 1997; Joron and Mallet, 1998). There are, however, some species of Müllerian mimic which appear polymorphic within any given locality. For example, *Laparus* (formerly *Heliconius*) *doris* is polymorphic throughout tropical America (Mallet 2001*a*)—according to Wickler (1968) there are no documented monomorphic populations of the species. This butterfly has up to four co-existing forms, some of which are probably mimetic while others appear non-mimetic (Joron, 2003). Similarly, the African Monarch *Danaus chrysippus* and its co-mimic *Acraea* spp. occur in several different yet sympatric forms within East Africa (Smith, D.A.S., 1973; Edmunds, 1974). As noted in Section 9.4.3, the bumble bee *Bombus rufocintus* is dimorphic with forms that resemble two distinct earlier emerging *Bombus* species (Plowright and Owen, 1980). Local polymorphism in Müllerian mimicry clearly does not render Müller's hypothesis untenable, but it does mean that Müllerian mimicry systems may sometimes be the outcome of complex interactions, involving more factors than are portrayed in the basic models.

The reasons for these different instances of polymorphism are likely to be complex and varied, and indeed there may be no single most important explanation. For instance: (1) habitat heterogeneity within any given locality may create conditions allowing a mimetic polymorphism to persist (a hypothesis advanced to explain polymorphism in *H. numata*, Brown and Benson, 1974; Joron et al., 2001); (2) weakly defended prey may gain advantage by resembling several different unpalatable models (a theory of 'quasi-Batesian' mimicry advanced to explain, for instance, the two forms of European ladybird *Adalia bipunctata* which may resemble different models—Brakefield, 1985; Speed, 1993*a*, and Chapter 11); (3) polymorphism may arise simply as a disequilibrium, such as the bringing together of previously isolated geographical races (as suggested for the African monarch, Owen et al., 1994). To complicate matters further, there is evidence of non-random mating in some polymorphic Müllerian mimics, such as the African swallowtail (e.g. Smith, D.A.S., 1973), which would be expected to perpetuate a polymorphism once it forms.

Mallet and Joron (1999) pointed out that if the alternative forms of a prey species are sufficiently common that the *per capita* death rate from predator education is small or negligible, then polymorphism of mimetic and other aposematic displays within a species may in fact be nearly stable. Thomas et al. (2003) found that novel conspicuous palatable prey (artificial pastry baits) could spread (via a simple evolutionary algorithm in which 'survivors' reproduce) within a small population of more familiar forms because adult robins (*Erithacus rubecula*) consistently avoided such prey. Hence, occasional establishment of new unpalatable forms in a community may be far from impossible. As we argue below (Section 9.7.7), Müllerian mimics could in theory benefit both from shared predator education and from their mutual poisoning of predators ('Müllerian mutualisms', Turner and Speed, 2001). Müllerian mutualisms between chemically defended species could therefore help maintain a low density threshold at which polymorphisms become possible, thereby enhancing the abundance of aposematic polymorphism.

### 9.7.7 Do Müllerian mutualists only benefit simply from shared predator education?

If chemically defended but non-mimetic species co-exist then they may share predators and thus both contribute to the toxin loads that predators carry. By co-existing and adding to the toxin loads of their shared predators then two chemically defended species may benefit merely from each others' presence, provided an increased burden of toxicity makes a predator less likely to attack those prey species. Turner and Speed (2001) called this situation a 'Müllerian mutualism'.

In the case of mimicry among unpalatable prey species, Müllerian mimics may benefit not only because they share out the cost of educating naive predators, but also because they both poison the same predators. Even if they carry completely different toxins mimetic prey species may still benefit one another because predators with a high burden of one of the toxins, may be unwilling to take the gamble and attack either (Sherratt et al., 2004*b*). More generally, if there are differences in levels of chemical defences between species then predators may periodically wish to sample the least well defended form as an emergency food supply, in which case mimicry can in theory arise as a form of parasitism of the less well defended form on the better defended form. Thus, while we have documented field and laboratory evidence of anti-apostatic selection and shared learning (see Section 9.5), there may be several other, not necessarily mutually exclusive reasons beyond Müller's 'classical' theory that can explain the mutual resemblance of unprofitable prey items. We return to these issues in Chapter 11.

## 9.8 Overview

The once controversial hypothesis of Müllerian mimicry is now a much better understood phenomenon. There is good comparative evidence that defended species, sometimes only distantly related, have evolved a common form through natural selection. There is also no shortage of mathematical and computer models showing how Müller's hypothesis could work. Yet there are surprisingly few experimental studies to elucidate the details of predator behaviour that might lead to selection for a common phenotype. Furthermore, the experiments that have been done all tend to indicate that Müller's theory may be wrong in specific detail (e.g. there may not always be $n$ of each distinct unpalatable prey type consumed, independent of density, before predators learn to avoid them). This does not mean that Müller's general argument is wrong: indeed, the few field studies that have been conducted have all produced results that are consistent with his theory. Much of the work conducted on Müllerian mimicry has so far centred on neotropical butterflies, despite the fact that it is a taxonomically widespread phenomenon. We believe that this field can be fruitfully advanced in many different directions in the following decades. Included on the research community's collective 'to do' list should be: (1) more experiments to understand if and how predator generalization plays a role in the development of Müllerian mimicry; (2) more experiments and theory on multi-species mimetic interactions, particularly how it relates to the evolution and maintenance of mimicry rings; (3) a better elucidation of how behaviours leading to Müllerian mimicry can maintain geographical heterogeneity; and (4) more experimental field studies, especially with non-lepidopteran groups.

PART III

# Deceiving predators

# CHAPTER 10

# The evolution and maintenance of Batesian mimicry

## 10.1 Scope

The fact that some species resemble their backgrounds, or components of their backgrounds was recognized long before the publication of Darwin's theory of evolution by natural selection (e.g. Kirby and Spence, 1823). Indeed, the observation that certain unrelated species resemble one another had also long been reported (e.g. Burchell, 1811, in Poulton, 1909*a*; Boisduval, 1836, in Poulton, 1890). Darwin himself described at least one instance of superficial resemblance (letter to J. S. Henslow 15 August 1832, Darwin, 1903) 'Amongst the lower animals nothing has so much interested me as finding two species of elegantly coloured true Planaria inhabiting the dewy forest! The false relation they bear to snails is the most extraordinary thing of the kind I have ever seen.'[5] Yet the possibility that individuals of one species may actually gain selective advantage by mimicking members of another species was not recognized until the work of Bates (1862) who developed his theory while reflecting on earlier observations of species resemblances among Amazonian butterflies. In this classic paper, Bates (1862) described several different types of mimicry, but the form that he is now most closely associated with ('Batesian mimicry') occurs when members of a palatable species (the 'mimic') gain a degree of protection from predators by resembling an unpalatable or otherwise defended species (the 'model'). Charles Darwin was clearly much impressed by Bates article (letter to Bates 20 November 1862, Darwin, 1887) 'You have most clearly stated and solved a most wonderful problem. Your paper is too good to be largely appreciated by the mob of naturalists without souls; but rely on it that it will have *lasting* value.' Indeed, it has.

Since the publication of Bates (1862) and Müller (1878)—see chapter 9—there have been an enormous number of papers on the subject of mimicry. There have also been numerous reviews including Wickler (1968), Rettenmeyer (1970), Edmunds (1974), Turner (1984, 2000), Waldbauer (1988), Malcolm (1990), Endler (1991), Mallet and Joron (1999), Joron (2003), and Gilbert (2004). The subject of mimicry has been variously described as 'a problem in ecology and genetics' (Sheppard, 1959) and a phenomenon at the 'interface between psychology and evolution' (Mallet, 2001*b*). Given that the use of the term mimicry has also extended to resemblance of non-living components or parts of individuals (see Chapter 1), then it is not surprising that there has been some semantic debate, and various classifications proposed (VaneWright, 1980; Edmunds, 1981; Robinson, 1981; Endler, 1981; Pasteur, 1982). Mimicry of living individuals is thought to arise for several different reasons including gaining access to either prey or hosts ('aggressive mimicry', Eisner et al., 1978; Hafernik and Saul-Gershenz, 2000), nuptial gifts (Thornhill, 1979; Saetre and Slagsvold, 1996), mates (Forsyth and Alcock, 1990), pollinators (Roy and Widmer, 1999; Johnson, 1994, 2000) or parental care (Davies et al., 1998). In keeping with our chosen topic ('avoiding attack') we will restrict much of our discussion of mimicry in this book to a consideration of selection on potential prey species to resemble other species, thereby gaining protection from predators

[5] On 31 July 1867, Darwin wrote to Fritz Müller proposing that the resemblance of a British Planarian to a slug might be an instance of mimicry (Darwin, 1887).

('defensive' mimicry). We have already considered one form of defensive mimicry—Müllerian mimicry. In this chapter we consider Batesian defensive mimicry, and in the following chapter we consider the controversial topic of the relationship between Batesian and Müllerian mimicry. For those readers interested in the phenomenon of mimicry in general, we briefly review some other forms of mimicry in Chapter 12.

Although Batesian and Müllerian mimicry are generally treated as distinct phenomena, it is important to note at the outset that it is sometimes difficult to distinguish between the two forms of mimicry in practice, and mistakes can be made. For instance, the viceroy butterfly (*Limenitis archippus*) has traditionally been regarded as a palatable Batesian mimic of the unpalatable monarch (*Danaus plexippus*) yet Ritland and Brower (1991) presented red-winged blackbirds with abdomens of these two species and found that the viceroy and monarch were approximately equally as unpalatable, a finding more consistent with Müllerian than Batesian mimicry. This particular finding is not entirely surprising—Poulton (1909*b*: 222) also suggested, on the basis of the unpalatability of related *Limenitis* species, that the viceroy was likely to be a Müllerian mimic, and Brower (1958*a*,*c*) showed it even more directly through palatability experiments. Yet the fact remains that some of the examples that we are about to describe as Batesian mimics may have a more complex relationship with their models than the form we are implying. Our emphasis in this chapter will not be on taxonomic coverage (Edmunds, 1974, for example gives some excellent descriptions of intensively studied mimetic associations in butterflies), but rather on highlighting general issues. We begin by briefly reviewing some examples of Batesian mimicry, and phylogenetic evidence that such similarities are unlikely to have arisen simply through common decent. We then compare and contrast the various theories that have been proposed to understand the phenomenon. Next, we examine the evidence for Batesian mimicry and its predicted properties, and finally we address several important questions and controversies, many of which remain only partly resolved.

## 10.2 Taxonomic distribution of Batesian mimicry

### 10.2.1 Examples of Batesian mimicry

Perhaps the classic examples of Batesian mimicry occur in adult butterflies and hoverflies, and we will deal in more depth with several specific cases of these in later sections. However, Batesian mimicry appears to be an extremely widespread taxonomic phenomenon and there is no shortage of examples in other groups. For example, the Bornean grasshopper *Condylodera tricondyloides* closely resembles tiger beetles, even in its mode of running, and may gain selective advantage by imitating these better-defended species (Wickler, 1968). Many spider species mimic ants in morphology and behaviour (McIver and Stonedahl, 1993), but also a range of other organisms including snails, beetles, wasps, and millipedes (Oxford and Gillespie, 1998). The relatively benign Australian ant *Camponotus bendigensis* may gain advantage by resembling the sympatric ant *Myrmecia fulvipes* in colour, size, and foraging behaviour (Merrill and Elgar, 2000). The ant *Camponotus planatus* appears to be mimicked by at least four arthropods in Honduras (Jackson and Drummond, 1974). One of these, the mantid *Mantoida maya* not only appears like the ant but also behaves like one (so much so that several intelligent but taxonomically naive children believed them to be ants). Unusually, the later instars of *M. maya* are strikingly different in colour and resemble wasps rather than ants, presumably because they are too big to pass themselves off as ants.

Several convincing examples of Batesian mimicry are known in fish (Smith, 1997). For instance, the plesiopid fish *Calloplesiops altivelis* exhibits colour patterns which make the caudal portion of the fish resemble the head of a moray eel, *Gymnothorax meleagris* (McCosker, 1977), a species which has skin toxins. When approached by a predator the fish flees to a crevice and leaves its tail exposed, thereby resembling the moray eel in several ways, including an eyespot. Similarly, the non-venomous blennies *Ecsinius bicolor* and *Runula laudanus* resemble the blenny *Meiacanthus atrodorsalis* which has a venomous bite (Losey, 1972).

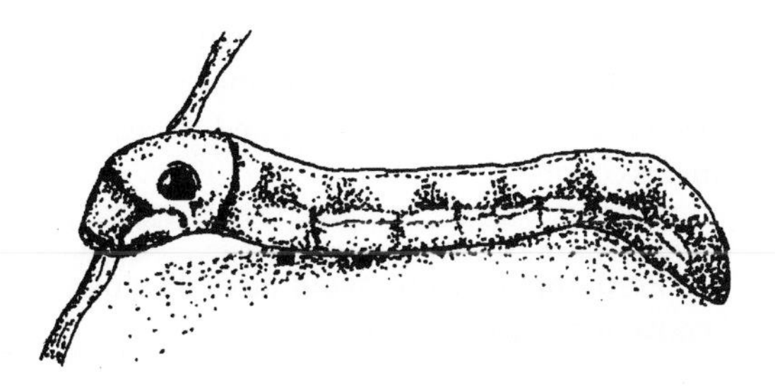

**Figure 10.1** Some hawk moth caterpillars appear to resemble snakes.

Some bizarre examples of Batesian mimicry have also been reported. Bates (1862: 509) describes 'the most extraordinary instance of imitation I ever met' of a large caterpillar which startled him by its resemblance (including eyespots and scales) to a small viperine (venomous) snake (it was so effective he used it to alarm everyone in the village where he was living). Indeed a number of lepidopteran caterpillars resemble snakes (Robinson, 1969, see Fig. 10.1), although the effect on non-human predators have not been determined. Individual fish of the species *Siphamia argentea* sometimes group together and collectively resemble the spiny sea urchin *Astropyga radiata* (Fricke, 1970, as cited in Edmunds, 1974). Juvenile *Eremias lugubris* lizards in southern Africa resemble 'oogpister' beetles (Carabidae: *Anthia*) in colour, gait, and size (Huey and Pianka, 1977). These beetles are considered noxious models because they spray an acid fluid when molested. Similarly, the anguid lizard, *Diploglossus lessonae* in Brazil is considered a Batesian mimic of the noxious millipede, *Rhinocricus albidolimbatus* (Vitt, 1992).

Batesian mimics sometimes fit the classical didactic model–mimic scenario, but they often form part of more complex mimetic communities, often involving mimicry rings (Mallet and Gilbert, 1995; Gilbert, 2004). For example, edible longicorns beetles in Arizona appear to participate as Batesian mimics, not of any specific species but of a more general Müllerian complex consisting of lycid beetles and arctiid moths (Linsley et al., 1961). Similarly, neotropical butterfly communities often consist of mimicry rings that support both Müllerian and Batesian co-mimics (DeVries, 1999; Chapter 9). More work is needed to understand the evolution and maintenance of Batesian mimicry in these complex communities.

### 10.2.2 Comparative evidence for Batesian mimicry

Examples of caterpillars resembling snakes, spiders resembling ants, and lizards resembling beetles support the contention that the shared resemblance does not depend on shared ancestry. Indeed, as Cott (1940) noted, similar appearances in putative mimics and models are often produced by widely different mechanisms. For instance, the transparent wings in mimetic neotropical Lepidoptera (consisting of both Müllerian and Batesian mimics) have been achieved by an array of different structural and optical mechanisms in different species (Cott, 1940). Similarly, the red spots near the base of the wing in the Great Mormon butterfly *Papilio memnon* mimic the body spots of their models (Joron, 2003). However, we must be prepared to accept the possibility that some instances of shared appearance are due simply to shared ancestry. To quote Fisher (1930: 156) 'Obviously the more closely allied are the organisms which show resemblance, the more frequently are homologous parts utilized in its elaboration, and the more care is needed to demonstrate that a superficial resemblance has been imposed upon or has prevented an initial divergence in appearance.'

Additional comparative evidence also supports the contention that mimicry has arisen as a specific adaptation to avoid attack, rather than as a general response to the environment. As Cott (1940) emphasized, mimic species often differ markedly from closely related non-mimic species. For instance, those hymenoptera that resemble the orange and black of unpalatable lycid beetles have coloured wings, rather than the typical colourless wings of the majority of hymenoptera. Of course it is possible that a common environment generates similar selection pressures on appearance (Nicholson, 1927, e.g. discusses the case of two species which are both selected to be cryptic, and how a common environment such as under bark or in a water body may require a similar body structure). Yet it is worth noting that while models and mimics often share the same geographical and ecological environment, only selected palatable species resemble the model. As Nicholson (1927) remarked 'Actually, we find that in many cases closely related mimics resemble a series of unrelated models which differ from one another greatly in appearance and which have only one factor in common, that they are all found in the same environment.' Likewise, instances of sympatric polymorphic Batesian mimicry (multiple forms co-existing of the same mimetic species—see Section 10.3.9) are difficult to explain on the basis of a shared environment. Finally, it is important to note that mimetic resemblances invariably affect only the visible characters—resemblances are only 'skin deep' (see Cott, 1940 for examples). As Fisher (1930) noted, 'Mimetic resemblances are not accompanied by such additional similarities as do not aid in the production, or strengthening, of a superficial likeness.' This supports the view that the observed similarity has arisen primarily due to selection on visual appearance.

Formal phylogenetic analyses of the evolution of defensive signals are relatively rare (Härlin and Härlin, 2003) and while basic questions relating to Müllerian mimicry have been addressed using modern phylogenetic tools (see Chapter 9) questions relating to Batesian mimicry have not. We hope that more phylogenetic work can be conducted in the future to address historical questions relating to the evolution of Batesian mimicry, such as how many times it has arisen in certain groups, and how it evolves in response to diversification in the unpalatable model species.

## 10.3 Experimental evidence for Batesian mimicry and its characteristics

Below we describe some of the evidence for Batesian mimicry and its properties. These experiments provide general support for the phenomenon of Batesian mimicry, and match predictions of theoretical models (see Section 10.4 and Appendix B) relatively well. However, they also tell us more about the underlying nature of Batesian mimicry. Most of the experiments we describe are multifaceted (we could equally cite the same paper under several different headings), but for parsimony we have had to be very selective in what we report. Interested readers are recommended to read the original accounts.

### 10.3.1 Predators learn to avoid noxious models and consequently their palatable mimics

Perhaps the most direct evidence that predators can be duped into avoiding palatable prey after experience with similar-looking but more noxious prey, comes from laboratory studies. Classic research of this kind includes the work of Mostler (1935).[6] Mostler showed that, in addition to their propensity to sting, the abdomens of wasps and honeybees (but not bumble bees) were unpalatable to a wide variety of species of bird. When young birds were offered wasp-like flies (hoverflies) or other diptera for the first time then they were readily eaten. However, mimetic hoverflies tended to be attacked less often when birds had been given experience with their noxious models (to which they bore either a particularly close resemblance, or only an approximate similarity). Better known work includes the studies of Brower (1958a–c; see also Platt et al., 1971) who found that captive blue jays learned to avoid attacking the pipe-vine swallowtail (*Battus philenor*), monarch (*Danaus plexippus*) and queen (*Danaus gilippus berencice*) butterflies after they have tasted them and subsequently tended to reject species (which were either moderately unpalatable or palatable) that resembled them.

While birds have been widely used as predators in these types of study, it is clear that predators other than birds can similarly learn to avoid distasteful prey and thereby be deceived into rejecting palatable prey. For example, Brower et al. (1960) showed that toads (*Buffo terrestris*) learn to avoid attacking bumble bees and that these toads subsequently had a greater tendency to reject edible asilid fly mimics. In an experiment that any fly-fisherman would enjoy, Boyden (1976) cast palatable *Anartia fatima* and unpalatable *Heliconius* spp.

[6] We are grateful to Francis Gilbert, University of Nottingham, UK for providing an English translation of this work.

butterflies to free ranging *Ameiva ameiva* lizards as well as their reconstructed mimics (wings glued on bodies of different species). The degree of protection afforded to mimics was greater with lizards known to have had prior experience with the unpalatable model.

Far fewer studies have been conducted with invertebrate predators (such as dragonflies, spiders, bugs, and wasps), despite the fact that many species are abundant and visual hunters. Bates (1862: 510) himself observed that 'I never saw the flocks of slow-flying Heliconidae in the woods persecuted by birds or Dragon-flies, to which they would have been easy prey; nor, when at rest on leaves, did they appear to be molested by Lizards or the predacious Flies of the family Asilidae, which were very often seen pouncing on Butterflies of other families,' indicating that he had observed that invertebrate predators refrain from attacking certain types of prey, and had appreciated its potential significance. It has been shown, for example, that praying mantids learn to avoid food associated with electric shocks (Gelperin, 1968) and distasteful prey (e.g. Bowdish and Bultman, 1993), and limited observational evidence indicates that dragonflies avoid attacking wasps on encounter (O'Donnell, 1996). Berenbaum and Miliczky (1984) found that the Chinese mantid *Tenodera ardifolia sinensis* rapidly learned to avoid unpalatable milkweed bugs (*Oncopeltus fasciatus*) that had been reared on milkweed seeds. Furthermore, of six mantids that had learned to reject milkweed-fed bugs, none of these individuals attacked palatable sunflower-fed bugs (auto-mimics, see Chapter 12) when they were offered. In a recent paper, Kauppinen and Mappes (2003) evaluated the responses of the dragonfly *Aeshna grandis* on encounter with black and warningly coloured natural prey (painted flies and wasps), and artificial prey in the field. After classifying behavioural responses of the dragonflies towards the prey (e.g. pause but do not touch, touch but not grabbed), the authors concluded that black-and-yellow striped flies were 'avoided more' than black ones, which seems to suggest that this form of coloration confers a selective advantage against odonate predators. While this was a fascinating study, the experiments did not directly control for the frequency of encounters, but simply classified behaviours (largely tactile response rather than those involving life and death) of odonates *following* encounter. This is an important limitation because it is conceivable that black and yellow prey are more visible than black prey. Nevertheless, this interesting paper serves to demonstrate that the role of invertebrate predators in shaping Batesian mimicry merits much more serious scientific investigation.

### 10.3.2 Palatable prey altered to resemble an unpalatable species sometimes survive better than mock controls

One way to test whether mimicry enhances the survival of edible prey is to see if artificial mimics benefit from their resemblance to real models. In an ingenious series of experiments, Brower and colleagues (Brower et al., 1964, 1967; Cook et al., 1969) released palatable day-flying male moth *Callosamia promethea* (a species native to eastern North America) that had been painted in different ways, into areas of Trinidad. A major advantage of using this species was that the moths could be recaptured using traps baited with pheromone-releasing females. Experimental moths painted to mimic the unpalatable *Heliconius erato*, and the unpalatable *Parides neophilus*, were not recaught at significantly higher rates than control-painted (similar amounts of paint but no change in colour or pattern) controls (Brower et al., 1964). Follow-up experiments with different unpalatable models also generated mixed results, leaving Cook et al. (1969: 344) to conclude that 'under wild conditions no clear selective differential can be demonstrated with the promethean moth mimicry system'. Of course, failure to reject the null hypothesis does not prove the null hypothesis. It is possible that birds became aquainted with feeding on the moths where they were released, in a fashion similar to the 'feeding stations' that may have incidentally been formed in Kettlewell's classic experiments on melanism in *Biston betularia* (Grant and Clarke, 2000; Hooper, 2002, see Section 1.2). In mimicry field experiments, there may be a kind of 'uncertainty rule' operating—in the process of measuring the effect of mimicry, one has to disturb the densities of mimics

**Table 10.1** Total recaptures of black-painted, yellow-painted and orange-painted male promethea moths. Adapted from Jeffords et al. (1979)

| | Black | Orange | Yellow | Total |
|---|---|---|---|---|
| Recaptured | 57 | 52 | 35 | 144 |
| Not recaptured | 245 | 250 | 267 | 762 |
| Total released | 302 | 302 | 302 | 906 |

and models, which in turn affects the advantages of mimicry.[7]

Half a decade later, Waldbauer and Sternburg (1975) proposed an alternative explanation for the results of Brower and his colleagues, noting that the black non-mimetic controls of *C. promethea* may have gained protection through their resemblance to unpalatable *Battus* spp. which occur in Trinidad (the fact that *C. promethea* resembled the pipe vine swallowtail *Battus philenor* had already been noted, Brower et al., 1964). Thus, the survivorship of artificial mimics may have been compared against mimics, rather than non-mimics (Waldbauer 1988). Impressed with the elegance of earlier work and motivated by this reinterpretation, Sternburg et al. (1977) and Jeffords et al. (1979) used painted promethea males to evaluate the advantage of mimicry in the pipe vine swallowtail complex in Illinois. This time, some of the *promethea* moths were painted with yellow bars to resemble the non-mimetic forms of the palatable tiger swallowtail butterfly—others (in some experiments) were painted orange to resemble the unpalatable monarch, while other male moths of this species were again mock painted (considering them natural mimics of the pipe vine swallowtail). In these experiments, the yellow painted moths were significantly less likely to be recaptured than orange and black-painted moths (and surviving yellow-painted moths also showed more evidence of wing injury), providing perhaps 'the first field demonstration of the efficacy of Batesian mimicry' (Table 10.1, Jeffords et al., 1979). By contrast, in similar work conducted in Michigan (Waldbauer and Sternburg, 1987) male *C. promethea* moths painted to resemble the palatable tiger swallowtail, the monarch and the pipe vine swallowtail all had similar recapture rates to one another. One explanation may be that the pipe vine swallowtail does not occur locally, while the local monarchs feed on plant species that are most likely not sufficient to provide an emetic dose of cardenolides. Taken together, these works suggest that the selective advantage of Batesian mimicry may be dependent on the availability of suitable noxious models (see Section 10.3.3). The study of Jeffords et al. (1979) therefore remains the most convincing study of its kind. We need more work to establish whether the theories put forward to explain the failure of related experiments to reject the null hypothesis, are indeed justified.

[7] We are grateful to J. Mallet for this analogy.

### 10.3.3 Batesian mimics generally require the presence of the model to gain significant protection

One might expect that Batesian mimicry can only evolve and be maintained in the presence of a noxious model. Indeed, Waldbauer and Sternburg's interpretation of their 1987 study (see Section 10.3.2) was based on this belief. This view has been supported by several experiments that have demonstrated a greater advantage to mimics when they are sympatric with models. In a recent paper Pfennig et al. (2001) put out hundreds of plasticine models of snakes in triplicates. These triads included a replica of the putative mimetic snake in the area, a distinct striped version with the same proportions of colours, and a plain brown version. The replica mimics were either based on the scarlet kingsnake or the sonorian mountain kingsnake which resemble the venomous eastern and western coral snake, respectively. For both model–mimic pairs, the proportions of total attacks by carnivores on the plasticine mimics were significantly higher in locations where the model was absent compared to where it was present. Furthermore, coral snakes become increasingly rare with increasing latitude and elevation in the United States, and in this study the proportions of total attacks on mimic replicas increased with both latitude and elevation, consistent with the view that the protective benefits of mimicry are dependent on the presence of the model (Fig. 10.2a and b).

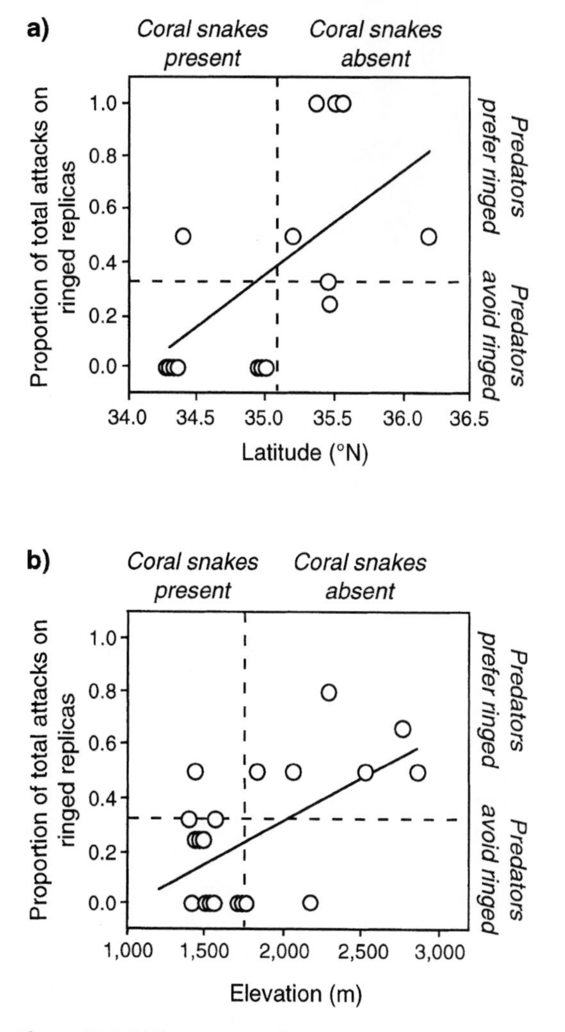


**Figure 10.2** (a) The proportion of carnivore attacks on ringed plasticine replicas of scarlet kingsnakes (a mimic of eastern coral snakes) was higher at those latitudes where coral snakes are absent. Similarly, (b) the proportion of attacks on replicas of the sonoran mountain kingsnakes (a mimic of western coral snakes) was higher at those elevations where the model is absent. If attacks were random then 1/3 of each presented type should be the first to be attacked. Redrawn with permission from Pfennig et al. (2001).

In general, the mimetic forms of a given species do tend to co-occur with their models. Bates (1862) drew much from the fact that models and mimics co-occurred, but also from the observation that their colour patterns changed in a parallel way from area to area. F. A. Dixey in his Presidential Address to the British Association for the Advancement of Science (1919) considered the phenomenon in detail, noting that members of the Oriental butterfly genus *Prioneris* appear to mimic members of the genus *Delias* (although it remains unclear whether this is a case of Müllerian or Batesian mimicry). Surveying their distributions, he reported that in all cases where pairs of species resemble one another, they inhabit the same area. Data for a similar phenomenon in snakes have been collated by Greene and McDiarmid (1981) (Fig. 10.3). Colubrid snakes of the genus *Pliocercus* are rear fanged (non-venomous) and probably consist of at most two species in Central America (Greene and McDiarmid, 1981). These snakes bear a remarkable resemblance to the forms of the front fanged venomous *Micrurus* coral snakes in the area (Fig. 10.3).

Perhaps the most enlightening evidence of the importance of the presence of the model is the forms (sometimes sub-species) a putative mimic species tends to adopt in areas where the model is not present. For example, in butterflies, the mimetic black female form of the eastern tiger swallowtail *Papilio glaucus* tends to occur where the pipe vine swallowtail *Battus philenor* is abundant (Platt and Brower, 1968; Edmunds, 1974). However, in southern Canada where *Battus philenor* is rare, the black female form of *Papilio glaucus* is rarely encountered and the genetically distinct yellow non-mimetic female form predominates (Layberry et al., 1998: 87). Another of the pipe vine swallowtail's mimics, the red spotted purple (*Limenitis arthemis astyananx*) is common in the United States, but it is the non-mimetic (and possibly disruptively patterned) subspecies *L. a. arthemis* that predominates in Canada (Platt and Brower, 1968; Layberry et al., 1998: 208). Similarly, the harmless snake *Lampropeltis doliata* appears to mimic the coral snake *Erythrolamprus*

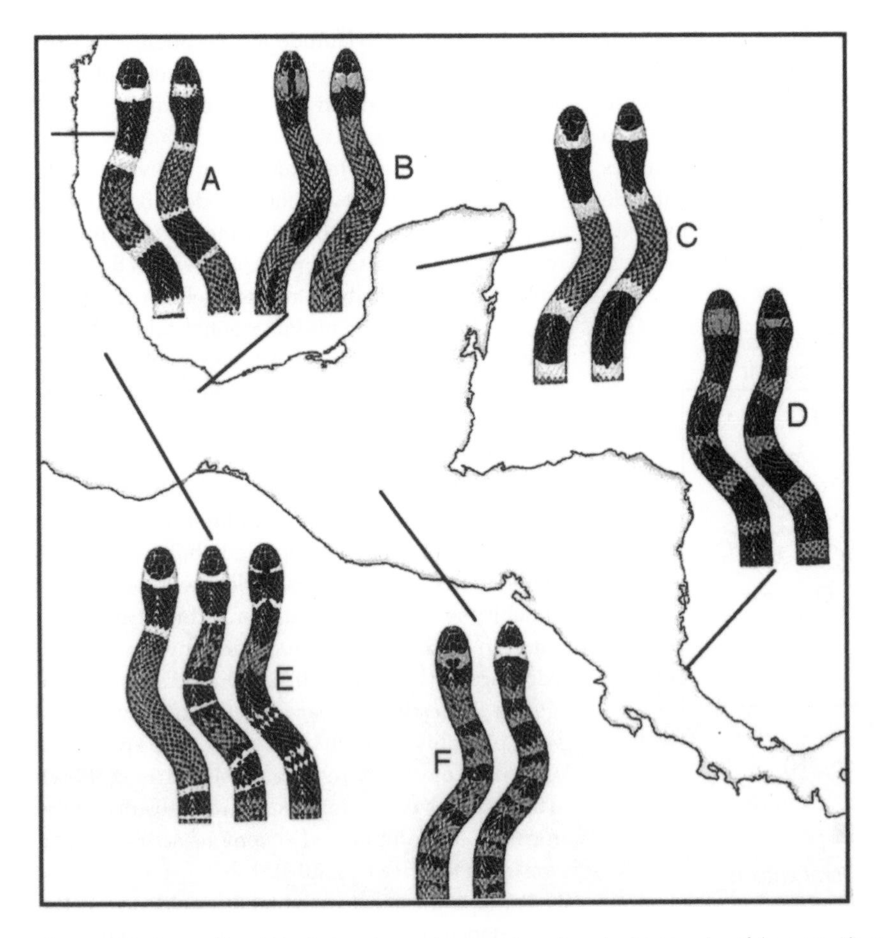


**Figure 10.3** Geographic variation in the colour and pattern of rear-fanged colubrid snakes of the genus *Pliocercus* in central America in relation to colour pattern variation in the more dangerous front-fanged snakes of the genus *Micrurus*. The presumed dangerous models, on the left in each set, are: (A) *M. fulvius*, (B) *M. limbatus*, (C) *M. diastema*, (D) *M. mipartitus*, (E) *M. diastema*, and (F) *M. diastema*. In (E), the centre snake is *Pliocercus*, and the snake on the right is *M. elegans*. Reprinted with permission from Greene and McDiarmind (1981), Coral snake mimicry: does it occur? *Science*, **23**, 177–199. Copyright 1981 AAAS.

*aesculapii* in the southern United States but further north where no venomous models are present the snake takes a more blotchy appearance, more consistent with disruptive colouration (Hecht and Marien, 1956; Edmunds, 1974).

While mimetic protection has long been assumed to disappear when the model is absent, it must be borne in mind that mimetic advantage in the absence of the model may *in theory* be possible when predators migrate from areas that contain the model (e.g. Poulton 1909*a*: 217; Carpenter and Ford, 1933: 43; Waldbauer, 1988). For instance herons and egrets show unlearned avoidance responses towards the highly venomous yellow-bellied sea snakes, yet naive green-backed herons from the Atlantic coast of Panama, where the sea snake does not occur, also showed avoidance reactions (Caldwell and Rubinoff, 1983). While there is clearly potential for the evolution of Batesian mimicry in the absence of the model, we do not know of any study that has formally demonstrated mimetic advantage to a given prey species on the basis of learnt (or innate) aversion to models in another geographical area.

Other puzzling examples of mimics persisting in the absence of a model have been observed. For instance, Clarke and Sheppard (1975) reported that females of the Nymphalid butterfly *Hypolimnas bolina* [which gave 'no convincing evidence' of unpalatability to birds (Clarke and Sheppard, 1975) although it may contain cardioactive substances obtained from certain food plants (Marsh et al., 1977) ] occurred entirely in their mimetic form in Madagascar even though its putative model (*Euploea* butterflies) was absent (a 'rule-breaking' mimic, Clarke et al., 1989). Rather than being maintained by the presence of migrating birds, or regular immigration of the butterfly from India, the authors argued that the resemblance was likely to persist either because of a lack of non-mimetic variants (especially likely if the species had recently immigrated to the area), or due to some selection unassociated with mimicry (it is also possible that the model once existed but is now extinct in Madagascar, Clarke et al., 1989).

### 10.3.4 The relative (and absolute) abundances of the model and mimic affects the rate of predation on these species

It has been widely predicted (see Section 10.4 and Appendix B) that the attack rates of predators on encounter with Batesian mimics and their models should depend not just on the presence of the model, but on the relative abundances of mimics and models. Bates (1862: 514) himself noted: 'It may be remarked that a mimetic species need not always be a rare one, although this is very generally the case.' Several studies have examined the nature of the quantitative relationship between predator attack rates and the relative proportion of models and mimics. As noted in Section 10.3.3, when predators have time-dependent memories, or when prey take time to handle (for instance) then one would expect that the absolute densities of models and mimics as well as their relative densities would affect their attack probabilities on encounter (Turner, 1978; Owen and Owen, 1984; Getty, 1985; Speed and Turner, 1999). Very little experimental work has been done to examine the direct effects of density dependence—for instance, we can find no experimental study that has examined the effects of increasing mimic and model density while keeping their ratio constant, or increasing mimic density while keeping model density constant.

In a classic study, Brower (1960) showed that the rate of attack of caged starlings on palatable (the mimic) and unpalatable mealworms (the model, dipped in quinine dihydrochloride) tended to increase as the percentage of models offered in a sequential mixture decreased, but that the relationship was non-linear. Thus, even when models comprised only 40 per cent of the models/mimics, then mimics often received similar protection compared to trials where there was a greater percentage of models. This is consistent with a non-linear 'zone of protection' (Turner, 1984; Sherratt, 2002*b*, see Section 10.5.5).

In a similar experiment, Huheey (1980) presented model honeybees (with wings removed) and mimic honeybees (wings and sting removed) to toads and tree frogs. Models and mimics were presented in a random sequence to the predators with a particular percentage of models, although in contrast to Brower's (1960) study, the same individual predator was given several different percentage model treatment regimes (to compensate for this problem, the first day's results of a new treatment were ignored). Overall the percentages of models and mimics attacked decreased as the percentage of models offered to both toads and tree frogs increased (it is difficult to assess the linearity of the relationships in this instance as only three model/mimic frequencies were presented). Other experiments with analogous results include the work of Avery (1985) who investigated the behaviour of captive house finches feeding on treated and untreated seeds, and Nonacs (1985) who investigated diet selection by captive long-eared chipmunks feeding on palatable and unpalatable dough balls (see Section 10.3.7).

Several other experiments have shown similar frequency-dependent effects when the mimics were not perfect replicas. For instance, Pilecki and O'Donald (1971) showed that neither poor mimics (artificial palatable pastry mealworms, paler in colour than models) nor their models experienced high attack rates by sparrows when the poor mimics were relatively rare (Experiment III). By contrast, significantly more mimics than models were taken when mimics and models were presented in a 1 : 1 ratio (Experiment IV). In this case the predators became more discriminatory when mimics were more abundant, although even here the poor mimics suffered less predation than expected compared to the palatable controls. In a more recent experiment, Lindström et al. (1997) sequentially presented captive great tits with a combination of model and mimic mealworm larvae (six in total). Models were dipped (and injected) with chloroquine and three small light blue 'nonpareils' (small round balls of brightly colored sugar used in food decorations) were placed close to their heads. Imperfect mimics were simply mealworm larvae with the nonpareils in the middle of their bodies. As might be expected, they found that imperfect Batesian mimics were attacked least frequently by the birds on a per capita basis when the models were more common. When the distastefulness of the models increased, the attack rates of birds on models decreased, but the corresponding attack rates on mimics did not change in a statistically significant manner. This final result may be a consequence of low sample size or reflect some ability of the birds to distinguish imperfect mimics from their models.

In sum, a number of experiments have demonstrated that the relative abundances of models and mimics influence the mean rates of predation on these types—as the ratio models/mimics increases then the attack rates on both models and mimics on encounter tends to decrease.

### 10.3.5 The distastefulness of the model affects the rate of predation on the model and mimic

Common sense suggests that, other things being equal, the protection an edible mimic gains from mimicry will vary with the intensity of the model's defences. This prediction has been supported on a number of occasions. As described above, Lindström et al. (1997) found the distastefulness of the model (or at least the concentration of noxious chemicals) affected the rate of attack of predators on the model. This result had previously been observed in a number of different studies. For instance, in a field experiment with artificial prey Goodale and Sneddon (1977) found that birds in gardens where highly distasteful pastry baits had previously been offered attacked none of their mimics (ranging from perfect to imperfect) in the first day of presentation, yet attacked 75 per cent of the readily distinguishable palatable controls (75 per cent triggering the completion of the trial). By contrast, birds in gardens that had earlier been offered less distasteful baits (flavoured with 20 per cent dihydrochloride solution) ate several of the mimetic baits and 75 per cent of the readily distinguishable palatable baits.

In another study, Alcock (1970) trained captive white-throated sparrows to overturn seed shells (painted with dots) under which was hidden half a mealworm (or no mealworm at all). The birds were then presented with a binary choice of a model and mimic. The mealworm models were either weakly unpalatable (salted) or highly unpalatable (smeared with tartar emetic) and hidden under shells decorated with specific markings. The mimetic mealworms were consistently palatable and were either good or poor mimics of the models (manipulated by changing the dots on the shells). Just as Pilecki and O'Donald (1971) were to theorize in the following year, when the mimic was poor and the model weakly aversive, then the mimic was attacked at a relatively high rate (and significantly more frequently than the model). However, when the model was emetic neither the model nor the mimic was frequently attacked, even when the degree of mimicry was poor.

### 10.3.6 The model can be simply difficult to catch rather than noxious on capture

Can a prey species that is easy for its predators to catch gain protection by mimicking a species that is sufficiently difficult to catch that predators sometimes decline the opportunity to pursue them? We will call this form of resemblance 'evasive Batesian mimicry'; although it has also been called locomotor mimicry (e.g. Brower, 1995), this term has also been used to mean mimicry of the movement behaviour of unpalatable prey (Srygley, 2001).

Van Someren and Jackson (1959) were amongst the first to suggest that a species could gain selective advantage simply by resembling another species that was difficult to catch. In his monograph, Holling (1965) also noted that 'effective escape behaviours are common features that inhibit predators, quite independently of edibility . . . Thus mimicry among edible species might simply reflect a technique of repelling predators that do not involve taste.' Rettenmeyer (1970) similarly included an effective escape mechanism in his list of features worth mimicking. Lindroth (1971) gave an interesting example when he proposed that the relatively slow moving ground beetle *Lebia* sp. (Carabidae) might mimic flea beetles (Alticinae, Chrysmelidae), which appear palatable to avian predators, but jump away when approached. Similarly, Balgooyen (1997) proposed that the short-horned grasshopper (*Arphia consperca*) mimicked the alfalfa butterfly (*Colias eurytheme*) in shape and colour because the latter was difficult to catch. In this instance, direct observations indicated that American kestrels were unsuccessful in pursuing the alfalfa butterfly, and occasionally appeared to hesitate or abandon attacks when preying on the short-horned grasshopper. To crudely estimate the relative catchability of the two species (and no doubt give his offspring some exercise) Balgooyen employed his 13-year-old son to attempt to catch the two species—as expected the butterfly took longer to capture and was captured less frequently. Caged experiments with young American kestrels indicated that the butterfly species was palatable, overall making a reasonable case that here the putative Batesian mimicry was based on evasiveness.

Experimental evidence that mimicry of a more evasive species can protect a slower-moving mimic from predation was provided by Gibson (1974, 1980) who found that Australian star finches (*Neochmia ruficauda*, Gibson, 1974) and robins (*Erithacus rubecula*, Gibson, 1980) continued to avoid attacking prey types (seeds and mealworms) of particular colours that had previously been difficult to catch (by means of a hinged platform). Interestingly, in both experiments two evasive models were used (conspicuous red and cryptic blue, Gibson, 1974; conspicuous orange and cryptic green, Gibson, 1980—they are referred to as cryptic because the platforms had green and blue painted marks). Interestingly, the more conspicuous evasive model appeared to provide higher and longer levels of protection to the mimic. In a similar experimental design, Hancox and Allen (1991) presented red and yellow pastry baits on a specially constructed bird table, which allowed one colour of bait to drop out of reach when attacked. In two sets of 20-day sessions (yellow evasive, red non-evasive followed by yellow non-evasive, red evasive) it was clear that birds rapidly learned to avoid attacking those pastry baits with the 'evasive' colour.

The notion that models can be unprofitable to pursue rather than costly to attack may also extend to aposematism (Chapter 8) and Müllerian mimicry (Chapter 9). For instance, Thompson (1973) argued that some of the forms of the polymorphic spittlebug *Philaneus spumarius* may be conspicuously coloured as a means of advertising their difficulty of capture (although this does not in itself explain why the polymorphism persists; see also Hebert, 1974 for an explanation based on resembling avian excrement and Harper and Whitaker, 1976 for an explanation based on frequency-dependent attack by dipteran parasitoids). A more likely candidate for evasion-based aposematism, first proposed by Young (1971) are conspicuous Morpho butterflies. These butterflies appear relatively palatable when presented to tropical birds (Chai 1986), but kingbirds (and most likely, many other bird species) find them exceedingly difficult to catch (Pinheiro, 1996). Similarly, the co-occurrence of several agile scarab beetles with orange elytra in the Namib desert has been explained as convergence on a common way of advertising ability to escape when pursued (Holm and Kirsten, 1979).

Brower (1995) suggests that the studies of Gibson and Hancox and Allen actually provide evidence against evasive Batesian mimicry: 'These results showing quick loss of aversion to prey unprofitability by reason of effective escape are in marked contrast to reports of long-term aversion to aposematic prey.' We agree that mimicry requires that predators remember previous encounters with similar looking prey, and that all three experiments suggest that in the absence of negative reinforcement (through experiencing evasive models) memories fade and with them any protection afforded to non-evasive mimics. Unlike Brower, we see no evidence in these studies that evasive mimicry cannot be sustained, only that the mimic would soon lose protection if the predator stops encountering models. However, this may be no different than what would occur if aversion were driven by unpalatably rather than evasiveness.

So far the evolution of escape mimicry has not been formally modelled. One might intuitively expect it to be a rare phenomenon if the cost of ascertaining whether a prey individual is easily caught (e.g. by attempting to catch it) were low: in these instances both models and mimics would be pursued, and mimicry would not evolve. In cases where it is costly to attempt to ascertain whether a given prey is a mimic (e.g. pursuit may be energetically demanding) then such mimicry should be entirely plausible (of course this same logic extends to other forms of Batesian mimicry).

### 10.3.7 The success of mimicry is dependent on the availability of alternative prey

Carpenter and Ford (1933: 28) noted that, since the motivation of predators to distinguish model from mimic depends on how hungry they are then 'the success of Mimicry depends on the abundance of other palatable food'. These conclusions were later supported in the specific quantitative models of Holling (1965), Emlen (1968), Dill (1975), Luedeman et al. (1981), Getty (1985), Kokko et al. (2003), and Sherratt (2003) who all showed that predators maximizing their rate of energy gain per unit time (or simply acting to avoid starvation) would be expected to more readily attack models and mimics when alternative palatable prey were in short supply. Indeed, Holling (1965: 36) proposed that '... the marked effects produced by removal of alternate prey or changes in their palatability are great enough that alternate prey should be included as an essential feature of mimicry theory.'

Several authors have attributed low predation pressure on imperfect mimics at least in part to the presence of alternative prey. For example, Hetz and Slobodchikoff (1988) presented the tenebrionid beetle *Eleodes obscura* (which produces defensive secretions when disturbed) with its sympatric, palatable mimic *Stenomorpha marginata* (also a tenebrionid beetle, without a defensive secretion) and palatable house crickets (*Acheta domesticus*). Mixtures of these prey (16 models, 7 mimics, and 7 crickets—a similar ratio to that found naturally) were placed out nightly in the field, thereby exposing them to predators that included bats, skunks, and ringtails.

Although (i) no trials without alternative prey were conducted, and (ii) the experiments were conducted at night where visual mimicry might be anticipated to be less important, the authors argued that the presence of alternative prey may have made predators more risk averse, such that they consumed imperfect mimics at a lower rate than palatable controls. More directly, Nonacs (1985) reported that the attack rates of predators (long-eared chipmunks) on mimics and models (palatable and unpalatable dough balls) decreased when alternative palatable prey were offered in abundance. In a simple but highly informative set of experiments, Jeffries (1988) found that less vulnerable freshwater invertebrate species tended to have higher survivorship in the presence of more vulnerable prey species.

### 10.3.8 Mimics do not always have to be perfect replicas to gain protection, particularly when the model is relatively common or highly noxious

The degree of similarity between mimic and model that is necessary to confer protection from predators has been investigated on several occasions. For instance, Schmidt (1958) trained chicks to avoid drawings of a particular butterfly associating these images with distasteful cornmeal. Schmidt found that drawings which only approximately resembled the model also tended to be avoided, whereas drawings of non-mimetic insects did not tend to be avoided. Other quantifications of the extent of model–mimic similarity necessary to confer protection include the work of Dittrich et al. (1993), who trained pigeons to distinguish images of flies (including hoverflies not considered to resemble wasps) and wasps. With some notable exceptions, pigeons ranked hoverfly wasp mimics according to their similarity to a wasp model in a comparable fashion to the way in which humans rank their

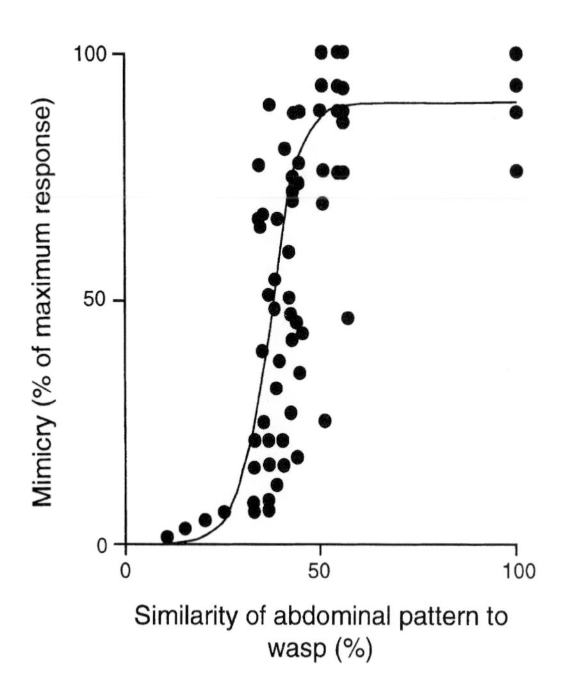


**Figure 10.4** The relationship between the image similarity of hoverfly specimens to a single wasp exemplar and the response of pigeons to these specimens, when they had been trained to distinguish wasps from flies. Overall there was a non-linear relationship between morphological similarity and perceived mimicry. Adapted from Dittrich et al. (1993).

similarity. Furthermore, the relationship between similarity (percentage of shared abdominal pattern) and the extent of mimicry (percentage of maximum response) was sigmoidal, in that all similarity scores beyond 50 per cent tended to generate similar responses in the birds (Fig. 10.4). Most recently, Caley and Schluter (2003) found that good artificial replicas of the toxic pufferfish (*Canthigaster valentini*) were approached less frequently by piscivourous fish than poor replicas. Once again the relationship appeared highly non-linear—the degree of protection was only seriously reduced for very poor artificial mimics.

As noted above, the noxiousness of the model may affect the attack rate of predators on models and mimics, but it also has implications for the degree of mimetic resemblance that can confer significant protection (Alcock, 1970; Pilecki and O'Donald, 1971). Perhaps the classic example of this is the work of Duncan and Sheppard (1965), which found that the level of electric shock associated with a dark green drinking solution (the model) influenced the colour range of drinking solutions (mimics) that tended to be subsequently rejected by chicks. The fact that even poor mimics may gain protection from predators has important implications for the evolution of mimicry and we return to this issue in Section 10.5.5.

### 10.3.9 Frequency-dependent selection on Batesian mimics can lead to mimetic polymorphism

In general, if Batesian mimics increase in frequency then predators will learn of their presence and tend to increase their attacks. Since rarity compared to the model maximizes the survival of a Batesian mimic, then one might expect selection to promote polymorphisms in which some conspecifics resemble one noxious model and other conspecifics resemble another. In addition, theoretical considerations suggest that the equilibrium frequency of different mimetic forms, if influenced solely by predation, should be such that each morph gains as much protection as other forms.

Mimetic polymorphism frequently occurs in Batesian mimicry complexes, but is by no means a diagnostic characteristic (Joron and Mallet, 1998). The palatable salamander *Desmognathus ochrophaeus* is considered to have colour morphs that resemble several species of unpalatable salamander (Brodie and Howard, 1973; Brodie and Brodie, 1980; Brodie, 1981). Other potential examples of polymorphic Batesian mimics include the non-aggressive social wasp *Mischocyttarus mastigophorus* that exhibits dark and pale colour morphs, thought to resemble the aggressive swarm-founding wasps *Agelaia xanthopus* and *Agelaia yepocapa* (O'Donnell, 1999). In these and other cases it should be borne in mind that morphs often co-occur locally but often the models only partially overlap in distribution. Hence polymorphism is likely to be maintained by a complex interplay of local balancing selection and the mixing of morphs from different areas.

Perhaps the best-known example of mimetic polymorphism occurs in the African mocker swallowtail butterflies (*Papilio dardanus*) whose females are Batesian mimics of different model species in Africa, and are often polymorphic in a single region, resembling several different models (Vane-Wright et al., 1999). Why mimicry should be restricted to one sex is a separate question as to why one or both sexes are polymorphic (see Section 10.5.3), but the two are often related. For instance, in south and central America, the kite swallowtail butterflies of the genus *Eurytides* consist of 15 recognized species, 14 of which appear to be Batesian mimics, largely of papilionid *Parides* species (West 1994). Of these Batesian mimics, six species are poly- or dimorphic without sex limitation and the remainder have sex-limited polymorphism (West 1994).

Some of the most fascinating examples of polymorphic mimicry include cases where the model is sexually dimorphic and the two sexes of mimic resemble each of the sexes of the model. Typically the male and female mimic resemble their respective sexes (Carpenter and Ford, 1933, describe the nymphalid butterfly *Pseudacraea eurytus* which mimics the Acraeine butterfly *Planema epea*), however, sometimes the male resembles the female and vice versa (Carpenter and Ford, 1933: 91). Perhaps the most impressive example of a species varying with the model it resembles (although it is clearly not a genetic polymorphism) is the recently recorded species of Indo-Malayan octopus in which individuals impersonate a range of venomous animals that co-occur in its habitat (Norman et al., 2001). There is already some evidence that this mimicry can be employed facultatively, with the octopus adopting a form best suited to the perceived threat at any given time, although this proposal needs further testing. The tortoise beetle (*Metrona bicolor*) may also engage in facultative mimicry, with its bright gold appearance resembling a water drop when at rest (Bates, 1862), but when in motion it transforms into a red-and-black ladybird mimic.

Formal experimental evidence for the prediction that Batesian mimicry is capable of generating and maintaining polymorphisms is lacking. Perhaps the most explicit study to test this property was reported by O'Donald and Pilecki (1970) who presented palatable and unpalatable (dipped in 1 or 3 per cent quinine hydrochloride) pastry 'mealworms' to garden house sparrows. There were two forms of distasteful model (blue and green) and three forms of palatable prey (blue mimic, green mimic and non-mimetic yellow). They found from comparing separate experiments that if the models were only slightly distasteful then the rare mimic was at an advantage over the common one. However, if the models were highly distasteful then a rare mimic had no advantage over a more common one. There has been some question of the appropriate interpretation of this experiment (Edmunds and Edmunds, 1974; O'Donald and Pilecki, 1974) due primarily to the differences in the consumption rate of the two unpalatable models (blue and green) used, and statistical questions relating to the appropriate population from which the observations were drawn. While we understand both sets of arguments, we have to conclude that O'Donald and Pilecki (1970) does not provide unequivocal evidence of frequency-dependent predation risk in mimics that it purports to, particularly in the light of the fact that time is a confounder and the experiments that were contrasted were not replicated. However, the basic experimental approach is sound, and we believe that their original methodology could be further developed to help investigate the evolution and maintenance of mimetic polymorphisms.

## 10.4 The theory of Batesian mimicry

Appendix B summarizes the mathematical and computer models that have been developed to understand Batesian mimicry and its properties. Models that deal predominantly with the putative spectrum between the extremes of classical Batesian and Müllerian mimicry, are discussed in Chapter 11, while models dealing with automimicry are considered in Chapter 12.

As might be anticipated, the various models of Batesian mimicry have taken several different approaches. While the vast majority of papers considered simply phenotypes (usually with perfect resemblance between model and mimic), several papers have considered the genetical basis of mimicry explicitly (Matessi and Cori, 1972; Barrett, 1976; Charlesworth and Charlesworth, 1975*a–c*; Turner, 1978). Two main types of modeling have

**Box 10.1 Mathematical models of Batesian mimicry**

*Huheey (1964) model*

1 A 2 3 4 B C 5 6 7 D 8 9 10 11 E 12 F G H 13 14 I J K 15 L M N O

This approach assumes that following the attack of a predator on a model, both mimics and models are avoided for $n - 1$ future encounters before the predator samples them again. In the encounter sequence above, $n = 3$ while numerals represent models and letters represent identical mimics (from Huheey 1988). Attacks on models are highlighted as (**!**) while attacks on mimics are highlighted as unstippled areas. Given these assumptions, the probability that an individual model or mimic will be eaten on encounter ($P$) is given by:

$$P = (1/(p + n\,q))$$

where $p$ and $q$ are the relative frequencies of mimics and models respectively ($p + q = 1$).
Note that $P \rightarrow 1$ as $p \rightarrow 1$, so if the mimic is relatively common then the attack probability on models and mimics on encounter approaches 1. Less intuitively, if there are no mimics ($q \rightarrow 1$) then the attack rate on models approaches $1/n$.

*Oaten* et al. *(1975) model*
A predator will eat a prey item with signal $s$ if

$$b\,f_1(s)\,p > c\,f_2(s)\,q$$

where $b$ and $c$ are the benefit and cost of attacking a mimic and model respectively, $p$ and $q$ are the relative frequencies of mimics and models (reflecting *a priori* probabilities of encountering a mimic or model) and $f_1(s)$ and $f_2(s)$ are the probability density of mimics and models associated with signal $s$. The distributions of $f_1$ and $f_2$ can be assumed normal if the signal comprises a number of independent sub-signals. Perhaps the most counterintuitive result of Oaten et al. (1975) is that, while there will always be individual selection on mimics to resemble their models, there exist solutions in which improving the mean similarity of mimics to models increases the overall attack rates of predators on these mimics. This property was rediscovered by Johnstone (2002) who used a kin selection model to show how imperfect mimicry might sometimes be maintained.

been particularly influential: an encounter-memory approach (Huheey, 1964) and a signal detection (Oaten et al., 1975) approach (see Box 10.1).

Huheey's (1964) widely cited work considered a predator that avoids sampling models and mimics (which are identical in appearance) for $n - 1$ future encounters after attacking a model. The study inspired a number of related papers that used Markov chain analysis to translate memory-based predator foraging rules into attack rates on models and mimics (Estabrook and Jespersen, 1974; Bobisud and Portraz, 1976; Arnold, 1978; Luedeman et al., 1981; Kannan, 1983—see also Brower et al., 1970; Pough et al., 1973 described in Chapter 12). Each of these papers attempted to analyse the consequences of particular rules of how predators might be expected to behave (often based on optimizing a memory parameter), typically in increasingly more realistic settings (e.g. allowing aggregation of the model and mimic or alternative prey). While Huheey's model helped focus debate and clarify arguments, our overall feeling is that the basic model was rather too prescriptive in its 'avoid for $n$ encounters' rule. By not considering time as an explicit parameter, and by using encounters to mark the passage of time, it is not immediately clear how $n$ relates to a time-based memory and how simple parameters such as total density (which will affect encounter rates) might influence the outcome (see Huheey, 1988 for a discussion).

By contrast, Oaten et al.'s (1975) signal detection approach did not consider a memory parameter or learning at all, but focused on the ability of predators to distinguish between models and mimics, and the consequences of making right and wrong decisions. Oaten et al.'s (1975) paper has not been widely cited, but despite its extreme psychological simplifications it is elegant and intuitive. A similar approach has since been adopted by several other related studies of Batesian mimicry, most notably Getty (1985) who developed models that also incorporated handling times in the optimisation algorithm.

Other models of the evolution and maintenance of Batesian mimicry are more difficult to categorize. Relatively few studies have considered the population dynamical consequences of Batesian mimicry (but see Yamauchi 1993, Sherratt, 2002*b*), while only three theoretical models have explicitly considered the role of predator hunger (Holling, 1965; Kokko et al., 2003; Sherratt, 2003). One interesting model (Holmgren and Enquist, 1999) employed neural networks to represent how predators might respond to stimuli, but the potential of this approach to the study of mimicry remains largely unexplored.

In an important paper that dealt largely with sex-limited mimicry (see Section 10.5.3), Turner (1978) argued that while Batesian mimicry was typically seen as a frequency-dependent phenomenon (with protection dependent on the relative densities of mimics compared to models), there may be an element of density dependence ('number dependence'). To make his case, Turner (1978) described a scenario in which the protection to mimics was proportional to $(1-q^n)$ where $q$ is the proportion of mimics in a model–mimic population and $n$ is a constant. In another scenario, the protection to mimics was proportional to $p^n$ where $p$ is the proportion of models in the combined population. He then showed that as the density of mimics increased in the population while the density of models was fixed then both $(1-q^n)$ and $p^n$ declined. One might think that this analysis indicates that the protection from mimicry is density dependent but it does not. To disentangle the effects of frequency and density one needs to vary density while keeping relative density constant and *vice versa*. In fact it is easy to show that $(1-q^n)$ and $p^n$ is not at all dependent on the exact population densities of models and mimics, but only their ratio.

While Turner's specific argument for density dependent effects is therefore not convincing, this is not to say that density dependence does not arise. Indeed density-dependent effects may sometimes be important. For instance, predators that can metabolize or sequester toxins might not refrain from attacking models and mimics at all if they are sufficiently rare that they have recovered before they see the next one. Similarly, if memory is time-based, then absolute density is likely to have a direct role in affecting selection for mimicry. To take an extreme case, if models and mimics are collectively so rare that a predator forgets about models and mimics before it encounters a new individual, then there may be little advantage in being mimetic.

Despite their variability in structure, overall the predictions of the models are remarkably consistent. Almost all papers in Appendix B make the intuitive prediction that the attack rate by predators on mimics and models should increase the higher the ratio of mimics to models (Emlen's 1968 paper does not support this prediction in all instances). Furthermore, all of the studies that considered prey profitability explicitly found that sufficiently noxious models can protect even highly numerous mimics (e.g. Holling, 1965; Estabrook and Jespersen, 1974; Oaten et al., 1975; Getty, 1985; Sherratt, 2002*b*). Similarly, the presence of alternative palatable prey is predicted to enhance the degree of protection afforded to the model and mimic (e.g. Holling, 1965; Emlen, 1968; Luedeman et al., 1981; Getty, 1985; Kokko et al., 2003; Sherratt, 2003). Finally, it has now been shown several times that imperfect mimicry can in theory evolve and be maintained for a variety of reasons including genetical constraints (Charlesworth and Charlesworth, 1975*a*), lack of further selection (Sherratt, 2002*b*, 2003) or even kin selection (Johnstone, 2002).

## 10.5 Questions and controversies

### 10.5.1 Why are not all palatable prey Batesian mimics?

Carpenter and Ford's (1933: 28) response to this familiar 'ridiculous question' was a *reductio ad absurdum*: the success of mimicry depends upon abundance of other palatable food. In effect, if all species became mimics, then there would be no advantage to mimicry. This, however, does not in itself explain why only a limited subset of species become mimetic, and which species these are. For morphological protective mimicry to be selected for, we require the prey species to be exposed to visually hunting predators and a suitable noxious model to resemble. Many hoverfly species for instance, may be mimetic because they are regularly exposed on flowering heads and because these sites are also visited by wasps and bees. It almost goes without saying that some species (such as nocturnal flying and day resting moths) would not benefit from adopting the visual appearance of an unpalatable species, since they would not be clearly seen by predators. There may also be anatomical restrictions (size and shape) that may drastically restrict the range of traits that would confer any initial mimetic advantage at all, and at least have some influence on choice of model.

A related question is, why be a Batesian mimic rather than a cryptic species? Though the obvious answer to this is 'to avoid predation', it may be incomplete. As we have argued in Chapter 7, for some species there may be high opportunity costs of crypsis; by hiding from predators, prey may lose opportunities to forage, to thermoregulate and to find mates. Hence, mimicry may evolve in species that have much to gain from conspicuousness, but for whom the costs of secondary defences are too high. One of the interesting implications that follows is that it is not just the presence of potential model species that matters to a putative mimic, but the presence of the right kind of aposematic model, that inhabits the right kind of niche. If a cryptic species suffers high costs from a lack of foraging opportunities then it needs to mimic a model that gives it protection while it forages. Similarly, for other species thermoregulation may be highly important, in which case Batesian mimicry will only evolve if there is a model that inhabits an environment with the right kind of thermoregulatory regime.

Looked at from the other perspective the question 'why be a Batesian mimic' rather than an unprofitable aposematic species, suggests two further answers. First, some Batesian species may have originally been aposematic, but because of the costliness of secondary defences, became cheats. Second, for some species the costs of acquiring secondary defences in the first place may simply be too high so that Batesian mimicry is the next best method of removing the opportunity costs that follow from crypsis (see, e.g. Turner, 1984).

### 10.5.2 Is the spatio-temporal coincidence of the models and mimics necessary?

While the co-occurrence of models and mimics in the same general location seems an important (but not limiting) pre-requisite for the evolution and maintenance of mimicry (e.g. Pfenning et al., 2001, Section 10.3.3), the co-occurrence of models and mimics at the same time may be far less important (Rettenmeyer, 1970; see Waldauber, 1988 for review), indeed it may be unnecessary (see Gibson, 1974, 1980 for results from an experimental design which supports this). Anecdotal observations suggest that some vertebrate predators may remember unpleasant experiences for periods over a year (Mostler, 1935; Rothschild, 1964). This may especially be the case if models occur before mimics and if continued encounters with models strengthens learning (Huheey, 1980) and enhances memorability (Speed, 2000). Given such memories, it may sometimes be advantageous for a mimic to delay its appearance until after the appearance of the model (as formalized by Bobisud, 1978; see Appendix B). One might wonder if there might be counter selection on the model to delay its appearance until after the emergence of the mimic, but all else being equal it is likely that any mutant model would not tend to survive as well. After all, by appearing first, the model is 'uncontaminated' by palatable mimics.

The buff ermine moth (*Spilosoma lutea*) is considered a mimic of the distasteful white ermine moth (*S. lubricipeda*), although it may be a Müllerian rather than Batesian mimic because it is unpalatable to some predators (Rothschild, 1963). As might be expected in this situation, the white ermine moth reaches peak abundance before the buff ermine moth (Rothschild, 1963). Similarly, Brodie (1981) noted that the palatable salamander *Desmognathus ochrophaeus* is active later in the season (May–August) than its unpalatable salamander model *Plethodon cinereus*[8] (March–April, the time when

[8] Once again, some confusion about 'palatability' arises. In their *Science* paper Brodie and Brodie (1980) refer to *Plethodon cinereus* as edible and show how mimetic forms of the species are at advantage over non-mimetic forms. Here Brodie (1981) considers *P. cinereus* as an unpalatable model because of its distasteful skin secretions.

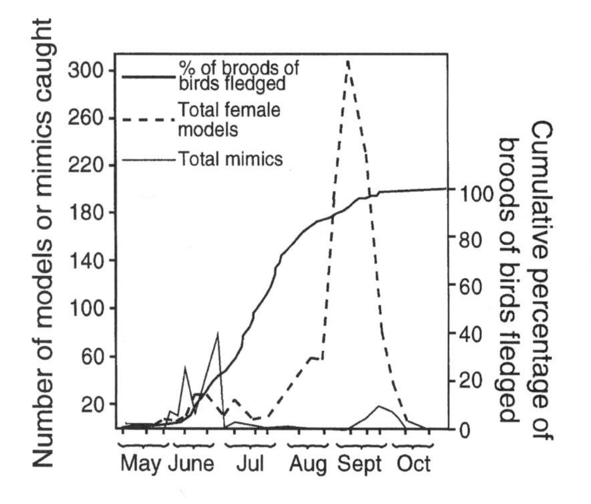


**Figure 10.5** The seasonal occurrence of high-fidelity hoverfly mimics in relation to the occurrence of their hymenopteran models, and the cumulative percentage of bird fledged. Data were collected in central Illinois by Waldauber and Sheldon, 1971.

migratory ground-feeding birds that prey on salamanders begin to arrive). By contrast, many good hoverfly mimics of stinging Hymenoptera in North America reach a peak in spring before their models are most abundant (Waldbauer and Sheldon, 1971; Waldbauer et al., 1977; Waldbauer and LaBerge, 1985; Fig. 10.5). One explanation for this puzzling phenomenon, for which there is some evidence (see Waldbauer, 1988) is that while adult birds may avoid mimics based on their previous seasons' experience with models, early emergence avoids the time when fledgling birds (which have not yet learned to avoid models) begin to capture their own insect prey. This is an intriguing explanation, but it cannot be universal. Working in the north-west of England, Howarth and Edmunds (2000) could find little evidence that hoverfly mimics were rare compared to models when fledgling birds are abundant, and generally found that hoverfly mimics were synchronous in their flight season with their hymenopteran models.

### 10.5.3 Why is Batesian mimicry often limited to one sex?

In some species, Batesian mimicry is limited to one sex. While it is a relatively common phenomenon in putative Batesian mimics, it is rare in species thought to be unpalatable Müllerian mimics (Turner, 1978). Interestingly, in butterfly species where sex-limited mimicry widely occurs it is invariably the female that exhibits mimicry, often as part of a female-limited polymorphism with both mimetic and male-like female forms (Turner, 1978, Section 10.3.9). Male-limited Batesian mimicry has however been reported in other taxonomic groups including the Buprestid beetle *Chrysobothris humilis* whose males resemble members of the chrysomelid subfamily Clytrinae (Hespenheide, 1975). Hespenheide (1975) explained this unusual case of male-limited mimicry as a consequence of sexual differences in behaviour: males often sit on legume twigs (waiting for ovipositing females), and thereby occur more frequently where the model occurs, and are more frequently exposed to predators. Another potential example is the promethea moth *Callosamia promethea* (see Section 10.3.2) where the males may gain advantage by resembling the pipe vine swallowtail butterfly (*Battus philenor*) yet the females appear non-mimetic. Males are almost certainly exposed more to visual predators, primarily through their need to fly towards pheromone-releasing females during the day (by contrast, females lay their eggs at night, Waldauber, 1996: 127). Many hoverfly mimics of bumble bees in the Holarctic also exhibit sex limited Batesian mimicry: 17 cases of male-limited and 11 female-limited mimicry have now been recognized (Gilbert, 2004).

If Batesian mimicry is advantageous, why do both sexes not always adopt it? So far two main theories have been put forward to explain the occurrence of sex-limited mimicry—one based on differences in the ecology of the two sexes, and the other based on sexual selection. In his classic paper on Malayan swallowtails, Wallace (1865) explained the widespread occurrence of female mimetic forms: 'with their slower flight, when laden with eggs, and their exposure to attack while in the act of depositing on leaves, render it especially advantageous for them to have some additional protection.' This same argument is reflected in our answer to why only some species are Batesian mimics (Section 10.5.1). Later, Belt (1874: 384–385) added the suggestion (not necessarily mutually exclusive) that sexual selection may help explain non-mimetic forms in males: mimicry does not arise in this sex because it would render them at a disadvantage during courtship. Turner (1978) took up the challenge of modelling this latter phenomenon, noting that sexual selection may resist colour changes more strongly in males than females. Using a population genetic framework, he showed how female-limited mimicry could readily arise when modifiers that suppress the expression of mimicry in males get selected for. The predicted outcomes include female-limited mimicry, or better mimicry in females than males and males retaining some of their colours that play a role in mate choice (see Appendix B).

Experimental evidence that male and females of the same species may experience different selective pressures with respect to predation, is widespread. Bowers et al. (1985) capitalized on an unusual finding in 1980 of large numbers of detached wings (matched for 309 individuals) of the checkerspot butterfly *Euphydryas chalcedona* in California, almost certainly the remains from being attacked and eaten by birds (coldness may have contributed to making the species more vulnerable that year). By comparing the coloration and sexual composition of this sample with a representative sample that had not been killed, Bowers et al. (1985) argued that birds had attacked significantly more females than males, and less of those males with high amounts of red on their forewing. The authors proposed that females may have been actively selected (they are larger and probably represent a more nutritious meal) while redder males may have benefited from warning coloration—however, this remains speculation and we need to bear in mind that 1980 may have been an unusual year for the checkerspot butterfly. More direct evidence that sexual differences in potential rates of predation can select for mimicry in just one sex was presented by Ohsaki (1995) who examined the frequency of beak marks on the wings of palatable Papilionidae and Pieridae and unpalatable Danaidae in Malaysia. The analysis of beak marks is fraught with problems, not least because it also indicates successful escape from a predator. However, females of each palatable family had a greater rate of wing damage than males. The numbers of recorded observations of *Papilio poytes* were relatively low, but non-mimetic females of this species had a higher per capita incidence of beak marks than mimetic females or males (which were themselves similar). Ohsaki (1995) points out that these observations indicate that 'females benefit greatly when they become mimetic, whereas males will benefit much less even if they become mimetic', but this rather leaves open the question why all females are not mimetic (Ohsaki, 1995 suggests this may possibly be a consequence of local disequilibrium, or some other cost to mimicry).

Evidence that sexual selection (either through female mate choice or male–male interactions) can maintain non-mimetic males, comes from several quarters. Krebs and West (1988) blackened the wings of male *Papilio glaucus* butterflies to look like the mimetic female form and examined their mating success compared to yellow-painted controls. No significant difference was found in the proportion of presentations to both mimetic and non-mimetic females, which resulted in mating. However, female solicitation of males that themselves had failed to court occurred at a significantly higher rate for yellow-painted than black-painted males. There is perhaps, however, slightly more evidence for non-mimetic male forms having some advantage in male–male competition. For instance, Lederhouse and Scriber (1996) investigated the benefits of male colouration in the eastern black swallowtail butterfly *Papilio polyxenes asterius*. The dorsal sides of males are less convincing mimics of the pipevine swallowtail *Battus philenor* (to human eyes at least) than the dorsal side of females, and the species is therefore generally considered as one exhibiting female-limited mimicry. Males of the species seek out and defend hilltop territories ('hilltopping') where the vast majority of courtship takes place. Lederhouse and Scriber (1996) found that males that had been altered using marker pens to have more female-like (and therefore mimetic) colouration, were less likely on release to be able to establish (and maintain) a local hilltop territory (at least close to the release site) than mock-treated controls. One reason for this may be that altered males were involved in significantly longer male–male encounters, through being miss-identified as females. By contrast, experimental creation of female-like males did not appear to affect female choice per se: while altered males did not as readily obtain territories the mean duration of courtship flights that progressed to copulations were not significantly different in the territories of males that were treated and mock-treated.

In summary, it is perhaps fair to say that while there are several good candidate explanations for why Batesian mimicry might sometimes be limited to one sex, we still do not know which explanation is the most general and important one (if indeed there is such an explanation). As ever, more experiments are needed. One might also ask if sexual selection acts to conserve male appearance in Batesian mimics, why does it not tend to do so in Müllerian mimics? One explanation may lie in the differing nature of the frequency dependence (Turner, 1978; Joron and Mallet, 1998). By its nature, Batesian mimicry may involve selection against common morphs (Section 10.3.9). In contrast to Batesian mimicry, the mutualistic nature of Müllerian mimicry ensures that the more individuals that adopt a common form of advertising, the higher the fitness of all.

### 10.5.4 How is mimicry controlled genetically and how can polymorphic mimicry be maintained?

Butterflies have become the central model for understanding the genetics of mimicry, not only because they provide some of the most celebrated examples of protective resemblance, but also because they can be crossed under laboratory conditions, and their colour patterns are often easy to classify. In an early study in Sri Lanka, Fryer (1913) conducted breeding experiments on the swallowtail *Papilio polytes* whose females are polymorphic (one form resembles the male and other forms resemble different unpalatable models). Somewhat surprisingly, no intermediate forms resulted from the crosses, and it was also deduced that males (all of which are non-mimetic) could transmit genes relating to mimicry in their female offspring. Perhaps understandably, observations such as this fuelled Punnett's (1915) belief that mimetic resemblance was more likely to have arisen by a sudden 'sport or mutation', rather than a gradual accumulation of mutations providing closer resemblances. If true, then this would place more emphasis on mutation rather than selection as a creative force in generating mimetic resemblance. We now know that the underlying genetics are far more complex than at first believed. Here in addressing the above question we briefly discuss some illustrative examples of the genetics of Batesian mimicry and consider their implications, largely from the perspective of mimetic polymorphism. Most of our knowledge of the genetics underlying Batesian (and Müllerian) mimicry in Lepidoptera comes from the dedicated and detailed studies of E. B. Ford, C. A. Clarke, P. M. Sheppard, and J. R. G. Turner. Besides reference to these studies, a helpful review of the subject area can be found in Nijhout (1991) and short summaries of our state of knowledge are given in Mallet and Joron (1999) and Turner (2000).

In Section 10.3.9 we discussed the case of polymorphic mimicry in the African mocker swallowtail butterfly *Papilio dardanus* in which females have evolved colour patterns that mimic several different species of unpalatable danaid butterfly. Detailed heritability studies of Ford (1936) and Clarke and Sheppard (1959, 1960*a*) have revealed that this particular example of female-limited polymorphism is controlled at a single genetic locus with up to 10 alleles, only four of which are involved in mimicry (not all female forms of *P. dardanus* are known as mimics). While there are as many as six alleles in any geographical race, breeding experiments within and between different races have shown that crosses between races produce very variable, non-mimetic progeny. Thus, the nature of the expression of this gene depends on other co-evolved genetic 'modifiers' specific to that population (Nijhout, 1991). As Turner (2000) notes, high-quality mimicry is therefore produced not as an innate property of the original mutation, but as the result of the fine-tuning of the genome by further modifiers. In the populations where the mimetic form is found, but not in other populations, natural selection has favoured genes that improved the level of resemblance beyond a coarse similarity to the model. More intriguingly, many of these modifiers have no obvious phenotypic effect in the absence of mimicry. For instance, modifier genes in a population of *Papilio dardanus* in Ethiopia shorten the tails of mimetic forms (so that they resemble their tailless models) but do not shorten the tails of non-mimetic forms (Turner, 2000).

The fact that the main pattern gene has such a complex and diverse set of simultaneous effects on the phenotype (colour and pattern) has also led to the suggestion that the gene is a complex locus (a 'supergene'), consisting of several closely linked genes each with a specific effect on the pattern (Clarke and Sheppard, 1960*b*). Supergenes have long been discussed in the context of mimicry but formal experimental evidence of supergenes in *P. dardanus* remains relatively tentative (Nijhout, 1991). Slightly more convincing evidence for a supergene complex (Nijhout, 1991) comes from the rare unusual morphs observed in breeding of *Paplio memnon* (the likely outcome of rare crossover events where close genes become unlinked), through which Clarke and Sheppard (1971) inferred the presence of several separate closely linked genes for traits such as wing shape and colour. The different polymorphic forms of *P. dardanus* and *P. memnon* readily interbreed, and on a theoretical basis, the combination of supergenes and modifiers readily explains how several different mimetic forms of the same species can be selected and maintained within a population (Charlesworth and Charlesworth, 1975*a–c*).

Of course, the fact that supergenes and modifiers may help maintain mimetic polymorphisms does not in itself explain how such genetic structures may arise. This is in essence a problem of the 'evolution of evolvability' (Kaufmann, 1993). It seems unlikely that selection could gradually bring together a number of loosely linked loci (Turner, 1984). However, it must be remembered that polymorphic mimetic butterflies are at most a subset of the Lepidoptera found in any given area. Hence, it is possible that polymorphic mimicry has tended to evolve only in those species in which many of the relevant loci already happened to be linked—a preadaptation. This 'sieve' working on particular predisposed forms of genetic system is analogous to Haldane's earlier argument that selectively advantageous alleles are more likely to spread in populations when they are dominant. It also helps explain why some groups of species are able to exhibit polymorphic mimicry, such as the Papilionidae, while other groups are rarely mimetic. Furthermore, this sieving effect may not be as unlikely as may first appear: if traits such as pattern and colour can be achieved by several different genes, then it is quite possible that some of them will be linked in some species, allowing selection to operate a phenomenon that has been referred to as 'the largesse of the genome' (Turner, 1977).

Overall, it is clear that polymorphic Batesian mimic species may require a special form of genetic architecture if its members are to simultaneously reach optima on different adaptive peaks. Researchers do not currently know enough about the genetics of mimicry in monomorphic mimetic species to discuss the various ways in which mimicry is achieved, but intuitively it may not require supergenes. We are still at an early stage of combining our knowledge of molecular and developmental biology and how it relates to mimicry. This may add a whole new layer of complexity. For instance, while the characteristics of the different eyespots are developmentally coupled in non-mimetic *Bicyclus anynana*, Beldale et al. (2002) found that artificial selection for increased size of one individual eyespot tended to proceed in a manner largely independent from selection imposed on another eyespot. The authors argued that this flexibility was probably related to the compartmentalization of the wing, possibly mediated compartment-specific genetic components that regulate the expression of the eyespot-forming genes. Clearly, there is more to learn.

### 10.5.5 Why are imperfect mimics not improved by natural selection?

One widely held belief is that there should always be strong selection pressure on mimics to resemble their models as closely as possible. Thus, it has been argued that close Batesian mimicry will tend to evolve when the mimic evolves faster toward the model phenotype than the model evolves away from the mimic (Fisher, 1930), a requirement that will almost inevitably be met (e.g. Nur, 1970; Holmgren and Enquist, 1999). Yet these views are somewhat at odds with the fact that there are many cases in nature in which potential Batesian mimics do not appear to resemble their models particularly closely. For example, many species of hoverfly are generally regarded as Batesian mimics of wasps and bees, yet to the human eye at least, they do not resemble their models particularly well (e.g. Dittrich et al., 1993; Edmunds, 2000). There is something of further interest here too. In some cases the similarity of mimics to their potential models is considered very good ('high fidelity'), whereas in other species the resemblance is poor. What is the reason for these differences?

The widespread occurrence of apparently imperfect Batesian mimics has so far been explained in a variety of different ways (Edmunds, 2000). For instance, mimics that seem imperfect to humans may actually appear as good mimics to predators (Cuthill and Bennett, 1993; Dittrich et al., 1993). Alternatively prey may be under no further selection to improve mimetic similarity, an outcome that is particularly likely when the model is extremely numerous or noxious (Section 10.3.5, Schmidt, 1958; Duncan and Sheppard, 1965), or there are plenty of alternative prey so that risk-taking by the predator is inappropriate no matter how low the risk (Dill, 1975; Sherratt, 2003). Another possibility to bear in mind is that the imperfect mimicry may arise as a consequence of a 'breakdown' in mimicry (Brower, 1960) as the mimic/model ratio increases. For instance, Sheppard (1959) investigating the field distributions of mimetic African butterflies found that the proportion of individuals of a given species with a poor resemblance to the model was higher when mimics were relatively common. Similarly, Azmeh et al. (1998) pointed out that the larvae of many hoverfly mimics that are judged imperfect are aphidophagous and may have increased dramatically in numbers following agricultural development. They found that the median similarity of hoverfly species to the wasp *Vespula vulgaris* was higher in areas considered less disturbed. Of course it is possible that certain imperfect mimics (such as particular species of hoverfly) may not be Batesian mimics at all but may be signaling their own unprofitability (e.g. their ability to escape predation, see Pinheiro, 1996; Azmeh et al., 1998; Edmunds, 2000). This is an interesting theory, but it should be noted that some species of hoverflies also engage in behavioral mimicry, for instance, by waving their legs to resemble hymenopteran antennae (see Waldbauer, 1988; Golding and Edmunds, 2000).

One recent intriguing explanation for imperfect mimicry proposed by Edmunds (2000) is that it may arise as a consequence of selection to resemble simultaneously more than one model species living in separate subareas. Similar explanations have been proposed to explain the occurrence of imperfect Müllerian mimics (Sbordini et al., 1979, Chapter 9) The idea does not work well when the models and mimics are sympatric—the jack-of-all-trades will typically do worse than a specialist and at best gain the same high level of protection (Sherratt, 2002*b*). However, it is a plausible explanation when there are many different model species (a complex mimetic community) and the mimic distribution overlaps either temporally or spatially with a mixture of the models which occur at different times or areas (Sherratt, 2002*b*). Edmunds (2000) proposed that the multi-model hypothesis might explain why imperfect mimics appear more numerous than more perfect mimics, although it is now becoming clear that this observation is predicted by several very different theories (Azmeh et al., 1998; Edmunds, 2000; Johnstone, 2002).

Other theories for imperfect mimicry which we feel are less likely to hold true include the idea that imperfect mimics may temporarily confuse predators (the 'satyric mimicry' hypothesis of Howse and Allen, 1994). Howse and Allen (1994: 113) suggest that imperfect mimicry may be explained because it is optimally confusing (rather than optimally deceptive) and that 'small departures from optimal ambiguity can destroy the paradoxical nature of the image.' Yet the experimental results of Dittrich et al. (1993) that motivated this theory indicate that hoverflies with a closer resemblance to their model are at best selectively neutral (as judged by pigeon responses), and are never selectively disadvantageous. Johnstone (2002) extended a signal detection model to include kin selection, and argued that an optimal level of mimetic imperfection may occur since any relatives of a more perfect mimic would be subject to greater predation. The possibility that kin selection could play a role in maintaining mimetic imperfection is an intriguing one. However, this theory may suffer from a similar problem as Fisher's 'kin-selection' explanation for the evolution of secondary defences in prey

(Chapter 5); edible prey are often too vulnerable to predation to risk grouping together, even if they are Batesian mimics. Classic imperfect mimic species such as hoverflies disperse widely as adults and do not maintain family groupings, thereby rendering Johnstone's (2002) theory unlikely.

In sum, there are many explanations for imperfect Batesian mimicry, some more plausible than others. Note however that several factors could come together to influence the perfection of mimicry and few of the explanations above are mutually exclusive.

### 10.5.6 How does Batesian mimicry evolve, and why do models simply not evolve away from their mimics?

Batesian mimicry may evolve in several different ways. One possibility is the evolution of close Batesian mimicry by a single mutational step (Punnett, 1915; Goldschmidt, 1945). However, good mimicry is unlikely to evolve by these means, even if the species share many similar genes and/or developmental pathways (Turner, 1984). Alternatively, Batesian mimicry may evolve as a series of gradual improvements in similarity, which has become the traditional way of portraying evolutionary change (Fisher, 1927, 1930). This may well be possible, but the fact that the first mutants begin with only a vague resemblance to the model *may* mean that they gain little overall advantage compared to conspecifics, such that there may be counter selection on appearance (e.g. thermoregulation) (Turner, 1977). The most widely accepted theory for the evolution of Batesian mimicry today is the two-step process [Nicholson, 1927, also attributed to E. B. Poulton (Turner, 2000)], effectively a compromise between the two earlier perspectives. First, a mutation with a major phenotypic effect places the mimic phenotype relatively close to the model. We have already seen that even imperfect mimics may gain substantial protection from predation, particularly if the model is highly noxious. Next, this approximate but beneficial similarity can be further enhanced by selection on 'modifier genes' (see Section 10.5.4) to improve the accuracy of mimicry.

It is apparent from the genetical studies of polymorphic Batesian mimicry (Section 10.5.4) that Nicholson's scenario qualitatively fit the facts—while Batesian mimicry often involves the products of a relatively small number of genes each with major effects (e.g. wing shape and colour), further modifiers could help improve the general similarity to the model. However, there is currently some debate over the role of mutations with small and large effects in the evolutionary process. As Charlesworth (1994) noted after exploring a refinement of his earlier mathematical model (Charlesworth and Charlesworth, 1975*b*): 'the evolutionary dynamics of Batesian mimicry is consistent with the view that mutations with relatively large effects on mimetic resemblance are favoured by selection in both stages of the evolution of mimicry.' Thus, which scenario fits best is still under debate, and the two-stage process should not be taken as fact.

In section 10.5.5 we asked how imperfect mimicry can persist, but it is equally important to ask why the models do not simply evolve away from the mimics. Put another way, why cannot the model evolve as quickly away from the mimic as the mimic evolves towards it? One explanation is based on the relative success of rare mimic and rare model mutants: while any change in the mimic phenotype towards the model will provide selective advantage, major mutants of the model species away from the mimic will not spread as rapidly because they are rare and not recognized as distasteful (Nur, 1970; Turner, 1977). Even if models could readily evolve away from mimics, it is unlikely that models could ever 'shake off' mimicry completely since selection to avoid mimicry depends on the presence of a high mimetic burden in the first place. In essence, mimicry may be a race that cannot be won by models unless they adopt forms that mimics cannot readily evolve towards. When we observe Batesian mimicry systems we may not be looking at outcomes of a race that the mimic has won. In these cases, the systems may be better characterized by a non-equilibrium co-evolutionary chase in which frequencies of phenotypic forms of mimics and models are continually changing over time. Gavrilets and Hastings (1998) and Holmgren and Enquist (1999) described this chase quantitatively—however, we note that the form of frequency dependence that is assumed in certain mathematical models, for example, linear frequency dependence (Gavrilets and Hastings, 1998), may generate atypical instability.[9]

### 10.5.7 What selective factors influence behavioural mimicry?

So far we have dealt almost exclusively with morphological mimicry, yet many Batesian mimics also behave like their models or produce similar chemicals. For instance, burrowing owls (*Athene cunicularia*) defend themselves against mammalian predators by mimicking the sound of a rattlesnake (Rowe et al., 1986; Owings et al., 2002) while the Indian rat snake *Ptyas mucosus* may be an acoustic Batesian mimic of the king cobra *Ophiophagus hannah* (Young et al., 1999). Similarly, it is well known that many species of mimetic hoverfly wave their front legs before them to resemble hymenopteran antennae, and make even sounds similar to the quark of a disturbed wasp (and pretend to sting) when grasped (see, e.g. Waldbauer, 1970, 1996: 124). Certain spiders that mimic ants also hold their front legs before them to resemble antennae (Edmunds, 1974; McIver and Stonedahl, 1993). Many questions remain unanswered. For instance, is behavioural mimicry more likely to evolve in species that are relatively poor morphological mimics? Furthermore, what aspects of behaviour are most likely to be copied?

We have already reviewed the evidence that chemically defended species tend to move more sluggishly and deliberately than other species that do not possess such defences (see Chapter 7). An important question to ask therefore is whether morphological Batesian mimics also adopt this aspect of their models. Srygely and Chai (1990*b*) report that members of the butterfly subfamily Dismorphiinae, which mimic highly unpalatable Ithomiinae butterflies, have slow regular flights like their models.

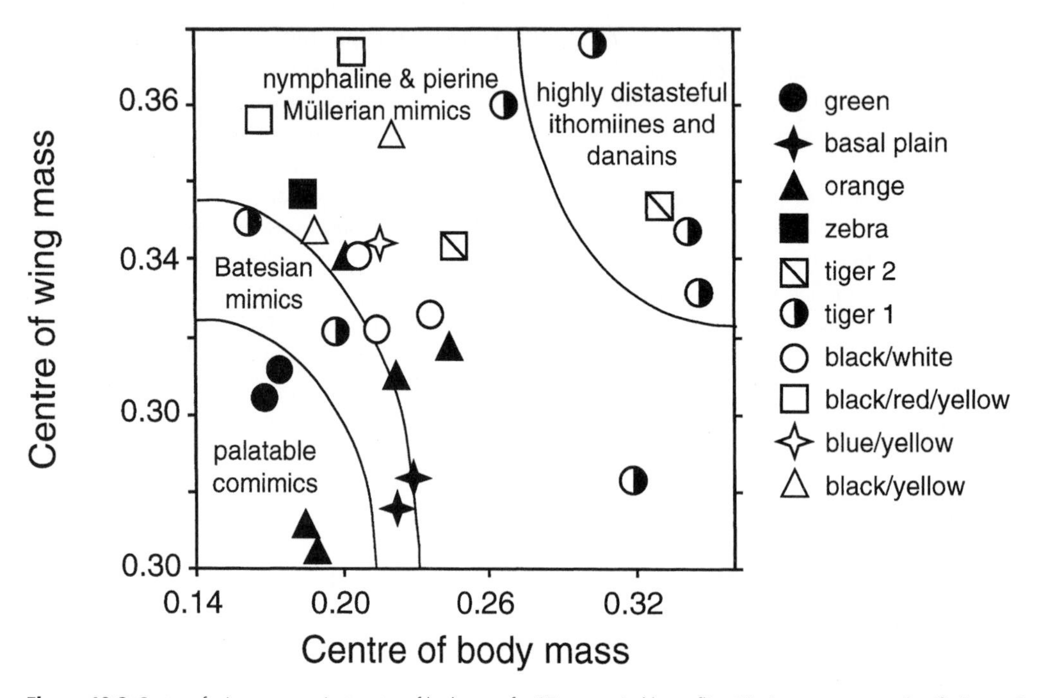


**Figure 10.6** Centre of wing mass against centre of body mass for 27 neotropical butterflies. Mimicry groups were classified according to the colours and patterns of the upperside of the wing. Palatable butterflies species tend to have centres of body mass positioned near to the wing base while unpalatable species have centres of body mass furthest from the wing base. Batesian mimics tend to have centres of mass which are just outside the zone for non-mimetic palatable species. Redrawn from Srygley (1994).

[9] We thank J. Mallet for pointing this out.

However selection in Batesian mimics towards a body shape (and consequent locomotory behaviour) that resembles a model may be opposed by selection to avoid capture when chased by predators (Srygley, 1994). For instance, the charaxine butterfly *Consul fabius* is a Batesian mimic of ithomiine and danaine models yet has morphological locomotory attributes (centres of wing and body mass) closer to evasive palatable non-mimics (Fig. 10.6, Srygley, 1994). Sherratt et al. (2004*a*) have recently supported this view, showing that computer-generated profitable prey only evolve locomotor similarity to morphologically identical unprofitable prey when the Batesian mimics are prevented from adopting fast movements that would allow them to avoid predators. As Srygely (1994) argues, 'Batesian mimicry should be a fertile field for future investigation of the evolution of flight-related morphology and kinematics.'

## 10.6 Overview

Batesian mimicry is now a well-studied phenomenon. While field demonstrations that Batesian mimicry provides a degree of protection from predators are remarkably rare, there is ample evidence that predators given prior experience with noxious models subsequently tend to avoid the palatable mimics. It has also been repeatedly argued, and shown, that the efficacy of Batesian mimicry is dependent on the abundances of the models and mimics, the noxiousness of the model and the abundance of alternative prey. Key areas for further work include: more phylogenetic studies to evaluate why, and how frequently, Batesian mimicry arises within particular groups; more work on the role of invertebrate predators as selective agents for mimicry; more work to disentangle frequency-dependent effects from density-dependent effects and more studies to test the intriguing phenomenon of evasive mimicry. We also need more experimental work to elucidate the selective advantages of Batesian mimicry in more complex and realistic mimetic communities—which unprofitable models are chosen and why, and under what conditions (if any) will a mimic evolve an appearance intermediate between two or more models?

# CHAPTER 11

# The relationship between Batesian and Müllerian mimicry

## 11.1 Context

The textbook distinction between Batesian (Bates, 1862) and Müllerian (Müller, 1878) mimicry is almost always based on the notion of edibility. Thus, Batesian mimicry is 'the resemblance of an edible species, the mimic, to an inedible one, the model', while Müllerian mimicry occurs when 'an unpalatable or venomous species resembles another' (Smith and Smith, 2001). Technically, these two forms of mimicry are perhaps better distinguished by the selective mechanisms thought to generate them. Thus, as we have seen (Chapters 9 and 10), Batesian mimicry is widely considered to have evolved in palatable prey as a consequence of selection to deceive predators into believing they are unpalatable, while Müllerian mimicry—if it is to be thought of as mimicry at all—as a consequence of selection to spread the burden of predator education (Fisher, 1930). One important consequence of these different mechanisms is the contrasting forms of frequency dependence they generate: Batesian mimics are considered parasites, while Müllerian mimics are considered mutualists (Turner, 1987, Chapter 9).

While a simple dichotomous view of palatability and unpalatability is central to the traditional distinction between Batesian and Müllerian mimicry, it is also widely appreciated that 'edibility' is not an absolute term (Cott, 1940): individual predators may vary in their motivation to attack particular prey types according to the predator's nutritional state (Pulliam, 1975), and dietary preferences may differ both within species (Sherratt and MacDougall, 1995) and among species (Malcolm, 1990; Endler and Mappes, 2004). Taken together, this variation raises the possibility that there may be a 'mimicry spectrum' (Turner, 1984; Huheey, 1988; Speed and Turner, 1999), in which the simple distinction between these two well-known forms of protective mimicry becomes blurred.

Even before the publication of the theory of Müllerian mimicry, Wallace (1871: 85) proposed that 'distasteful secretion is not produced alike by all members of the family and that where it is deficient, protective imitation comes into play.' Marshall (1908) similarly argued that hungry birds would consume weakly unpalatable prey, and that this could mean that weakly unpalatable species become parasitic, rather than mutualistic mimics. Nicholson (1927: 34–35) was another proponent of this view: '. . . no sharp demarcation can be drawn between Batesian and Müllerian mimicry. It is probable that no insect is wholly distasteful to all kinds of predaceous animals, and every intergrade appears to exist between the most distasteful species and those which are palatable to all predators. Batesian and Müllerian mimicry are therefore to be considered as the extreme types of deceptive resemblance, and not as two separate phenomena.' Indeed Nicholson (1927) went on to echo Wallace (p. 89): 'The incipient mimic need not therefore be palatable; it need only be less distasteful than its model, other things being equal . . . The model, however, must always be the form which is least liable to attack, whether this is due to its special distastefulness or its numerical superiority.' De Ruiter (1959: 353) was even more extreme in his views: 'Müllerian mimicry . . . is very unlikely to be realized except when predators live in the presence of such a superabundance of food that they never have to resort to relatively distasteful prey, a situation that will hardly persist for long enough to be of evolutionary importance'.

Perhaps the majority of contemporary researchers have maintained that there may be an important difference between these classical forms of mimicry, but that this difference may be somehow blurred by predators foraging on weakly distasteful prey in times of energetic need. Thus, Sheppard (1975: 183) stated: 'the edibility of an object is determined in part by the degree of starvation of the predator . . . It follows that it is not always possible to determine with any certainty whether every particular association is Müllerian or Batesian'. Benson (1977: 455) took a similar view, arguing that while the acceptability of a prey species could fluctuate substantially depending on ecological circumstances, any hunger-dependent change in mimetic status would be short-lived: 'With a variable environment of predators being more or less hungry or selective, the status of a mimic may change, but it will be intermediate only instantaneously as it passes over the knife-edge separation between the two categories.' Most recently, Srygley and Kingsolver (1998: 53) appear to have adopted a pluralistic approach, implying that both Müllerian and Batesian mechanisms can act in a given system to maintain mimetic similarity: '. . . incipient mimics may have a survival advantage over non-mimetic forms that are more readily sampled by predators early in the breeding season when demand is low. As the demand for resources increases with reproductive output, predators sample distasteful butterflies as well as those that are palatable, and sampling of mimetic forms would also increase . . . so that selection for mimetic perfection would ensue.'

Given such long history of debate on the relationship between Batesian and Müllerian mimicry we feel it is appropriate to discuss this controversial subject, and give our own views. As much of the debate centres on whether weakly defended mimetic prey species would have a parasitic, neutral, or mutually beneficial effect on a predator's attack probability on encounter with the better defended model, we begin by assessing the evidence that different species of defended prey can indeed differ in their overall level of defence. We then ask under what conditions should the presence of weakly defended mimic species increase the likelihood that a more highly defended model is attacked on encounter. Finally, we assess the experimental and observational evidence that some Müllerian mimetic relationships are actually parasitic.

## 11.2 Evidence of interspecific differences in levels of secondary defence

Here we restrict ourselves to a few key examples. As we will see in Chapter 12, the efficacy of a prey item's defence can vary substantially within species. It is also well known that the same general form of prey defence can vary in effectiveness among prey species. For example, while many wasps have stings, the sting of certain wasp species (such as the hornet) are particularly painful. Similarly, while some snakes have sub-lethal venoms, other species have deadly venoms. Summers and Clough (2001) collated data from Daly et al. (1987) which clearly showed that the diversity, quantity and toxicity of the toxins found in 21 species of poison frogs (Dendrobatidae) varied considerably. Of course, different prey species may protect themselves from predation in qualitatively different ways including spines, toxins and unpalatability (Chapter 5). Finally, some chemicals that are aversive to birds may in fact be primarily present for some other reason, such as antiparasite defence or as wetting agents, in which case we should not be surprised if some prey appear to have weak or moderate chemical defences. Given all of these observations, it would be surprising if defended prey species did not vary in their overall unprofitability to a particular predator species. To quote Dixey (1909: 564) 'every upholder of Müllerian mimicry, so far as I am aware, is not only ready to admit, but is prepared positively to assert that distastefulness is relative; that it exists, like other means of defence, in degrees that may vary indefinitely from species to species'.

In an early paper to address this question, Brower et al. (1963) presented laboratory-reared adult males (more than a thousand in total) of seven species of heliconiine butterflies (*Heliconius numata, H. melpomene, H. erato, H. sara, H.* [now known as *Laparus*] *doris, Dryas iulia* and *Agraulis vanillae*) to 62 individually caged birds (silverbeak tanagers). Palatable satyrid butterflies were used as controls. From examining the relative frequencies of reactions of these birds to the heliconiinae during their learning process, they concluded that while all were unpalatable, certain species (such as *H. numata*, *H. melpomene*, and *H. erato*) were significantly more unpalatable than other species (such as *H. [Laparus] doris*, *Dryas iulia*, and *Agraulis vanillae*). Brower (1958*a*) had earlier reported that the Viceroy butterfly (now generally classed as a Müllerian mimic of the Monarch—see Ritland and Brower, 1991) was more edible than the Monarch butterfly to Florida scrub jays (*Cyanocitta coerulescens coerulescens*), but less edible to these birds than other non-mimetic species. It may be that variation in the dietary regime of different prey species contributes to variation in their levels of toxicity, just as it does on an intraspecific basis (Chapter 12). A good example is a pair of geometrid moth species of the genus *Arichanna* described by Nishida (1994/1995). These species are visually similar and presumed Müllerian mimics. However, *Arichanna gaschkevitchii* is a monophagous specialist that collects relatively large quantities of gryanotoxins from its host plant. By contrast, *Arichanna melanaria* is an oligophagous generalist that collects smaller quantities of gryanotoxins and is presumably rather less harmful to its predators.

Overall, there appears a reasonable level of support for the contention that there can be between-species differences in the effectiveness with which a prey's secondary defences deter predators. As the above examples illustrate, this observation extends to collections of species such as heliconid butterflies and dendrobatid frogs, which have been associated with Müllerian mimicry (e.g. Kapan, 2001; Symula et al., 2001).

## 11.3 Why should weakly defended mimics increase the likelihood that more highly defended models are attacked?

So far, there have been three main reasons for questioning the distinction between Batesian and Müllerian mimicry. As we have seen (Section 11.1), numerous authors have pointed to the fact that edibility is not fixed, so that weakly defended mimics can act as parasites of better defended models in times of energetic need. We have also seen that some authors (e.g. Nicholson, 1927; Thompson, 1984; Malcolm, 1990) have argued that differences in abilities of predators to overcome defences may effectively produce a hybrid between Batesian mimicry (if one series of predators views the more weakly defended prey as profitable) and Müllerian mimicry (if another series of predators views the same weakly defended prey as unprofitable). Finally, there has been considerable debate over the role of predator psychology—if avoidance learning were more effective for highly defended prey, then weakly defended mimics may take a free ride, and sometimes parasitise, the more effective educational properties of the better defended model.

At the heart of all three of the above phenomena are the expectations that: (i) the long-term probability of attacking some defended prey on encounter is non-zero and (ii) outside mimicry, weakly defended prey have a higher long-term probability of being attacked by predators on encounter than highly defended prey. If these key conditions hold in real systems, then there may be selection on weakly defended prey to resemble the more highly defended prey in a manner that is beneficial for the former (in terms of attack probability on encounter) but deleterious for the latter. Although such a pair of defended species would generally be regarded as Müllerian mimics, the parasitic nature of the predicted relationship is more like Batesian mimicry and has therefore been called 'quasi-Batesian' (Speed, 1993*a*). In the following subsections we examine in more detail the theory underlying these phenomena and their assumptions. We then look at observational evidence for their general predictions.

### 11.3.1 Predator hunger

There is now a substantial body of work to support the contention that predators are more prepared to attack defended prey items in times of nutritive need (Poulton, 1890; Swynnerton, 1915; Sexton et al., 1966; Gelperin, 1968; Williamson, 1980; Chai, 1986; Hileman et al., 1995; Gillette et al., 2000). For instance, Srygley and Kingsolver (1998) argued that when demand for resources increased at the height of red-winged blackbirds' breeding season, then more individuals from three moderately distasteful species of butterfly that were tethered on platforms were taken by adults.

While questions over the role of hunger in influencing the nature of relationship between Müllerian and Batesian mimicry have been repeatedly raised, there has been surprisingly little work to pin these issues down in a formal way. Important contributions of Holling (1965) and Dill (1975) both included hunger effects in the context of their simulations of the evolution of mimicry, arguing that the alternative prey made predators more risk averse, but these formulations were mechanistic and not based on optimization or risk-taking criteria. In a call to re-evaluate the nature of the relationships in Müllerian mimicry systems, Speed (1993*b*) pointed out that defended prey may not only contain toxins, but also nutrients. Therefore, under times of hunger, weakly defended prey may be profitable to attack: (p. 1246): 'in periods of actual (or anticipated) hunger, and of scarcity of alternative prey, eating a species that contains some toxin and surviving is a better strategy than eating none at all and starving.' If one accepts that even highly unpalatable prey may act as 'an emergency food supply' (Brower et al., 1968), then one might expect that there are at least some ecological conditions under which a weakly defended prey would benefit from resembling a better-defended prey species, while the latter would not necessarily benefit from the association.

In a recent paper, Sherratt et al. (2004*b*) modelled just such a framework. Their analysis was limited to considering selection on the morphology of prey species in the potentially lengthy period of time when predators have become generally aware of the noxious qualities, and nutritive value, of their prey. Using a pair of stochastic dynamic programming equations (Mangel and Clark, 1988; Clark and Mangel, 2000) they identified the optimal state-dependent foraging rules that would maximise a predator's long term survivorship by simultaneously balancing its need to avoid poisoning and its need to avoid starvation. They then examined the implications of this behaviour for the evolution of prey morphologies.

The authors found that when palatable prey were rare then, as Speed (1993*b*) had anticipated, those prey species which contain relatively low doses of toxins become profitable to consume by hungry predators. Under these conditions, a weakly defended prey could indeed gain selective advantage by resembling a prey species which contained a higher dose of the same or different toxins. However, the precise nature of the ecological relationship between well-defended model and weakly defended mimic could either be mutualistic or parasitic depending on how mimic density increases when favoured by selection. Thus, increasing the density of the weakly defended mimic in the system while reducing the probability of encountering nothing actually decreased the attack rate of predators on all defended prey types because more food was being added to the system (see also Sherratt, 2003). By contrast, if the weakly defended, well-recognized controls were gradually converted to mimics then the attack rates of predators on encounter with the model/mimic complex invariably increased as the mimic burden increased. Perhaps even more surprisingly, when two prey species contained high levels of different toxins then they could gain mutual advantage by resembling one another by increasing predator uncertainty as to the specific kind of toxin a given prey item contains (see Speed and Turner, 1999, Section 9.7.7).

Of course, one could argue the above state-dependent mechanisms of generating similarity among distasteful prey, while broadly correct in theory, never operate in practice because: (a) all natural distasteful prey are so highly noxious that once learning is complete they are never attacked by predators and/or that (b) alternative palatable prey are always so abundant that predators are never hungry enough to need to eat these prey. Reassuringly, the dynamic programming equations predicted both outcomes: if prey are highly toxic and/or there is an abundance of alternative non-toxic food then there may be no need to utilize defended prey as 'an emergency food supply'. In these cases, and from a state-dependent perspective alone, the mimics would have no effect on their better-defended models.

Kokko et al. (2003) came to very similar conclusions, again using a dynamic programming model, arguing that defended prey species would tend to become quasi-Batesian mimics if food deprivation was a more serious threat to predator fitness than poisoning. Here they also included an opportunity cost—if predators spent time handling models and mimics then their foraging efficiency on alternative non-toxic prey would diminish. Nevertheless, their central conclusions were almost identical to Sherratt et al.—the abundance of alternative prey determines whether weakly toxic mimics are attacked or not.

Here then, the central issue as far as it pertains to selection on the morphology of Müllerian mimics is: (a) how well defended the species are and (b) how often predators are hungry. We suspect that at least some natural defended prey are not always so overwhelmingly noxious that predators will avoid them whatever their nutritional state. Clearly, the widespread observation that predators are more prepared to attack familiar distasteful prey species when they are hungry, suggests that defended prey can be tolerated. More specifically, Sargent (1995) assayed the acceptabilities of a range of species of lepidoptera, in which only 11 per cent fell into the range of moderate-high unpalatability and 33 per cent were slightly acceptable-unacceptable. Field observations of black-backed orioles and black-headed grosbeaks feeding on monarch butterflies at overwintering sites (Fink and Brower, 1981; Brower and Calvert, 1985) similarly indicate that even classical aposematic prey involved in mimetic relationships, are not always avoided by all predators.

How frequently such extreme bouts of hunger will occur in a natural setting is a much more difficult question to address. However, natural populations, including those of arthropods, often exhibit extreme fluctuations in density from year to year (see e.g. Hassell, 1978). Under these conditions one would expect periods of time in which predators are forced to consume distasteful prey. McAtee's (1932) data on the gut contents of birds points to non-trivial levels of attacks on aposematic forms (see Chapter 7). It is worth noting that even established common Müllerian butterfly models also show signs of bird attack (e.g. Benson, 1972; Mallet and Barton, 1989), but this may reflect a continuing need to educate the community of predators rather than state-dependent willingness to attack unpalatable models. As so often, we are without the appropriate data to know for sure whether natural predators frequently get extremely hungry, but we would not be surprised if they do.

Müller's hypothesis explicitly centres on sharing the costs of predator education, and it is important to note that the above hunger-based arguments (both in terms of verbal arguments and formal models) do not consider learning. Conversely, Müller's hypothesis ignores the possibility of selection on unpalatable prey after learning is complete. In fact, it is possible to combine an initial stage of learning (for instance, following Müller's 1879 original algorithm) with subsequent state-dependent optimal foraging (Sherratt, Ruxton, and Speed, unpublished). As might be anticipated, when learning is the primary source of mortality in distasteful prey then there is mutually beneficial (but unidirectional, see Mallet, 2001*a* and Chapter 9, Section 9.7.1) selection on the more weakly (or less numerous) unpalatable species to resemble the more unpalatable (or more numerous) species (although it does not always have to be so—see 11.3.3). By contrast, when the primary source of mortality in unpalatable prey items is mediated by hunger rather than learning, there is selection on the more weakly defended species to resemble the better-defended species. If the subsequent rise in mimicry brings no new food into the system, then such selection would be deleterious to the better-defended species.

We are left with the view that state-dependent foraging behaviour by predators can *in theory* turn some mimetic relationships among unpalatable species quasi-Batesian in nature. This result is not only intuitive (Speed, 1993*b*), but has also been supported by two recent models (Kokko et al., 2003; Sherratt et al., 2004*b*). Whether some relationships are Müllerian or quasi-Batesian should depend on the relative mortality experienced by unpalatable prey during the course of predator education, compared with mortality inflicted by state-dependent foraging behaviour. As will be argued in Section 11.4, we do not know enough to say for sure which process tends to be most important in nature, but there are some indirect signs.

### 11.3.2 Differences in predatory abilities: the 'Jack Sprat' effect.

As the nursery rhyme goes, 'Jack Sprat could eat no fat, his wife could eat no lean.' When predators differ in their abilities to catch, subjugate, and deal with prey then the outcome of selection on prey morphology can be complex (Malcolm, 1990). It stands to reason that if a weakly defended prey species can be tolerated by one species of predator but not another then the pre-requisites for quasi-Batesian mimicry may hold: weakly defended prey will have a higher long-term probability of attack on encounter by predators than highly defended prey. There is ample evidence that predators can differ in their abilities to overcome prey defences—one has only to think of mongoose which are prepared to attack venomous snakes that many other predators of equal size would not. Edmunds (1974) reported that 'cuckoos regularly take hairy, brightly coloured caterpillars that are rejected by most other species of bird'. Similarly, the black redstart may feed cinnabar caterpillars to its young while most other birds avoid them (Hosking, 1970). In essence, when predators differ widely in their preferences then one would expect mimicry systems to represent some form of hybrid between classical Batesian mimicry and Müllerian mimicry. We can find no empirical or theoretical work that has attempted to assess the effects of interspecific (or intra-specific) variability in predators' abilities to overcome prey defences, on the nature of *mimetic* relationships among defended prey.

### 11.3.3 Psychological models

Following the work of numerous authors (Owen and Owen, 1984, Huheey, 1988; Speed, 1993*a,b*; Speed and Turner, 1999) it is now clear that several models of predator learning predict quasi-Batesian mimicry. As noted above, the key reason for the generation of quasi-Batesian mimicry in the above models is a non-zero long-term attack rate on the weakly unpalatable prey (an 'asymptotic attack rate' Speed, 1993*a* or 'partial preference', Pulliam 1975). When this condition holds, then increasing the density of weakly unpalatable mimic typically increases the attack rate of predators on the even more unpalatable model, with predation rates invariably converging to that on the weakly unpalatable mimic alone.

Unfortunately, it remains unclear whether the various forgetting and memory jogging assumptions made in the above studies (e.g. Turner and Speed, 2001) represent accurate reflections of the ways in which real predators might be expected to behave (Joron and Mallet, 1998; Mallet and Joron, 1999; Mallet, 2001*a*). For example, in their analysis Owen and Owen (1984: 228) necessarily assume that 'the predator has only a short-term memory so that it always samples at an encounter immediately subsequent to a non-sampling encounter'. This extreme amnesia not only appears highly unrealistic, but also results in a minimum attack rate on even the most unpalatable prey of 50 per cent. More importantly, if two similar-looking species are both *always* costly to attack then one might expect that both should be eventually avoided, even if one is more costly to attack than the other. Turner and Speed (2001) made a similar argument when they discussed the possibility that continued exposure to negative stimuli might eventually lead to total aversion. Yet , if prey are defended but, for reasons of state dependence, not always costly to attack then learning may act in concert to bring attack rates to some non-zero levels (see Speed, 1993*a,b*; Turner and Speed, 2001).

Despite continued debate (e.g. Mallet, 2001*a*; Speed, 2001*a,b*), very little research has been done to evaluate whether predators do indeed behave in ways assumed by the above models. It is clearly difficult to know what assumptions for predator behaviour are, and are not, realistic without good experiments to guide us (Speed, 2001*a,b*). Mallet (2001*b*) similarly lamented (page 8929) 'It seems as though it should be simple to design experiments to test these ideas about predator learning. However, most experiments on the psychology of learning use highly standardized tests, and rarely assay varying densities of items to be memorized.' The only published experiment so far conducted with prey types which varied in their distastefulness (artificial pastry baits), found that increasing the density of mildly inedible mimics (at the expense of mildly inedible controls) tended to increase the attack rates on the more inedible type which they resembled (Speed et al., 2000). There is also evidence in controlled laboratory environments that the addition of 'moderately defended' mimetic prey increases the willingness of birds to attack highly defended individuals (Simon Hannah, unpublished PhD thesis). These studies were not designed to disentangle the state-dependent motivational forces (Section 11.3.1) from the psychological mechanisms of learning and forgetting. However, they do clearly indicate that more weakly defended mimetic prey may have a parasitic effect on better-defended species and give cause to take the above arguments even more seriously.

## 11.4 Observational data on the nature of the relationship between Batesian and Müllerian mimicry

If quasi-Batesian mimicry were widespread then one might expect there to be selection on weakly defended species to resemble several different well-defended models. Yet the fact that local polymorphisms are more typical (although not entirely common) in Batesian than Müllerian systems is suggestive of qualitatively different processes at work. Simmons and Weller (2002) for example examined the phylogeny of mimicry in tiger moths (Arctiidae : Euchromiini), which are generally considered unpalatable Müllerian mimics of wasps. These moths, some of which exhibit extraordinary mimicry in terms of narrow waists and transparent wings, may be more weakly defended than wasps because they cannot sting. If the evolutionary dynamic was Batesian or quasi-Batesian in nature one might expect multiple shifts in mimetic type as the taxon evolves, yet if it were Müllerian one might expect convergence to one visual form (Simmons and Weller, 2002). Overall their analysis indicated the phylogeny was more consistent with the predicted Müllerian distribution of mimetic traits than with that of a quasi-Batesian scenario.

Nevertheless, polymorphism may not be a definitive trait of quasi-Batesian mimicry. Some mimetic species may conceivably vary between Batesian and Müllerian states perhaps because of seasonal changes in predator demand for prey (e.g. Srygley and Kingsolver, 1998). When this is the case, it could be that mimetic monomorphism is the evolutionarily stable form that a prey can take, even if for part of a season, mimetic polymorphism would be favoured. Furthermore, one might expect less intense selection for polymorphism in quasi-Batesian mimics because it will take more weakly defended mimics than classical Batesian mimics to reduce the protection afforded to a better-defended model.

In Chapter 9 (Section 9.7.6) we presented specific arguments as to why several species traditionally regarded as Müllerian mimics, such as the burnet moth and *H. numata*, are polymorphic, and these explanations appear satisfactory without the need to invoke the quasi-Batesian effects, especially since these species are known to be highly defended. Similarly, the density-dependent effect observed by Kapan (2001) in which the per capita survivorship of a rare unpalatable polymorphic form increased with density (Chapter 9, Section 9.5.1), was more consistent with the positive frequency dependence predicted by Müllerian mimicry than the negative frequency dependence predicted by quasi-Batesian mimicry. Examples of highly toxic species evolving to resemble less toxic species (Symula et al., 2001) are perhaps best explained by classical Müllerian advergent mimicry in which a rare (or later emerging) highly defended prey evolves to resemble a more common (or earlier emerging), weakly defended species (see Box 9.2). By contrast, the advergent evolution of highly defended prey to resemble less defended prey cannot be explained on the basis of state-dependent predatory behaviour, and it may be challenging to explain on the basis of differential rates of learning and forgetting.

DeRuiter (1959) proposed that, to human eyes at least, there was a much closer resemblance among species believed to be distasteful (e.g. Danaine and Ithomiine butterflies) than he would expect from selection to simply have a common form of advertisement. On face value, such observations might indicate a parasitic relationship in which it paid one species, or both, to be indistinguishable in appearance. However, it is important to be extremely cautious: close Müllerian mimics such as *H. erato* and *H. melpomene* use the same pattern elements, and in some cases the same genes, to achieve a nearly identical appearance (Nijhout, 1991). Nevertheless, the very strong resemblances between Müllerian mimics that cannot be ascribed to close phylogeny point either to very high mortality through recognition errors, or variation in their utilization as a food source. In contrast to studies of mimetic imperfection in Batesian mimics (see Section 10.5.5), the degree of mimetic perfection in Müllerian mimics has received scant theoretical or empirical investigation.

## 11.5 Summary

We feel that Müller's mechanism of shared predator learning may admirably explain why any given species of defended species might spread faster from rarity by resembling an established defended species. We also acknowledge that it is qualitatively different in mechanism and implications from Batesian mimicry. Yet we feel that it is appropriate to proceed with caution. We do not know enough about the learning behaviour of real predators when faced with defended prey that differ in unprofitability to reject the psychology-based theories of quasi-Batesian mimicry. Furthermore, the state-dependent mechanism we have described will almost inevitably generate a form of quasi-Batesian mimicry, especially if predators are frequently hungry, if distasteful prey can be stomached in times of nutritive need and if mimics rise at the expense of non-mimics. Elucidating the relative roles of hunger and learning in influencing the selection on prey morphology will take time and effort. With the exception of Speed et al. (2000), we know of no published experiment that has investigated selection on distasteful mimetic prey that differ in their overall unpalatability. Clearly we need more experiments using unpalatable prey that differ in their unpalatability, and predators differing in their hunger levels, before we can finally settle the important issue of the relation between Müllerian and Batesian mimicry.

Finally, it is important to echo the sentiments of many mimicry workers over the years including M. Rothschild and M. Edmunds. Each example of mimicry may have its own unique attributes. Many mimetic systems involve a complex interplay of species which vary in their defensive attributes and predators which vary in their abilities to deal with them. Trying to shoehorn a mimetic system into a simple dichotomy of Batesian or Müllerian mimicry, which were themselves concepts developed predominantly with a single predator and two interacting prey species in mind, may not always be appropriate.

# CHAPTER 12

# Other forms of adaptive resemblance

## 12.1 Overview

The aim of this chapter is to briefly describe several further forms of mimicry in the natural world, namely aggressive mimicry, floral mimicry, intraspecific sexual mimicry, and automimicry. As noted in Chapter 10, there are several reasons why a resemblance to a given species might evolve, or be maintained, by natural selection. Not all of these explanations relate directly to the gaining of protection from predators. However, some forms of mimicry (such as pollinator mimicry) have close parallels to Batesian and Müllerian mimicry, while other forms of mimicry (such as aggressive mimicry) may, in theory, confer advantages both in terms of access to resources, as well as protection from predators. Indeed, as will be seen, there has been heated debate as to whether certain putative examples of aggressive mimics are actually Batesian mimics and vice versa. Given these close interrelationships between the different types of mimicry, we feel it is appropriate to briefly review these other forms even if they are not always related to avoiding attack. We have placed special emphasis in this chapter on the phenomenon of automimicry (in which members of the same species gain protection from predators by resembling better defended conspecifics), not only because the subject genuinely relates to 'avoiding attack', but also because it is has rarely been reviewed, it is likely to be common, and its presence raises some important issues.

## 12.2 Aggressive mimicry

In aggressive mimicry[10], a predator or parasite species resembles another non-threatening (or even inviting) species (or object) in order to gain access to prey or hosts (Cott, 1940; Wickler, 1968). Thus, the primary benefit to an aggressive mimic is seen as one of gaining access to resources rather than receiving protection from predators. Human hunters sometimes attempt to disguise themselves in order to get closer to their prey (see Cott, 1940 for examples). Similarly, larvae of the green lacewing *Chrysopa slossonae* cover their spiny backs with tufts of wax from the aphid *Prociphilus* when moving around its colonies feeding on the aphids—a 'wolf in sheep's clothing' (Eisner et al., 1978). Indeed, Tom Eisner and his colleagues showed that without these shields the lacewing larvae tend to be ejected by the attendant ants (yes, these proverbial sheep even have shepherds).

Examples of aggressive mimicry are abundant and diverse. For example, female fireflies of the genus *Photuris* (a Lampyrid beetle) attract and then consume male *Photinus* fireflies by mimicking the flash behaviour of *Photinus* females (Lloyd, 1965, 1975). It turns out that *Photuris* females also acquire defensive steroidal chemicals called lucibufagins from their *Photinus* prey, which confer protection against jumping spiders (Eisner et al., 1997). One might wonder how conspecific males fare with such sirens: in response to males of her own

[10] Sometimes referred to as 'Peckhammian mimicry' (Wickler, 1968) after E. G. Peckham and G. W. Peckham's work in 1889 and 1891 on mimicry in jumping spiders (Salticidae). Pasteur (1982) proposed that this term is unsuitable because there are several kinds of aggressive mimicry, and because other authors had earlier discussed similar possibilities, notably Kirby and Spence (1823) and Bates (1862).

species, the female *Photuris* gives a rather different flash response to that directed towards *Photinus* prey, and does not tend to eat her suitors.

Bolas spiders of the genus *Mastophora* are so called because they hunt with sticky balls on the end of a line, reminiscent of south American gauchos throwing their bolas (although in bolas spiders they are not thrown, but dangled; Yeargan, 1994). Eberhard (1977) observed that of 165 prey collected by a certain species of bolas spider, all were male noctuid moths of just two species. This was a far from random sample, as Eberhard confirmed by cataloguing the diverse range of taxa caught in sticky traps. Given that the male moths also tended to approach from downwind, it was proposed that bolas spiders produce compounds that act as sex attractants to these species. This conclusion has subsequently been supported by chemical analysis (e.g. Stowe et al., 1987; Haynes et al., 2002).

Wickler (1968) describes a fascinating case of aggressive mimicry seen in some North American freshwater clams whose larvae are parasites on fish gills. Adult females of *Lampsilis ovata ventricosa* bear a fleshy protusion on their upper mantle that is a convincing mimic of a small fish (with tail, fins and eye). When predatory fish approach this lure, the clam puffs larvae into its face where they attach to gills and begin their parasitic stage of their life cycle (Fig. 12.1).

In a highly unusual case of cooperative aggressive mimicry, larvae ('triungulins') of the blister beetle *Meloe franciscanus* climb to the top of stems where they form aggregations which resemble bees (Hafernik and Saul-Gershenz, 2000, Fig. 12.2). These aggregations attract (through visual, but most likely also chemical, cues) males of the bee *Habropoda pallida*, to which they attach during pseudo-copulation and are eventually transported back to the bee's colony which they parasitize.

Other potential examples of aggressive mimicry are controversial because it is not always easy to identify the most significant benefit arising from the similarity. For instance, one reason why certain Asilid flies may resemble hymenoptera is that it allows them to approach their victims more closely without alarming them. Yet asilids rarely approach their prey in a 'casual' manner as if disguised, and classical Batesian mimicry may be a better explanation for this

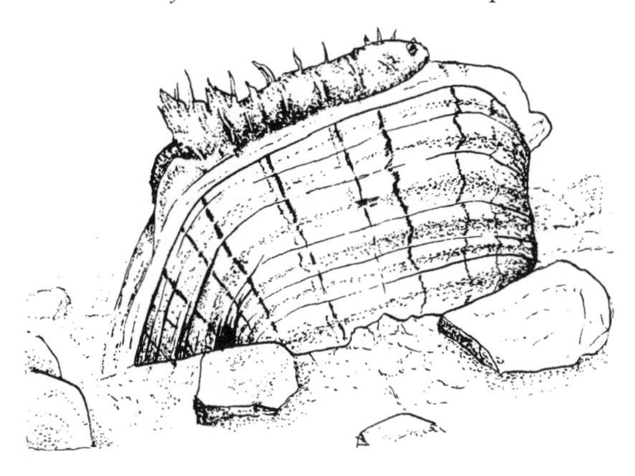

**Figure 12.1** A clam which may lure its hosts with a fleshy protuberance that looks like a fish.

**Figure 12.2** Cooperative mimicry of a bee by larvae of the blister beetle *Meloe franciscanus*.

resemblance (Poulton, 1904; Evans and Eberhard, 1970; Rettenmeyer, 1970). Similarly, the syrphid fly *Volucella bombylans* lives in the nests of bumble bees (and occasionally social wasps) in its larval stages where it feeds on nest debris (Spradberry, 1973). The adults have two main colour forms resembling white-tailed and buff-tailed bumble bees and enter nests to lay their eggs. Yet female *Volucella* enter nests independent of the colour pattern type of the bumblebee species, and there is no clear evidence that mimetic adults gain through access to nests rather than protection from predators (Rupp, 1989).[11]

## 12.3 Pollinator (floral) mimicry

Many flowers produce rewards to pollinators (nectar, pollen, or both), but some flowers that do not provide rewards may nevertheless attract pollinators by resembling rewarding flowers or other objects (Dafni, 1984; Roy and Widmer, 1999; Johnson, 2000). In Batesian floral mimicry, the mimic produces no nectar reward but the model does (Roy and Widmer, 1999). For instance, Brown and Kodric-Brown (1979) found a population of *Lobelia cardinarlis* that secreted no nectar but appeared to attract hummingbirds by resembling more abundant nectar-producing species (Williamson and Black, 1981 subsequently suggested alternative interpretations for this observation, including automimicry—see Section 12.5). Similarly, the orchid *Orchis israelitica* appears to mimic the lilly *Bellevalia flexuosa* to gain access to pollinators (Dafni and Ivri, 1981). In a delightful paper, Johnson (1994) described a study of the orchid *Disa ferruginea* in South Africa which appears to be dependent on a single butterfly species for pollination. The flowers of *D. ferruginea* produce no food reward, but appear to be visited by the butterfly because of their close visual similarity to other reward-producing species. This is a polymorphic mimic because in some locations a red-flowered form resembles one flower (a lily), while elsewhere an orange-flowered form resembles another species (an asphodel). Populations that are sympatric with their reward-producing models tended to have relatively high levels of pollination and fruit production compared to populations where the orchid grows alone. We should ask why the nectarless species do not evolve some reward, in order to gain their own pollinators without deception. In species where access to resources rather than pollinators limit fitness (and there are suitable rewarding models) then deception could save energy that can be allocated to the reproductive process. The same general argument opens up the possibility that automimicry (Section 12.5) could readily be selected for in plant–pollinator systems.

Of course it is quite possible that flowering plants might engage in a form of 'Müllerian' mimicry. Thus, two or more rewarding flower species might gain a mutual advantage by evolving a common advertising display (Roy and Widmer, 1999). Many cases of Müllerian floral mimicry have been proposed (see Dafni, 1984; Roy and Widmer, 1999 and references therin). For instance, the neotropical understory herbs *Costus allenii* and *Costus laevis* (Zingiberaceae) share the same pollinator (the bee *Euglossa imperialis*) and while the plants differ in vegetative characteristics (they are not closely related) their flowers are almost identical in colour. Schemske (1981) argued that low flower density has selected for floral similarity in these species. However, as Roy and Widmer (1999) emphasize, although these ideas have been around for some time, no one has performed the crucial tests to determine whether their similarity is adaptive.

Other forms of floral mimicry are more akin to aggressive mimicry. For instance, pseudo-copulation occurs when flowers mimic the female insect to attract mates, and thereby pollinate the plant. *Ophrys* (Orchidaceae) flowers produce a combination of visual, tactile and olfactory stimuli to attract hymenopteran males (e.g. bees) to attempt to mate with them (Fig. 12.3). Interestingly, any given male has to be duped twice to transfer pollen between individuals. Of course *Ophrys* flowers may compete for attention with real females and these flowers appear most attractive in the short time after the emergence of males and before the emergence of females (Dafni, 1984; Roy and Widmer, 1999). One

[11] Translation kindly provided by Dr Francis Gilbert, http://www.nottingham.ac.uk/~plzfg/syrphweb/Rupp1989.doc

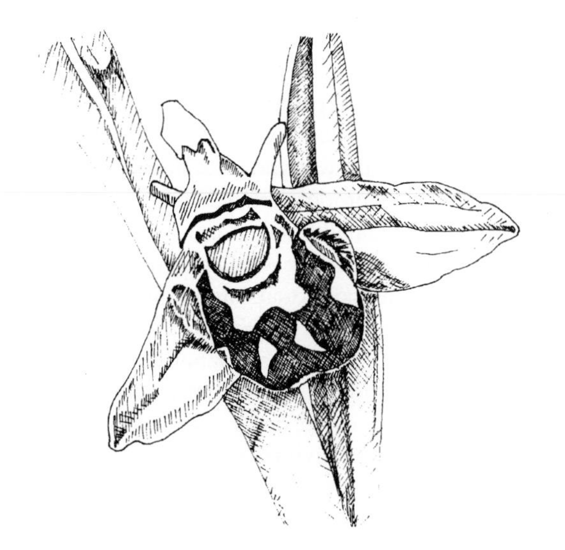

**Figure 12.3** A bee orchid which attracts pollinators without providing any nutritive reward.

should also not forget the effect of the deceit on the duped. Wong and Schiestl (2002) investigated the effect that the orchid *Chiloglottis trapeziformis*, which mimics the sex pheremones and appearance of its pollinating wasp *Neozeleboria cryptoides*, had on this pollinator. They showed that by producing these signals the plant made attraction of mates more difficult for female wasps, providing the basis of a co-evolutionary response.

Finally, there is scope for classical Batesian (and Müllerian) protective mimicry against herbivores in plants as a whole, in which a poorly defended species is selected to resemble a better defended species. While several potential examples have been suggested (e.g. Eisner and Grant, 1981), Augner and Bernays (1998) argue that no study has yet conclusively shown that this form of mimicry exists in terrestrial systems. As Roy and Widmer (1999) and many other researchers have emphasized, more work is needed to understand the evolution of mimetic relationships in plants.

## 12.4 Intraspecific sexual mimicry

In *intraspecific* sexual mimicry, a proportion of males resemble females of the same species, or vice versa. Males often mimic females in order to obtain mating opportunities, and hence many forms of intraspecific mimicry can be thought of as special cases of aggressive mimicry. For instance, while some parental male bluegill sunfish (*Lepomis macrochirus*) adopt a courting and guarding strategy, a proportion of male sunfish use female mimicry to approach courting pairs undeterred and steal fertilizations from parentals (Neff et al., 2003). A similar strategy is adopted by a proportion of males of the tropical rove beetle *Leistotrophus versicolor*, which bluff their way into the territories of larger males, allowing them opportunities to mate with females that are attracted to the territory (Forsyth and Alcock, 1990).

Sometimes the advantages of males mimicking females are highly unusual. Males of the scorpionfly *Hylobittacus apicalis* (Mecoptera) offer small arthropods as nuptial gifts to females during courtship and copulation. Yet some males of this species mimic females, allowing them a greater opportunity to rob the nuptial gifts of conspecific males (Thornhill, 1979). As Shine et al. (2001) recently argued, the unusual male garter snakes that produce female-like pheremones on emerging from hibernation ('she-males') may benefit because the mating balls that males form around them help keep them warm and less exposed to predators.

In other cases, a proportion of females may resemble males. However, the advantages to females resembling males are often much less clear, and explanations for this intriguing phenomenon, which extend from butterflies (Clarke et al., 1985; Cook et al., 1994) to hummingbirds (Bleiweiss, 1992), remain controversial. Perhaps the best studied (and most hotly debated) example of females appearing to resemble males is found in odonates (damselflies and dragonflies). Female-limited colour variation ('polychromatism', Corbet, 1999) is a widespread trait in this group, particularly in damselflies, where it is found in over half of the European genera (Andres and Cordero, 1999). In all polychromatic species, one female chromotype (the 'androchrome') resembles the male, while one or sometimes more (all similar looking) chromotypes do not: these are known as 'the gynochrome(s)'. In some species, such as *Ischnura ramburi*, androchromes are differentiable from males only after close inspection (Robertson, 1985).

Furthermore, the androchromes of some species have been found to more frequently adopt behaviours associated with males (Robertson, 1985; Forbes et al., 1997). Recent theories for this curious phenomenon have all been based on the need of females to avoid excessive harassment by males. Thus, the androchrome may have evolved either because males confuse androchromes with males and thereby avoid them (Robertson, 1985; Sherratt, 2001) or because males are simply confused by the appearance of a given female form when that female form is rare (Miller and Fincke, 1999; Van Gossum et al., 2001).

## 12.5 Automimicry

### 12.5.1 The phenomenon of automimicry

In the cases of intraspecific sexual mimicry described above, a proportion of one sex resembles the other sex. As Section 12.4 argued, there are a variety of reasons why such mimicry might be selected for. However, there may be an even more pervasive form of intraspecific mimicry in which weakly defended members of a species gain protection from predators by resembling better defended conspecifics.

Although there is a general tendency to consider members of an unpalatable (or otherwise defended) species all approximately equally unpalatable, this may not always be the case (indeed it may be frequently untrue, Bowers, 1992, 1993). The possibility that individuals of the same species can vary markedly in their forms of defences was first reported by Brower et al., (1967) who reared a strain of monarch butterfly (*Danaus plexippus*) on cabbage (*Brassica oleraceae* L.) and found that their larvae, prepupae and adults were all palatable to blue jays (*Cyanocitta cristata bromia*). Yet monarchs reared on milkweed *Asclepias curassavica*, a natural food plant that contains heart poisons (cardenolides), caused the same birds to vomit. Understandably, the authors suggested that these findings were consistent with the view that palatability of the monarch was directly related to the species of plant ingested by the larvae. Who else but Poulton (1914)—cited in Malcolm and Brower (1989)—had anticipated this general result, in suggesting that chemistry may help explain why monarch butterflies feed exclusively on milkweeds.

Given the fact that palatability may be influenced by the larval food plant selected by the ovipositing female butterfly, Brower and his colleagues (Brower et al., 1967, 1968) proposed that there was a real possibility of a 'palatability polymorphism' in the field. The authors referred to such polymorphism, in which palatable individuals are perfect mimics of unpalatable members of the same species, as 'automimicry'.[12] Subsequent laboratory research on the cardenolide concentrations of individual adult monarchs reared on 11 species of *Asclepias* has revealed a positive (non-linear) relationship between plant cardenolide concentration and butterfly cardenolide concentration (Malcolm and Brower, 1989; see also Brower et al., 1984). Indeed, over the past three decades it has become increasingly clear that a number of defended species sequester unpalatable chemicals directly from food materials in both marine (Pawlik, 1993; McClintock et al., 2001) and terrestrial environments (Bowers, 1992) rather than (or as well as) synthesizing such substances *de novo* (see Chapter 5; Rothschild, 1972; Duffey, 1980; Nishida 2002 for reviews). As a further example, Ritland (1994) reared Florida queen (*Danaus gilippus*) caterpillars on four different asclepid food plants and found that the adult butterflies in these treatments subsequently differed in their

[12] Some researchers (e.g. Guthrie and Petocz, 1970; Lev-Yadun, 2003) have referred to cases of mimicking parts of oneself (e.g. false thorns, false heads or false weapons) or even making false-replicas of oneself (Purser, 2003) as automimicry, but here we restrict ourselves to Brower et al.'s (1967) definition based on gaining protection from predators by mimicking more unpalatable conspecifics. This definition rules out several fascinating examples of sexual mimicry (see Section 12.4). It also rules out examples such as *Argiope* spiders making replica's of themselves as decoys against predators (see Purser, 2003 for some wonderful photographs). It is tempting to treat automimicry as *any* case of intraspecific variation in defense. This is a pragmatic interpretation, but we feel that automimicry is best reserved to cases where there is *sufficient* intraspecific variation in defense that weakly defended individuals would only gain protection from predators in the presence of better defended conspecifics.

cardenolide content and palatability to red-winged blackbirds (*Agelaius phoeniceus*). McLain and Shure (1985) found that the Lygaeid seed bug *Neacoryphus bicrucis* became palatable to anole lizards when fed on seeds that lacked the pyrrolizidine alkaloids of its natural host plant.

Since Brower et al.'s 1967 seminal work, the extent of this intraspecific variation in palatability has been formally investigated in field-caught specimens, particularly in butterflies, on both macrogeographic (e.g. Brower and Moffitt, 1974; Cohen, 1985; Bowers and Williams, 1995) and microgeographic (e.g. Eggenberger and Rowell-Rahier, 1991; Pasteels et al., 1995) scales. For example, wild populations of the monarch from Florida, Massachusetts, Trinidad, and Costa Rica exhibit high variation in palatability, with anywhere from 10 to 90 per cent of adult butterflies being emetic (Brower, 1969). While Californian monarch butterflies contained significantly higher mean concentrations of cardenolide than Massachusetts monarchs, both populations were sufficiently diverse that they contained individuals that would be palatable (non-emetic) and others that would be unpalatable, to blue jays (Brower and Moffitt, 1974). Moranz and Brower (1998) sampled 280 Florida queen butterflies at 11 sites in three ecozones (coastal salt marshes, and xeric and hydric inland areas) in Florida. The mean cardenolide concentrations of individuals varied significantly among queen populations separated by only a few kilometres, but they also varied temporally within populations. Thin layer chromatography revealed characteristic 'fingerprints' of the species of host plant on which each individual butterfly had developed. The strong positive correlation between mean cardenolide concentration of a population and their frequency of feeding on *Asclepias perennis* and *A. longifolia* indicated that the available host plants had mediated much of the observed differences in butterfly cardenolide concentrations.

Automimicry has been most intensively investigated in the monarch–viceroy–queen mimicry complex in North America, but there is also evidence of automimicry in other butterflies. For instance, the closely related African monarch *Danaus chrysippus* also feeds on milkweeds of a variety of species throughout its broad geographical range, and local populations may contain over 80 per cent adults that are non-toxic to birds (Edmunds, 1974; Brower et al., 1975). Bowers (1980) found that adults of the checkerspot butterfly *Euphydryas phaeton* were both unpalatable and emetic to blue jays when their caterpillars were reared on their usual host plant *Chelone glabra* (Scrophulariaceae). By contrast, caterpillars of this species that were transferred to *Plantago lanceolata* upon emergence from their over-wintering phase produced palatable, or partially palatable, adults when presented to the same predator species. In the field, post-diapause larvae often over-exploit individual *C. glabra* plants, such that feeding on alternative host plants is not uncommon: under these conditions, automimicry would be expected.

Further studies have revealed many other likely cases of automimicry in animals other than butterflies. It is important to note however that most of these studies, like the above studies on butterflies, demonstrate the *potential* for automimicry (more palatable individuals gaining protection from predators through their similarity to less palatable individuals) *if* certain conditions hold, but only show variability and do not unequivocally establish its selective benefit. Millipedes, which produce hydrogen cyanide as a means of defense, may generate anything from 0 to 600 μg of this compound per individual (Eisner et al., 1967). Although not direct evidence for automimicry, it is noteworthy that Dendrobatid frogs lose their ability to produce highly toxic alkaloids when bred in captivity (Daly et al., 1987). Pasteels et al. (1995) investigated variation in autogenous and host-derived chemical defences within several species of *Oreina* leaf beetles (Coleoptera : Chrysomelidae). Populations of these species were found exclusively synthesizing cardenolides *de novo*, or exclusively sequestering pyrrolizidine alkaloids, or doing both, depending on the local availability of food plants, indicating significant intraspecific variation in defense. Tullberg et al. (2000) found that fourth instars of the milkweed seed bug *Lygaeus equestris* were more palatable to chicks when the earlier nymphal stages were reared on the sunflower *Helianthus annuus* seeds, than when its nymphs were reared on seeds from its natural asclepiadaceous host plant *Vincetoxicum hirundinaria*. Interestingly, the effect of the host plant was far less marked in a parallel study on the seed bug *Tropidothorax leucopterus*, suggesting that this particular species had alternative means of defense, not dependent on the host plant (Tullberg et al., 2000). Berenbaum and Miliczky (1984) found that the Chinese mantid *Tenodera ardifolia sinensis* tended to regurgitate after being fed on milkweed bugs (*Oncopeltus fasciatus*) that had been reared on milkweed seeds. After one to four encounters, the vast majority of experimental mantids avoided attacking such bugs altogether (although they still ate palatable flies). Of six mantids that had learned to reject milkweed-fed bugs, none attacked a sunflower-fed bug when it was offered. By contrast, of five mantids fed on an equivalent number of sunflower-fed bugs, four attacked these bugs when they were offered subsequently. This result, while poorly replicated, is one of the few studies to formally demonstrate a protective advantage to automimicry.

Many other factors besides host food plant may mediate intra-specific variation in palatability. For instance, Alonso-Mejia and Brower (1994) found that the concentration of defensive cardiac glycosides (cardenolides) decreased in individual monarch butterflies as they aged, and proposed that freshly emerged individual monarchs can thereby serve as noxious models for the older individuals. Similarly, sex may sometimes account for considerable variation in noxiousness to would-be predators. For instance, males of many bees and wasps do not carry stings, but may be protected from predators by their resemblance to conspecific females that do have stings. Stiles (1979) investigated the extent of sexual dimorphism in bumble bee species in the western hemisphere. Although one might expect that undefended male bees would closely resemble defended female bees of the same species, this is not always the case. He argued that in those species where sexual dimorphism occurred, it was likely to have arisen as a consequence of selection counter to automimicry, based on the greater need of males to thermoregulate. Pasteur (1982) reports a study in Gabon by Bigot and Jouventin (1974) in which a naive monkey that had tasted an *Anaphe* female butterfly would not touch further individuals of either sex, even though it routinely ate males before. Eggenberger et al. (1992) reported that the defensive secretions of the Chrysomelid beetle *Oreina gloriosa* varied significantly with both season (certain defensive compounds decreased, and certain compounds increased, most likely a function of beetle age) and sex (females tended to produce significantly higher amounts of certain defensive compounds). As Bowers (1992) notes, variation in the extent to which a given species can defend itself may also depend on how recently (and in what context) this defence has been previously used (see Chapter 5). For instance, many venomous snakes may no longer be venomous if they have recently bitten a prey item or potential threat. Insects such as beetles, grasshoppers, and phasmids that use sprays or bleeding to protect themselves may temporarily be without this means of defense if they have recently discharged (Eisner, 1965).

At least some of the observed intraspecific variation in both sequestered and synthesized defense may well have a genetic component. Edmunds (1974) reported that West African *Danaus chrysippus* accumulate less cardenolide than East African conspecifics when fed on the same plant. Similarly the South American monarch *D. plexippus megalippe* was more emetic than its close relative *D. p. erippus* when both were reared on the same milkweed species (Rothschild and Marsh, 1978). Similarly, Eggenberger and Rowell-Rahier (1992) found evidence that the composition and quantity of defensive secretions of the Chysomelid *Oreina gloriosa* was under some form of genetic control. Mebs and Kornalik (1984) analysed the venom content of four specimens of eastern diamondback rattlesnake (*Crotalus adamanteus*) from the same litter and found that the venoms of two snakes were essentially free of toxin. They also found that the individual absence or presence of the toxin in the venom was constant over time, suggesting the possibility that some of the observed variation may be genetically determined.

One fascinating *potential* example of sex-based automimicry has recently come to light

(Conner et al., 2000). Males of the strikingly colourful Arctiidae moth *Cosmosoma myrodora* acquire a certain pyrrolizidine alkaloid by selectively feeding on particular plants. The male allocates some of this acquired alkaloid to a cottony mass of filaments that he discharges onto females during mating. In laboratory trials, males that had imbibed alkaloids were cut lose from the webs of the orb spider *Nephila clavipes* uninjured, while the majority of males that had not imbibed the alkaloids were killed by the spiders. Similarly all of the females that had mated with alkaloid-bearing males survived, while the majority of females that had mated with alkaloid-free males were eaten by the spiders. As Conner et al. (2000) state, while the moth is clearly aposematic, it is an open question whether vertebrate predators are also deterred by the acquired alkaloids. Furthermore, while males and females are likely to differ in their alkaloid content, it is also unclear whether spiders can be duped into releasing a less repellent individual on the basis of prior experience with a more repellent one (cf. Berenbaum and Miliczky, 1984).

Many *Danaus* and related species, including monarchs, also actively seek out sources of pyrrolizidine alkaloids from plants to use in the production of sex phenemones, which may also render them more unpalatable (Edgar et al., 1976; Kelley et al., 1987; Gullan and Cranston, 1994). Brown (1984), for instance, found that orb spiders (*N. clavipes*) readily ate freshly emerged Ithomiine butterflies (despite the fact that the larvae of these butterflies are believed to gather poisons from their host plants, primarily Solanaceae). By contrast, both the palatable butterfly *Biblis hyperia* when smeared with pyrrolizidine alkaloids and wild-caught Ithomiines were cut lose by these spiders. Thus, Ithomiines may be protected in the wild against this abundant spider predator through pyrrolizidine alkaloids that are gathered from plants as adults. Brown (1984) reported high variation in total pyrrolizidine alkaloids between wild-caught adults in a given population, once again indicating a high level of intraspecific variation in defense. Whether this variation is sufficient to confer protection to the less well-defended individuals remains to be seen.

Finally, it is important to note that plants themselves may exhibit considerable intraspecific variation in their degree of defense (e.g. Dolinger et al., 1973; Lincoln and Langenheim, 1978, 1979; Louda and Rodman, 1983*a,b*; Lincoln and Mooney, 1984; Bowers, 1992; Bowers and Stamp, 1992) and it is not inconceivable that less well protected plants might gain protection from herbivory from resembling better-defended individuals (see Section 12.3; Augner and Bernays, 1998). Indeed, it is possible that better-defended plants may have evolved ways to resist exploitation by automimics; recent theory has suggested that plants produce bright autumn colours at some cost, and that better-defended (healthy) plants use these costly traits as honest signals of their ability to resist herbivores (Archetti, 2000; Hamilton and Brown, 2001). There is some circumstantial evidence supporting this provocative theory (Hamilton and Brown, 2001; Hagen et al., 2003), but there are also some important difficulties with the hypothesis, particularly in relation to the ecology of insect herbivores such as aphids (Holopainen and Peltonen, 2002; Wilkinson et al., 2002).

### 12.5.2 The challenge to theoreticians

There are two principal conceptual challenges relating to automimicry. First, how can it be maintained (since automimics are 'palatable', why do predators not simply attack both automimics and their less palatable models?) and second, how can it evolve? Of course, the first question is no more challenging to address than the question as to why perfect interspecific Batesian mimicry can persist (Chapter 10). The second question is potentially much more interesting because if defense carries a cost (see Chapter 5), then one can imagine a case where palatable 'cheats' are always at an advantage over unpalatable conspecifics. Somewhat surprisingly, this issue has only been raised and debated in the last decade (Guilford, 1994). We now consider both questions in turn.

*The maintenance of automimicry*

Fittingly, Brower et al. (1970) were the first to suggest a mathematical model of the maintenance of automimicry.[13] In their model, a single encounter with an unpalatable prey caused the predator to reject all other similar-looking individuals ('single trial learning') for $n$ encounters with such prey (including the current experience, cf. Huheey, 1964). If a predator did not encounter noxious prey it would eat $n$ prey in a comparable period of time. They defined variable $m$ as the number of prey available per predator in the population, such that if $n < m$ there are more prey available than can be eaten (similarly if $n \geq m$ and all prey are palatable, then survivorship will be zero). Finally, the authors let $k'$ represent the proportion of prey that were unpalatable. Under these conditions, the average number of prey eaten by one predator in $n$ encounters is given by the geometric series:

$$1 + (1-k') + (1-k')^2 + \cdots + (1-k')^{n-1}$$

The sum of any geometrical series of the form $(1 + r + r^2 + \cdots + r^{n-1})$ is simply $(r^n - 1)/(r - 1)$ such that the average number of prey eaten is:

$$\{(1 - k')^n - 1\}/\{(1 - k') - 1\}$$

which reduces to:

$$\{1 - (1 - k')^n\}/k'$$

and the fraction of prey surviving is therefore:

$$1 - \{1 - (1 - k')^n\}/mk' \quad [=j_{k'}]$$

When all prey are palatable, then proportion $1 - (n/m)$ $[= j_0]$ would survive. Strangely, the authors use the difference in the two rates $(j_{k'} - j_0)$ as a measure of 'automimetic advantage' and show this difference is highest when $n = m$ and when $k'$ is high. However, this simply compares the survivorship of a mixture of palatable and unpalatable prey with that of palatable prey. Perhaps a more enlightening approach is to simply look at the survivorship $(j_{k'})$ of automimics (and their models) when the majority of the population is unpalatable compared to the survivorship of automimics when the majority of the population are automimics (see Fig. 12.4). Individuals within a population of unpalatable prey always have a higher survivorship than individuals within a population with palatable prey. As the proportion of palatable automimics in a population increases then typically the survivorship of models and mimics remains high until the parasitic burden becomes too much and the survivorship of models and automimics declines steeply (Fig. 12.4).

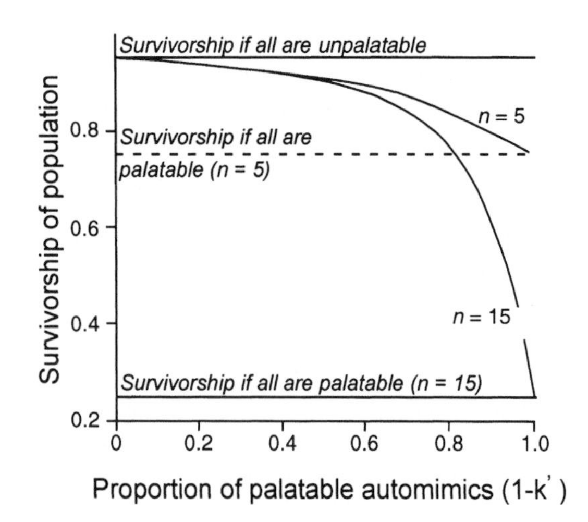


**Figure 12.4** A reformulation of the model of Brower et al. (1970) showing how the survivorship of models and palatable automimics changes as the proportion of palatable automimics increases, $m = 20$.

Several years later, Pough et al. (1973) revisited the same mathematical model, but this time they assumed the prey was less noxious in that two attacks were required (either consecutively or separated by any number of palatables) before rejection occurred. We will not review their specific findings here because they were qualitatively similar to that reported earlier: not surprisingly, automimetic advantage (compared to palatable non-mimics) is less when predators require two attacks on unpalatables before they reject prey with this appearance. Similarly, automimics always survive better than their theoretical counterparts that cannot exploit the appearance of unpalatable conspecifics.

*The evolution of automimicry*

As Guilford (1994) noted, 'if automimics are visually indistinguishable from their automodels, and if toxins are costly to the prey containing them, then automimicry tends to render aposematism

[13] Automimicry has sometimes been called 'Browerian' mimicry (Pasteur, 1982) in honour of its primary discoverer.

evolutionarily unstable.' How then can warning signals persist in the face of possible cheats that gain the benefits from signaling, without paying the cost? There are several potential solutions to this important question. First and foremost, being unpalatable may not always carry a cost (see Chapter 5). As Leimar et al. (1986) note, several experiments have failed to show a physiological cost (e.g. Dixon et al., 1978) and some have even shown that plants containing high concentrations of cardenolides may enhance larval growth (Erickson, 1973; Smith, 1978). Thus, it is quite possible that long periods of co-evolution of one species with its host plant render those individuals that have not managed to sequester toxins at no overall advantage compared to more toxic conspecifics. However, there are a growing number of studies that now show that chemical defences are indeed often costly (see Chapter 5). Furthermore, if defences were cost-free, one might wonder why there are automimics at all—perhaps in these cases it is simply a case of the automimics not obtaining sufficient levels of defensive compounds.

A second reason why automimetic 'cheats' do not take over the population may be that automimics are preferentially attacked over automodels. This simple mechanism might well explain the absence of automimics in defended prey with an obvious means of physical defense such as urticating hairs or spines, but it could also act in more subtle ways. In particular, Gibson (1984) proposed that automimicry in lepidoptera might be maintained by some complex interplay between vertebrate and parasitoid predation on their larvae, rather than adults. She proposed that birds might typically prefer to feed on larvae that are foraging on non-toxic food plants, while parasitoids might prefer larvae that are raised on toxic food plants. Changes in the densities of automodels or automimics might cause one or both predatory groups to switch preferences thereby maintaining a balanced polymorphism. Twenty-three years later, Losey et al. (1997) presented data (and a mathematical model) supporting the contention that colour polymorphism in the pea aphid, *Acyrthosiphon pisum* might be maintained by a similar type of balancing selection by a predator (the ladybird *Coccinella septempunctata* which prefers red morphs) and a parasitoid (*Aphidius ervi* which prefers green morphs). It is unclear, what (if any) data Gibson's (1984) interesting explanation was based on, but the possibility remains that if lack of defense can be detected by predators (either directly or by focusing on herbivores foraging on certain host plants), then automimics will be selected against.

Third, defended automodels may survive predatory attacks more frequently than undefended automimics. Clearly such a difference in survivorship can arise either as a direct consequence of the defense, and/or as a consequence of the predator appreciating that a prey item is defended and releasing it. There is now considerable evidence that defended species often have a higher probability of surviving attack by predators than undefended species (Jävi et al., 1981; Wiklund and Jävi, 1982), in part because defences such as toxins are often located in the outer parts of tough bodies (Brower and Glazier, 1975). For instance, Boyden (1976) found that avoidance of *Heliconius* butterflies by *Ameiva* lizards was not only based on wing colour and pattern, but also on a distasteful element in the body or wings. Thus, not all Heliconids were killed on attack by the lizards, but all of the palatable control butterflies were. Similarly, Brower (1958*a,b*) found that scrub jays that had been fed edible butterflies tended to seize all those offered (both palatable and unpalatable), eating the former and dropping the latter. If defense is costly but increases survivorship then the balance between surviving attack and minimizing physiological costs can in theory produce an evolutionarily stable combination of automodels and automimics (Speed and Ruxton, unpublished results). Till-Bottraud and Gouyon (1992) made a similar case when they proposed a mathematical model to explain why some leaves on a plant may produce cyanide and other leaves produce little. They argued that for individual plants there is an optimum frequency of cyanogenic leaves according to the cost and herbivory. However, when several individuals are present, the overall evolutionarily stable strategy (ESS) level of protection per plant is less.

As Guilford (1994) notes, if predators commonly react on attack towards potential prey in a way that helps them discriminate automimics from automodels, then aposematic signals may be interpreted as 'go-slow' signals rather than 'stay away' signals. Thus, it is possible that 'predators learn not to avoid aposematic prey, but to sample them more cautiously in order to determine their true palatability more accurately' (Guilford, 1994, abstract). Such cautious attacking (allowing automodels to survive predatory attacks more frequently than automimics) would clearly help filter out palatable cheats from unpalatable conspecifics. In support, Sherratt (2002*a*) found that under some conditions, the 'go-slow' behaviour analogous to that described by Guilford (1994) might evolve in some members of a predator population, especially if it reduced the probability of predators being harmed in the process of attacking potentially defended prey.

Unfortunately however, there remains very little direct evidence that automodels would survive an attack more frequently than automimics. In a fascinating paper, Fink and Brower (1981) observed flocks of black-backed orioles and black-headed grosbeaks feeding on monarch butterflies in their dense overwintering sites in central Mexico. They explicitly tested the hypothesis that the birds preferentially kill and eat individuals containing little or no cardenolide. Thus, they observed orioles attack 98 monarchs and release 75 without harm; grosbeaks similarly released 65 of 92. As the birds never eat the wings, Fink and Brower (1981) were able to estimate the cardenolide content of those killed and partially eaten and compare it with a sample of unattacked butterflies. The frequency distributions were very similar, indicating that a higher cardenolide content gives individual butterflies no additional protection from these birds. It turns out that while grosbeaks kill and continue to eat the individually captured butterflies without influence of differences in their cardenolide content, orioles eat fewer of the butterflies with higher cardenolide content. However, as the butterflies are dead, this will not affect selection per se. In follow-up work, Brower and Calvert (1985) found that black-backed orioles and black-headed grosbeaks killed proportionately more male than female monarchs than would be expected from their abundances in nets, and they rejected proportionately more female abdomens by dropping them. Males contained on average a lower concentration of cardenolides than females, and may have slightly less cardenolides in total (a result of marginal significance). The authors suggest that birds may be able to distinguish the sexes prior to capture (hence this may not be a case of perfect automimicry), and they may also release more noxious females on capture.

### 12.5.3 Summary

The possibility that less well protected individuals may gain protection from predators by resembling better-defended conspecifics, has only been widely appreciated and discussed in the last few decades. It seems reasonable to assume that almost all defended species will show some form of intraspecific variation in defence, and that this variation may sometimes be sufficient to confer protection to more weakly defended individuals that would not otherwise arise. Potential examples of this automimicry are seen in a wide variety of taxa ranging from plants to bugs, but it has been most thoroughly investigated in butterflies. The phenomenon raises a key issue—if defense carries a cost, then why are defended individuals not undermined by similar looking conspecifics that share the benefit of any defensive signal, but do not carry the cost? So far two main explanations have been offered, which may both be right. First, it is possible that some defenses do not carry costs compared to not having them and that not having the defence is a simple consequence of the environment rather than genes, such as the host plant that an individual happens to feed on. Second, predators may impose greater mortality on less well defended cheats than better-defended conspecifics, either mediated by higher attack rates or lower survivorship. The evidence for both sets of explanation is currently rather limited, and automimicry is a subject which deserves much more detailed investigation.

# CHAPTER 13

# Deflection and startling of predators

Chapter 10 focused on sensory mechanisms by which prey can gain advantage by increasing the likelihood that predators misidentify them as belonged to another (unattractive) species: Batesian mimicry. We now deal with another group of mechanisms that may lead to sensory deception of a predator. Whereas Batesian mimicry works to reduce the chance that an attack will be launched on a prey item, these mechanisms generally act to reduce the chance of an attack being successful.

The first of these mechanisms (which we call *deflection*) works by increasing the predator's propensity to strike at a region of the prey's body that is expendable or highly defended. The second mechanism induces a 'fright' reaction in the predator (quantified as a reduced propensity to carry through—or delay before carrying through—an attack) by producing an unexpected visual or auditory signal, and is called the *startle* mechanism. We finish by considering two other cases of sensory manipulation of predators: death feigning by prey and displays by adults that may be aimed at distracting predators from the search for their young.

## 13.1 Deflection defined

There is ample evidence (e.g. Blest, 1957; Smith, S.M. 1973*a,b*; Lyytinen et al., 2003) that predators preferentially bias their initial strikes to certain parts of prey individuals' bodies, and that this preference can be influenced by markings on the prey's body. Our interest here is in whether prey species have evolved markings that manipulate this aspect of predatory behaviour in a way that confers a fitness advantage to prey. We would expect this advantage to be observed as an increased likelihood of escaping from an attack.

Deflection of predatory attacks onto expendable or relatively invulnerable body parts is sometimes called *parasematism* or *diversion*. We first consider the empirical evidence for such deflection. We then consider how such a signal could evolve, especially if it involves a loss in crypsis to the bearer. Finally, we consider why predators have apparently not been selected to avoid this manipulation.

## 13.2 Empirical evidence for deflection

### 13.2.1 Lizard tails

The most quoted alleged example of deflection involves species of lizard that can shed and regrow their tails. Many different types of animal are considered to have body parts that can be shed when grabbed by a predator without being fatal, a process called autotomy (Cooper, 1998*b*). This body part can often be brightly coloured, contrasting strongly both with the substrate and with the cryptic coloration of the rest of the animal. This bright coloration is hypothesized to increase the likelihood of an attacking predator striking the bright body part (e.g. the lizard's tail) in preference to another, less expendable, body part. This sensory manipulation of the predator, in turn, is hypothesized to increase the prey's probability of surviving the attack.

Studies have demonstrated that reptiles that have already autotomized their tail are less likely to survive predatory attack than equivalent individuals with intact tails (e.g. Congdon et al., 1974). However, these results on their own shed little light on the effectiveness of tail coloration, since being tailless may affect the ability of prey individuals to resist and/or flee predators regardless of any deflective effect of the *coloration* of tails.

More convincing are the experimental manipulations of tail colour reported by Cooper and Vitt (1985, 1991). These studies used juvenile skinks of two species exhibiting brilliant blue tails on an otherwise black body (*Eumeces fasciates* and *E. laticeps*). These species also feature autotomy, with the tail being shed during struggles with a predator. In their first experiment, Cooper and Vitt compared unmanipulated skinks with those that had had their tail painted black. Skinks were placed singly in a small container with a predatory snake. First, Cooper and Vitt demonstrated that the blue colour influences the placement of predatory bites. All skinks were bitten by the snake on either the body or the tail. Only 50 per cent (nine of 18) of blue tailed individuals were bitten on the body; in contrast 94 per cent (15 out of 16) of black-tailed individuals, were bitten on the body. Bite placement was suggested to have important fitness consequences for the prey:

> Hatchlings bitten on the tail rolled and twisted their bodies briefly prior to autotomy, presumably attempting to escape with the tail intact. When the tail was bitten, autotomy invariably occurred immediately proximal to the position of the snake's grip. In each case, the lizard escaped while the snake held and swallowed the thrashing tail. Lizards bitten on any portion of the body other than the tail were unable to escape and were swallowed by the snake.

To separate the effect of colour manipulation from any other effect of painting (e.g. an unpleasant smell), Cooper and Vitt compared individuals with unpainted tails to two manipulated groups. One manipulation produced individuals with tails painted black; the other produced tails painted in a colour that matched the natural blue colour of the lizard's tail closely (at least to humans). All (nine of nine) black-painted individuals were bitten on the body and consumed, none escaped. Of those with blue-painted tails, 42 per cent (six of 14) were bitten on the tail and escaped while the snake ate the autotomized tail. Of the unpainted controls, which also had blue tails, 55 per cent (five of nine) were bitten on the tail and escaped. Hence, the prey's tail colour seems to have the more important effect on placement of bites than whether or not the tail is painted.

These experiments are very suggestive that bright coloration of a detachable body part can bring fitness benefits through deflection of predatory attacks, at least in a laboratory setting.

Castilla et al. (1999) created plasticine replicas of juvenile lizards of the genus *Podarcis*, placed them in typical basking positions and inspected them for beak imprints after 1 week. *Podarcis* is particularly suited for study, because different populations show quite different tail colorations. The plasticine models were of two types, one with a brown tail that blended with the rest of the animal's body, and the other with a bright green tail that provided a strong contrast to the body. Castilla et al. provide a concise summary of their results:

> Replicas with bright green tail color experienced the same rates of attack by birds as replicas with cryptic brown tails. However, the proportion of replicates that showed bill markings on the tail only was highest for green-tailed replicas. In contrast, the frequency of predatory attacks towards the head or body was similar in the two groups of replicas. Our experiment appears to support the classical prediction of the adaptive value of a green autotomic tail in lizards.

We feel that this final sentence is a little less cautious than we would prefer, as the 'support' for the theoretical prediction is not particularly strong. In truth, it would be difficult to get strong support from this methodology. The authors themselves acknowledge this weakness:

> Unfortunately, many replicas showed imprints all over their surface, such that it was impossible to determine whether the initial attack was directed towards the tail, body or head.

Behavioural displays as well as (or instead of) bright colouration may also act to deflect predators towards lizards' tails: see Section 13.5.1 for further discussion.

### 13.2.2 Tadpole tails

Caldwell (1982) suggested that dark tail coloration in tadpoles (which generally have dark heads) deflects predatory attacks away from the head to a less vulnerable region. She states that

> tadpoles are very fragile organisms and do not survive damage to the head and body or near the base of the tail. But tadpoles can survive the loss of the end of the tail and complete metamorphosis successfully.

Caldwell studied the species *Acris crepitans*, which she suggests exist as a morph with a dark tail and a morph with a mottled tail. She reports that ponds had a higher frequency of dark morphs than creeks or lakes. This is suggested by Caldwell to relate to one potential predator of tadpoles (dragonfly larvae) being common in creeks, but not the other habitats (where fish were the most abundant likely predators of tadpoles). Deflection is considered by Caldwell to be an ineffective tactic against fish, which can engulf entire tadpoles, but potentially effective against dragonfly larvae, which—being smaller—must strike at a localized part of tadpole.

The success of the dark tail was suggested (to Caldwell) by the higher frequency of tail damage detected in black morphs than in mottled ones. This represents, at best, very weak evidence for deflection: since tail damage reflects unsuccessful predation attempts, this methodology provides little information on the frequency of successful attacks on the two morphs (a similar objection has been raised to the use of beak marks in butterflies as a measure of their vulnerability to predation; see Edmunds, 1974). Setting this methodological problem to one side, one important potential confounding factor is age. Is it possible that tail coloration darkens with age, and that spawning times and/or development rates are different in lakes and creeks compared to ponds? If so, black-tailed individuals might be expected to have more tail damage just because they are older and so have been exposed to predators for longer. In support of this thesis, Figure 4 of Caldwell's paper shows a series of tadpoles selected to show the range to tail coloration in the sample; there is a clear trend in this picture of seven tadpoles towards the larger (and so perhaps, older) ones having darker tails.

Caldwell also reports on predation experiments, with equal numbers of the black-tailed and mottled-tailed tadpoles added to a container containing dragonfly larvae. Survivorship did not differ significantly between the two, although black-tailed tadpoles showed a significantly higher propensity to show tail damage at the end of the experiment. These results are difficult to interpret since, in at least some of them, tadpoles that were apparently uninjured at the end of one trial were returned to the holding tank from which tadpoles from the next trial were drawn. Since the sample size used in the statistical analysis was 100 of each 'morph', but only 'approximately 50 tadpoles of each morph were available' (Caldwell, 1982), this suggests that pseudo-replication and non-random selection are very substantial problems in this set of experiments. These experiments certainly show that dragonfly larvae will prey on tadpoles with both dark and mottled tails, but little beyond that.

In a later paper, Caldwell reports on tail coloration in the tadpoles of another species, *Hyla smithii*. 91 per cent (10 of 11) of black-tailed specimens found had damage on the black portion of the tail, whereas only 43 per cent (three of seven) of mottled-tailed specimens showed tail damage (Caldwell, 1986). This time, Caldwell does explore the possibility of the tail coloration being age dependent, and indeed suggests that 'based on this sample, pigmentation increases with age'. According to our argument above, this provides an explanation for higher damage to darker tails that does not require any differential predation, simply that older individuals are more likely to be damaged than younger ones.

We conclude that there is currently no evidence that the coloration of tadpoles' tails functions to deflect predators.

### 13.2.3 Eyespots on fish

Many tropical fish feature a dark spot at the posterior end of their bodies; these 'eyespots' have been suggested as having a deflective function. McPhail (1977) experimentally added dark spots to some members of a species that does not feature them (*Hyphessobrycon panamensis*). He then compared the fate of manipulated fish with those of unmanipulated controls in laboratory predation experiments:

> Fish with artificial caudal spots escape predators more often than the same species without spots, but the difference was not significant. However in fish captured, the place on the body where the fish were seized was shifted posteriorly by the presence of a spot.

Sadly, details of the experimental methods, data collection, and analysis are incomplete and

**Table 13.1** Results of Dale and Pappantoniou (1986)

| Experimental conditions and observations | *C. aureus (Goldfish)* | *B. conchiosus (Rosy Barb)* | *C. festivum (Flag Cichlid)* |
|---|---|---|---|
| Number of trials | 30 | 20 | 10 |
| Mean specimen size ± 2 SD standard length in mm | 41.1 ± 10.2 | 42.3 ± 9.0 | 37.4 ± 3.2 |
| Percentage strikes directed towards the head | 71 | 34 | 51 |
| * Average trial duration in minutes ± 2 SD | 1.07 ± 1.1 | 4.83 ± 0.50 | 4.12 ± 0.80 |
| percentage of test animals surviving with both eyes intact | 0 | 56 | 38 |

* Trial was stopped after 5 minutes or when both eyes had been removed

ambiguous, and so it is difficult to draw any firm conclusions from this work.

Dale and Pappantoniou (1986) explored the deflective function of eyespots using the cutlips minnow (*Exoglossum maxillingua*) as a predator. This species exhibits eye-picking behaviour, where it removes and consumes the eyes of attacked fish. Experimental details are not provided, but they compared 20 trials with single goldfishes (*Cichlasoma auratus*) as prey with 20 trials using single goldfishes with an artificial eyespot (created by injecting India ink under the scales in the tail region. For fish with eyespots 49 per cent of strikes were directed to the head; compared to 68 per cent of strikes to fish given a sham injection that did not alter their appearance. Further, the mean time taken to remove both of the prey fish's eyes was 1.2 min for unmanipulated fish and 2.3 min for fish with eyespots.

In a further series of experiments, unmanipulated goldfish were compared with two species that naturally feature eyespots (Rosy Barb, *Barbus conchonius* and flag cichlid, *Cichlasoma festivum*). The results shown in Table 13.1 are again suggestive of an anti-predatory role for eyespots. Finally the authors report 'A series of experiments with wooden models having various combinations of eye, eye camouflage, and eyespot markings have further corroborated these results', although no details are given.

Winemiller (1990) observed that

> *Astronotus ocellatus* and several other large cichlid fishes of South America exhibit bright ocelli, or eyespots, near the base of the caudal fin. *Astronotus ocellatus* sympatric with fin-nipping piranhas shows less extensive fin damage than sympatric cichlids of similar size that lack distinct caudal ocelli.

Winemiller suggests that this two-species comparison 'supports the hypothesis that eyespots reduce piranha attacks by confounding visual recognition of the prey's caudal region.' Since the two species involved differ quite considerably in behaviour and ecology, we find this argument unconvincing.

Meadows suggested an entirely novel anti-predatory function for eyespots (Meadows, 1993). His investigation of these spots on the four-eye butterflyfish (*Chaetodon capistratus*) found that they were generally oval in shape, being longer perpendicular to the main (head to tail) axis of the fish's body. Meadows suggests that predators are more successful when attacking perpendicular to their prey's longitudinal axis. Thus, a predator might expect to see a round eye. An oval 'eye' may indicate to the predator that its angle to the prey is other than perpendicular. As a result the predator may misdirect its strike or may delay its attack while it manoeuvres into the 'correct' position. This idea is currently without an empirical foundation, but is worthy of further investigation. Misleading viewers as to the prey's direction was one of the key objectives of many camouflage patterns painted on warships during the twentieth century (Williams, 2001). Alternatively, Karplus and Algom (1981) suggested that fish use distance between the eyes to evaluate whether a fish is a predator, and suggest that false eyes may function to make relatively benign fish appear to be more fearsome to potential competitors or predators. This idea remains untested.

### 13.2.4 False head marking on butterflies

Since the early nineteenth century, several naturalists have observed that the morphology and ventral colour pattern of the wings of some butterfly species create the clear impression (at least to humans) of a head at the posterior end of the resting butterfly. This impression is often strengthened by naturally occurring behaviours that tend to emphasise the false head. Tonner et al. (1993) explored the possible deflective function of false head wing patterns in the Burmese junglequeen butterfly, *Stichophthalma louisa*. This was done by scoring of wing damage on captured individuals. The outer edge of the wing was divided up into 13 sections (see Fig. 13.1), sections 11–13 were considered to be the false head area (FHA). Damage was scored as symmetrical (appearing on both wings) or asymmetrical (appearing on only one). They demonstrate convincingly that symmetrical damage occurs most frequently in the FHA, whereas asymmetric damage occurs outside this region. The interpretation of this result rests on the hypothesis that asymmetric wing damage is inflicted when the butterfly is in flight and the placement of such attacks is unaffected by wing markings, whereas symmetric damage is inflicted when the butterfly is at rest with its wings held together, and the placement of these attacks is potentially influenced by underside markings. This assumption is supported by the arguments of Robbins (1980):

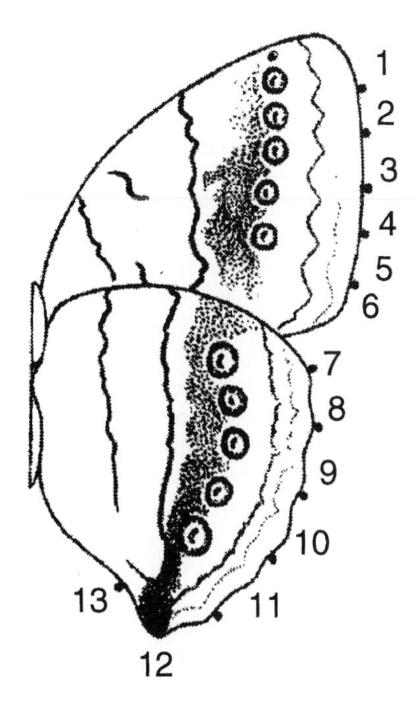


**Figure 13.1** Wing pattern on the underside of the Burmese Junglequeen, with 13 sections identified by Tonner et al. (1993).

> A number of authors consider lycaenid butterflies with the anal angle (or adjacent areas) of both hindwings broken off to be indirect evidence of a predator's unsuccessful attack directed at the 'false head', Three lines of evidence support this proposal. First, Van Someren (1922) confirmed that the unsuccessful attacks of lizards produce this kind of wing damage. Second, I marked individuals of *A. aetolus* using felt-tip markers, and monitored them under field conditions for several weeks to determine whether symmetrically missing pieces of hindwing can result from gradual wear. I found that hindwing margins gradually frayed with age, rather than breaking cleanly to produce symmetrical damage. Third, I confined six *A. aetolus* females in net bags (for an average of three days each) over plants with recurved spines on branches and both leaf surfaces (*Solanum lancaeifolim*) to determine whether sharp objects, such as thorns, might cause symmetric gaps in hind-wing margins. Although wing margins of these individuals frayed rapidly, as is usually the case with butterflies confined by net bags, I found no symmetrical damage. Thus, I conclude that the rate at which lycaenid butterflies experience symmetrical damage in their hindwings is a relative measure of the frequency of unsuccessful predator attacks.

On this basis, there seems to be some evidence suggesting that hindwings with a piece missing are suggestive of a failed attack on the butterfly, directed to the missing part of the wing. Unfortunately, another passage of Robbins (1980), directly undermines Tonner et al.'s assumption that distribution of damage on different parts of the wing can be taken as a good guide to the distribution of points of attack:

If false head wing patterns do deflect predator attacks, then the frequency of predator-inflicted damage should be greatest at the false head. Such a comparison assumes that the wings of lycaenid butterflies will break off wherever grabbed. To test this assumption, I measured the force needed to break different parts of lycaenid wings using an artificial beak apparatus. I found that the outer margins of both wings and hindwings adjacent to the anal angle break most easily, while the forewing costal vein and the area where all four wings overlap are the most resistant to breakage (more than four times stronger than the anal angle area). These results are corroborated by the incidence of beak marks (impressions of beaks on butterfly wing surfaces) on lycaenid butterflies. The majority of beak-marked individuals which I have seen had been grabbed by all four wings or across the forewing costal vein. This result indicates that wings do not break when grabbed in these areas. *Thus frequencies of wing damage to different areas of the wings cannot be used to test the false head hypothesis.* [our emphasis] However these results also indicate that, in terms of probability of escape, it is most advantageous for the butterfly to be grabbed by its false head.

On the basis of Robbin's observation above, we disagree with the conclusion of Tonner et al. (1993) that their work supports the false head hypothesis. Greater damage in a certain area is not convincing evidence that attacks are concentrated in that area. Further, siting of unsuccessful attacks may well be an unreliable indicator of the siting of successful attacks.

Robbins (1981) found that species that he considered 'classic examples of false-head butterflies' showed a higher incidence of symmetrical damage to the hindwings than species that he considered had fewer of the defining characteristics of a false head. One potential explanation of this observation is that the first group are more successful at deflecting attacks to the anal area, although—in view of his own observations quoted above—it is equally compatible with the hypothesis that false head marking do not influence attack placement but those species with false head marking are subject to stronger predation than those without such markings. Further, we must state again that we feel that inference about the siting of successful attacks on an animal from siting of damage from unsuccessful attacks is inherently unreliable.

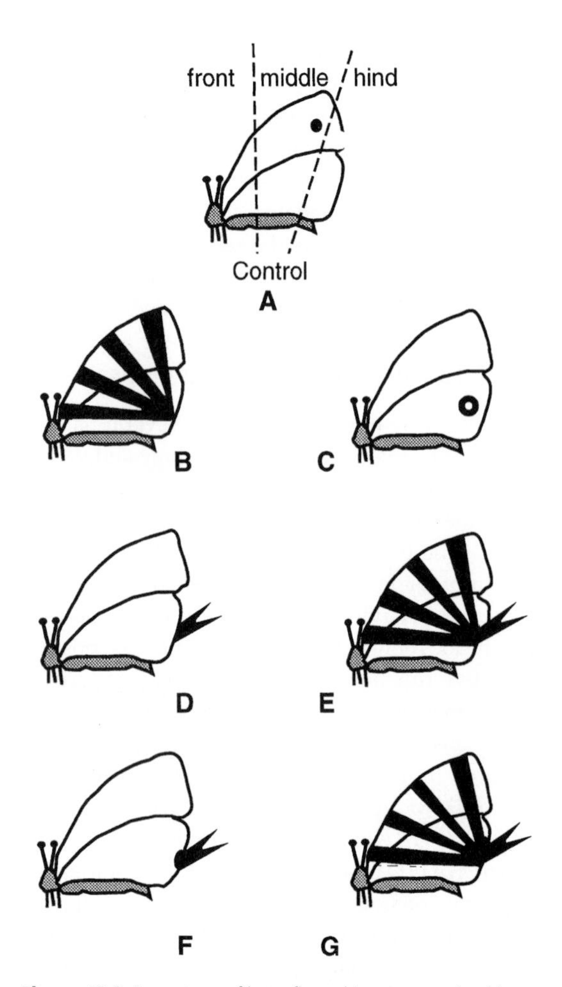


**Figure 13.2** Seven types of butterfly markings presented to blue jays by Wourms and Wasserman (1985). (a) Control showing front, middle, and hind regions used to identify strike points. (b) Lines converging in the anal angle of the hindwing. (c) Eyespot 0.5 cm in diameter. (d) Tails consisting of 1.0 cm wing segments painted and dried in place. (e) Convergent lines and tails. (f) Tails and anal spot. (g) Convergent lines, tails, and anal spot.

Wourms and Wasserman (1985) present results of an experiment that was aimed at providing quantitative information on the effects of prey colour patterns on rates of successful predation. They presented experimentally modified dead cabbage butterflies (*Pieris rapae*) to blue jays (*Cyanocitta cristata*). Six different modification types were considered, see Fig. 13.2, each representing components of the classical false head markings. Attacks by the birds were videotaped for later analysis. This analysis differentiated between the initial attack and subsequent handling of the prey item for consumption, often after the jay had carried it to a perch. For both phases, the area struck in each attack was identified as front, middle, or hind. For the initial attack phase, only the eyespot treatment (Fig. 13.2c) differed significantly in the spatial distribution of attacks from the control (Fig. 13.2a), showing a greater preponderance of strikes in the rear area (where the eyespot was marked). In contrast, during the handling phase, all six treatments differed significantly from the controls, with all showing a strong shift in strike positions towards the rear section (accounting for between 39 and 65 per cent of strikes) in comparison to only 14 per cent of strikes while handling control individuals.

A follow-up experiment introduced live butterflies singly into a cage with a blue jay. All butterflies presented were attacked. Of 79 individuals modified according to the scheme shown in Fig. 13.2g, 20 per cent escaped from the bird 'because of mishandling', compared to only 10 per cent of 104 controls. These experiments are certainly interesting, but have some significant limitations. The experiments have been criticised because the test species is very different in size, shape, behaviour, ecology, and phylogeny from the species that naturally show false head markings (Cordero, 2001). An interesting extension of the experiments with dead butterflies, would have been to have several versions of scheme (Fig. 13.2c) with the false eye in different positions, to explore if the placement of the eyespot, rather than simply its existence, influences strike placement. A control for painting, using a paint that visually matches natural coloration would also be worthwhile. The experiments with live prey are particularly interesting, but the sceptic would argue that the experimental process required to manipulate markings might have a strong influence on prey behaviour, or that the predators did not find them as palatable as the controls.

Recently, Cordero (2001) has suggested that birds actually preferentially attack the rear of butterflies and that a false head acts to deflect attacks towards the butterfly's real head. This is suggested to increase the chance that the butterfly will detect the oncoming strike and take evasive action. This hypothesis predicts that a false head should deflect attacks forward. This prediction means that Cordero's hypothesis should be experimentally separable from the traditional hypothesis. One novel experimental approach that might be useful is to use birds trained to peck at images on a computer screen, this would allow images to be changed without many of the confounders that painting introduces. However, if Wourms and Wasserman are correct, and the false head has its strongest influence during handling, rather than the initial attack, then this approach would be less useful.

It is also important to remember that there are alternative hypotheses for the functioning of false head marking. One is that false heads reduce capture success of predators by misinforming them as to the orientation of prey, and to the likely direction that prey are likely to flee in. Again, this hypothesis does not seem immediately amenable to testing using computer images.

The most thorough investigation of the possible deflective function of marking's on butterfly wings is that of Lyytinen et al. (2003). They used the species *Bicyclus anynana*, which has two forms in the wild: a wet-season form with prominent eyespots on the ventral surface, and a dry-season form with no or much reduced spots. Individuals of both types were presented singly in the laboratory to two types of predator: a lizard (*Anolis carolinensis*) and a bird (pied flycatcher: *Ficedula hypoleuca*). The lizards were also tested using a double-mutant line (*Bigeye-comet*) that features particularly large eyespots. Despite considerable statistical power, this investigation found no evidence of differential survival rates between the three types. Hence, they conclude that 'this study provides no support that marginal eyespot patterns can act as an effective deflection mechanism to avoid lizard or avian predation'. A follow-up study (Lyytinen et al., 2004) used naive hand-reared pied flycatchers, rather than the wild-caught individuals of the first study. In one of three trials like those of the first study, the birds successfully attacked and killed noticeably fewer spotted butterflies (three of 12) than unspotted ones (seven of 10). Although small sample sizes prevent this from being conclusive, this does hint that eyespots may have some anti-predatory function against naïve predators.

**Table 13.2** Results of Powell (1982)

Number of times a hawk missed a fake weasel of the designated size-spot morph

| Size-spot morph | Hawk 1 | Hawk 2 | Hawk 3 | Total |
|---|---|---|---|---|
| Long-tail no spot | 1 | 1 | 0 | 2 |
| Long-tail tail spot | 11 | 4 | 9 | 24 |
| Long-tail black spot | 2 | 0 | 2 | 4 |
| Short-tail no spot | 9 | 7 | 9 | 25 |
| Short-tail tail spot | 1 | 0 | 1 | 2 |
| Short-tail black spot | 3 | 0 | 0 | 3 |

Note: Each kind of fake weasel morph was presented 12 times to each hawk

### 13.2.5 Weasel tails

Powell (1982) hypothesized that the black tip to the tail of a stoat (*Mustela erminea*, sometimes called an ermine) acted to draw attacks from potential avian predators towards the tail, which is a smaller target than the stoat's body and more easily missed by the predator. He further hypothesized that weasels (*Mustela nivalis*, sometimes called 'least weasels'), being smaller than stoats and having shorter tails, do not have a black tip to their tail because this would be too close to the body to provide any advantage. These hypotheses were tested using three captive hawks trained to attack targets that moved rapidly across the floor of their enclosure. Six types of targets were used, three of similar size and tail length to a stoat and three to a weasel. One of each set of three was all white, one was white with a black tip to the tail and one white with a black band on the body (see Table 13.2 and Fig. 13.3). Each hawk was presented with each of the six models on 12 occasions, in random order. On every occasion the hawk attacked the moving target, but sometimes missed (see Table 13.2). Of the larger long-tailed targets, each hawk was much more likely to miss the model with the tail spot than the ones with no spot or a band on the back. Conversely, for the smaller, short-tailed models, they were each much more likely to miss the spotless model than either of the other two. These surprising but clear-cut results very much warrant further investigation, especially with regard to other mammals with black-tipped tails.

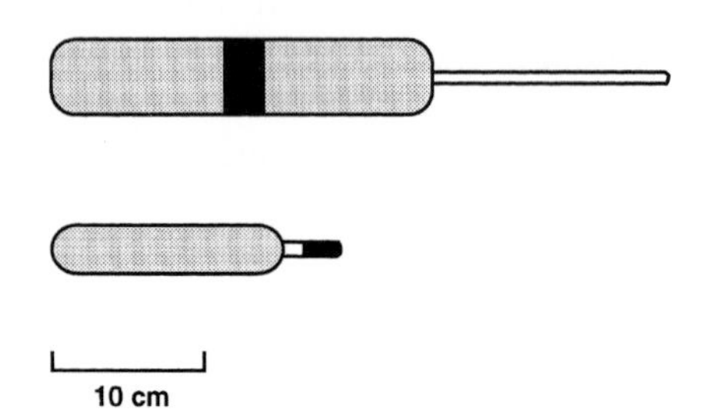


**Figure 13.3** Examples of two fake weasel six-spot morphs: long-tailed model with a black spot, short-tailed with a tail spot.

### 13.2.6 Summary of empirical evidence for deflective signals

There seems good evidence that brightly coloured autotomizable tails in reptiles can have a deflective effect, at least in laboratory experiments. No comparable evidence is available for any other prey type: although there are sufficiently tantalising results for eyespots in fish and in butterflies, and for tail coloration in weasels, to warrant further exploration. We will argue in the next two sections that deflective signals seem evolutionarily feasible to us. We suggest that the current paucity of empirical evidence for such signals should not be taken as indicative of their likely scarcity in nature. Rather, it reflects lack of scientific exploration of the phenomenon, and a lack of rigour in some of the studies that have been carried out. We hope that our discussion of the case studies above will help in designing more effective future studies.

## 13.3 How can deflective marking evolve if they make prey easier for predators to detect?

One potential drawback to brightly coloured deflective signals is increased detection by predators. Cooper and Vitt (1991) explored this with a simple model, reproduced below.

Imagine that an individual with a cryptically coloured tail is detected by a predator with probability $P_d$. After detection, the predator attacks and the probability of the prey escaping this attack is $P_e$. If these two probabilities are independent, then the prey individual's probability of being captured by a predator is

$$P_d(1 - P_e).$$

We assume that having a conspicuously coloured tail increases the probability of detection by $\alpha$, but also increases the probability of escaping by an amount $\beta$.

Conspicuous tail coloration will be favoured if the inconspicuous type is more likely to be captured by the predator: that is, if

$$P_d(1 - P_e) > (P_d + \alpha)(1 - P_e - \beta).$$

This rearranges to the condition

$$\beta > \frac{\alpha(1 - P_e)}{P_d + \alpha}.$$

Hence, conspicuousness will be favoured if $\beta$ (the advantage that is gained from conspicuousness) is large or $\alpha$ (the increased probability of detection due to conspicuousness) is small compared to $\beta$ and $P_d$ (the probability of being detected in the absence of the conspicuous signal). Also unsurprisingly, a large probability of detection in the absence of conspicuous coloration favours selection of the conspicuous signal. More interesting is that a high probability of escape in its absence (high $P_e$) favours the evolution of the conspicuous deflective coloration. This suggests that tail autotomy—and perhaps associated behaviours that draw attention to the tail—(increasing $P_e$) probably developed before the conspicuous coloration of these body parts in some species. More generally, this model demonstrates that if deflective markings cause an increase in the rate at which their bearer is attacked, this need not necessarily mean that such markings will not be selected for. They can still be selected for, providing their enhancement of probability of escape from an attack is sufficient to compensate for this potential cost.

## 13.4 Why do predators allow themselves to be deceived?

The empirical evidence above suggests that some prey may be able to reduce their risk of being captured in a predatory attack by inducing the predator to attack specific parts of their body. Why do predators allow such deflection to occur, if it costs them prey items? We hypothesize that deflection occurs because of lack of familiarity with the prey type. Hence, we would predict that deflective markings would be relatively unsuccessful when used by prey species that the predator attacks frequently, compared to prey items that it attacks infrequently. Specialist predators should not be fooled by deflective markings, whereas generalist predators have to accept such costs as a by-product of having evolved to be able to handle diverse prey types. The generalist predator may find deflective marking difficult to combat in one species encountered infrequently if similar visual cues are useful when attacking a different species that are encountered more frequently. This argument may provide a theoretical framework for consideration of why some styles of signal will be more effective at deflecting than others. It also raises the testable hypothesis that prey that use deflective signals will generally not be the main prey of predatory species that they successfully deflect. We might also expect to see predators habituating, such that their probability of being fooled by deflective marking declines with increased exposure. Whilst this has been demonstrated repeatedly for startle signals (see next section), it has not been explored for deflective signals. However, predators of reptiles that shed their tails upon attack may not be under strong selection pressure to stop 'falling for this trick', since they do end up with a substantial meal from the tail, particularly as tails are often used as fat stores.

## 13.5 Startle signals

We consider the general features of signals designed to make a predator hesitate in its attack, before considering a number of proposed examples of such startle signals.

### 13.5.1 General considerations

Edmunds (1974) defines startling signals (also referred to as frightening or deimatic signals) as follows.

> Deimatic behaviour produces mutually incompatible tendencies in a predator: it stimulates an attacking predator to withdraw and move away. This results in a period of indecision on the part of the predator (even though it may eventually attack), and this gives the displaying animal an increased chance of escaping.

The classic example of a startling signal is suggested to be the bright colours that some otherwise-cryptic moths can be induced to display by disturbing them whilst at rest. The moths have brightly coloured hind-wings that they generally cover with cryptically coloured forewings, but which can be voluntarily and suddenly revealed (Blest, 1957).

One key feature of startling signals is that they are induced by the proximity of a predator. It is generally postulated that the sudden appearance of a bright display or loud noise induces an element of fear or confusion in the predator, giving the prey individual an increased chance of fleeing before being attacked. Although often postulated, this possible mechanism has subjected to little rigorous experimental testing. Importantly, the fact that a signal is induced by proximity of a predator, is not of itself sufficient reason to identify it as startling. For example, the signal may function aposematically, but the prey may have evolved not to display its aposematic signal constantly if the trade-off between its costs and benefits change temporally. Thus, there may be an advantage to crypsis to the prey individual when faced with some predators, but to aposematic signalling when faced with other types of predator. Hence, the prey has evolved to be cryptic unless it encounters a predator of the latter type, when it switches to reveal an aposematic signal. Other factors that might lead to the evolution of an aposematic signal that is only used when an appropriate predator is detected are the expense of the signal production, the demands of social signalling and the demands of crypsis as an aid to capturing its own prey. Similarly, a deflective signal may only be used when a predator is detected if the deflection signal is expensive to produce or conflicts with other signalling.

The idea of aposematic signals that can be used only when a predator is detected, has been given little previous consideration (but see Guilford and Cuthill, 1989). The drawback to this potential adaptation (compared to a continuously deployed aposematic signal) would be reduced effectiveness of the aposematic signal, if the prey individual risks the predator attacking before it has had a chance to deploy its signal. Even if the prey is defended, these attacks could have a fitness cost. However, this cost may be more than compensated for by reduced detection rates by predators because a cryptic signal can be used most of the time. This idea warrants further exploration, particularly for prey faced by a suite of predators, with varying responses to an aposematic signal. The theory of Cooper and Vitt (1991), highlighted in Section 13.3, should be a useful tool for such explorations. Cooper (1998*a*,*b*) explores the use of displays of the autotomous tail of a lizard that he considered to enhance deflection, but which are sometimes used when no predator has been detected. Cooper argues that such 'anticipatory displays' can be advantageous in circumstances where the fitness benefits from deflecting ambushing predators that have already detected the prey individual (but which the prey item has not detected) outweighs the cost of alerting predators that had not previously detected the prey individual. This argument is backed up with a quantitative model and with empirical observation that the displays commonly occur immediately after travel to a new location (when risk of ambush by an undetected predator is high), but stop soon afterwards (as the probability that a predator is near enough to ambush but has not yet been detected by the prey declines).

Other expected features of startling signals are that they only have a transitory effect on the predator, and that the signaller must flee immediately after giving the signal. This gives a way to potentially differentiate between the alternatives based on aposematic signalling suggested here and startling signals. Signallers that do not flee very soon after signalling can much more readily be explained by our alternatives outlined in this paragraph than by a startling effect on the predator. Similarly, we would expect predators to habituate to a startling signal, but not to an aposematic one.

### 13.5.2 Distress calls as startle signals

Many vertebrates are known to emit vocalizations when in utmost danger (often called fear screams or distress calls). Many functions have been postulated for these, but the most comprehensive study of a possible startle function to these distress calls is that of Conover (1994). He allowed opossums and racoons to attack a caged bird (a starling) and noted their response after avian distress calling was broadcast through a loudspeaker. Of the five opossums, three made no noticeable response to the distress call, one jumped back and released the birdcage at the onset of the distress calling, the other released the bird cage and retreated after 15 s of distress calling. Of the 10 racoons, four showed no response to the calls, two jumped back and released the birdcage at the onset of the calling, two did so after over 10 s of broadcast calling, and two appeared to attack more vigorously at the onset of calling. Little can be concluded from these results except perhaps that predators may sometimes respond to distress calling by releasing the prey.

More interesting are observations of distress calling by birds held by Conover himself. For 10 of the 11 species tested, birds distress-called while simultaneously moving their limbs in an attempt to struggle free. This pairing of calling with struggling is more consistent with the startle hypothesis than with any other commonly cited explanation for distress calling (e.g. warning kin, calling for help to conspecifics, or seeking to attract other predators to challenge the one currently holding the caller) although they are not mutually exclusive. Unfortunately, these interesting observations are not backed up with quantitative data, but certainly warrant further investigation. Also consistent with the startle hypothesis is Conover's observation that calls were more frequent when the bird was held by the wing or leg compared to the head or body, and so might be more able to break free. This observation is backed up by quantitative details, which do show a convincing effect. However, before Conover's observations can be held up as evidence of a startle function to distress calling, a demonstration that birds are physically able to call when held by the head or body would be useful. Exploration of the use of distress calling by birds held by humans are interesting, but they do not hold the promise of being as conclusive a test of the startling function of such calling as experiments with natural (or near-natural) predation events.

Conover (1994) raises a concern about the hypothesis that distress calls have a startle effect. We quote the relevant passage below:

> The strongest evidence against the startle hypothesis is Högstedt's (1983) finding that avian species in open habitat distress called less than species from moderately-exposed habitat, and they in turn distress called less than species that lived in dense cover. I am unable to reconcile this with the startle hypothesis.

We do not understand Conover's concern. Whilst this observation does not follow naturally from the startle hypothesis, it is at least equally as challenging to reconcile with any of the other proposed functions of distress calling.

Wise et al. (1999) explored the effect of broadcasting starling distress calls on single coyotes that had just been presented with a chicken carcass. On first exposure, the majority of the 30 coyotes tested showed a startle response, resulting in less rapid and vigorous attacking of the carcass compared to control treatments without the call. The coyotes, however, quickly habituated to the calling and did not respond with a startle in further treatments with the call.

Neudorf and Sealy (2002) reported on the distress calling of birds caught in mist nets.

> If calls startle the predator, then we would expect a correlation between calling and struggling behaviour; however we found no such correlation. In attempting to startle a predator into releasing the caller, the caller is expected to struggle and call simultaneously . . . but most individuals that we handled struggled regardless of whether they called or not.

We do not find this convincing evidence against the startle hypothesis. Under the startle hypothesis, we might expect birds to struggle when they call, and it appears that Neudorf and Sealy's birds did just that. The hypothesis does not suggest that birds should not struggle when they are not calling. We also find their other conclusion less than convincing.

> We found a significant correlation between body size and calling frequency [i.e. rate of calling] that supports the predator startle hypothesis. Larger species may more effectively startle the predator because their calls are higher in amplitude.

They present no evidence that amplitude of call is related to either caller size or startling effect, and we do not see why either of these assumptions should be self-evidently true.

Conover argues that startling is only one of several tactics that a bird could adopt to free itself from a captor. Alternatives are feigning death in the hope that the captor will loosen its grip (see Section 13.6), or trying to inflict injury on the captor. These three tactics cannot be employed simultaneously, and use of one reduces the time available for using the others. Hence, these tactics should not be used indiscriminately, and we should not expect distress calling to be universal. However, in conclusion, there is some evidence that distress calls could have a startling effect on predators, although no study can be considered highly convincing.

### 13.5.3 Visual startle signals

The classical study of startle coloration is by Blest (1957). He used a species of Nymphalid butterfly (*Aglais urticae*) which, when disturbed, open their wings abruptly exposing a bright dorsal pattern that was previously hidden and contrasts strongly with their normally cryptic appearance. Blest compared the reaction of bird to unmanipulated butterflies and individuals that had been treated as follows.

> The scales may be easily removed from the upper surface of the wings by gentle friction without injuring the butterflies. Insects so treated displayed normally but the surface that they expose is a dull colour. Birds could be induced to show an escape response when the untreated butterflies displayed but not when the treated ones did.

Sadly, we are not able to draw very much from Blest's work, as full details of these experiments are not provided. However, he does seem to show that inexperienced birds showed a startle response to intact butterflies that they did not show to modified ones—that had had their spots removed. However, it is vital to demonstrate convincingly that this treatment did not affect the butterflies in any way other than in the visual appearance of their wings. As a first stage of this, comparison of manipulated butterflies with a control group that experienced the same process of capture and handling but that did not have their spots removed would be instructive.

Further work supportive of Blest's conclusions comes from the study of Cundy and Allen (1988). They produced a mechanical butterfly that could be triggered to reveal its hindwings under the experimenter's control. This was mounted on a bird table and triggered when a free-living bird approached. One version of the model had eyespots on the hindwings, the other was identical except for the lack of eyespots. Revealing of the hindwings on the eyespotted model induced a flight response by the target bird on four of eight occasions; whereas the control model induced flight on one of five occasions.

Cundy and Allen also performed similar experiments with a model where the experimenter could induce illumination of markings on a model butterfly when a bird approached: there were four model types: where the marking was an eyespot, a circle a cross or a non-illuminated eyespot control. Again, the response of approaching free-living birds were observed: the results were as in Table 13.3.

**Table 13.3** Birds taking to flight, or not, in response to the four stimuli of Cundy and Allen (1988)

| | Eyespot | Circle | Cross | Control |
|---|---|---|---|---|
| Flight | 5 | 2 | 2 | 0 |
| No flight | 16 | 18 | 17 | 12 |

Although the sample is too small for statistical testing, Cundy and Allen's results are suggestive of a startle function to eyespots. Their sample sizes were small due to mechanical difficulties with their equipment, but further development of their approach is warranted.

Convincing evidence of a startle effect comes from experiments reported by Ingalls (1993) with hand-reared blue jays. These birds, which were inexperienced with warning coloured aversive prey, took significantly longer to touch novel colours that possessed bold black bands compared to novel unbanded colours, when discs of these two types covered a food reward. From this, the author concludes that being conspicuous, as opposed to simply being novel, appears to enhance a startle reaction. The birds eventually habituated, such that they showed no delay in touching the initially startling disks.

Ingalls builds on earlier works by Vaughan (1983) and Schlenoff (1985). Schlenoff presented artificial moth models to blue jays. The models had patterned hindwings that were concealed behind grey forewings until a bird removed the moth models from a presentation board, when the hindwings were suddenly revealed. The underside of the models featured a food reward as well as a clear view of the hindwings. Birds that had been trained on models that had grey hindwings showed a startle response (delay before taking the food reward) when exposed to models with brightly coloured hindwings. This startle response declined with repeated experience. Interestingly, the reverse treatment of birds trained on models with brightly coloured hindwings, did not show a startle response when exposed to models with grey hingwings. This suggests that conspicuousness, rather than simply novelty or oddity is required to trigger a startle response, agreeing with Blest's observations reported earlier. In a further experiment, Schlenoff found that birds that had habituated to one conspicuous hindwing pattern still showed a strong startle preference when another pattern was presented to them. This suggests that birds habituate to a specific stimulus, rather than reducing a generalized startle response. Another of Schlenoff's experiments demonstrated that a familiar hindwing pattern that did not induce a startle response when associated with a one familiar forewing pattern, would produce a startle response when associated with another forewing pattern, which was familiar to the predator but which had previously been associated with a different hindwing pattern. This suggests that a familiar signal when presented in an unexpected context can produce a startle response.

In the experiments of Vaughan (1983), blue jays were trained to push flaps aside to reveal a coloured disc that had to be removed in order to obtain a food reward. Novel coloured discs or colours that the bird have experienced infrequently in the past produced a startle response (measured as a latency from flap removal to food consumption), compared to the time taken in reference trials using discs of the colour that the bird had been habituated to in training.

Together, these works demonstrate convincingly that startle responses can occur in a laboratory setting, the next challenge will be to demonstrate that they occur in a more natural setting, that prey are adapted to enhance the advantage that they get from startle responses and that prey do gain a fitness advantage from startle responses by their predators.

Experiments like those of Cundy and Allen (1988) suggest that eye-like patterns are particularly 'startling'. We agree with Guilford (1992) that 'perhaps the extreme danger associated with the sudden appearance of a pair of eyes forces animals to generalise widely to react to all sorts of eye-like stimuli even if they are not in their appropriate context.'

### 13.5.4 Sound generation by moths attacked by bats

It is now clear that several types of insects can detect imminent attack by echolocating bats and modify their behaviour accordingly (Miller and Surlykke, 2001), often by abrupt changes to their flight-path. Some moths and butterflies modify their behaviour by producing clicking sounds. Several (non-exclusive) mechanisms have been postulated to explain this: these clicks may interfere with the bat's sonar, reducing the chance that the bat can accurately find the prey, they may be an aposematic signal, and/or have a startling function. Our interest here is in whether there is any evidence for this last function. However, we note that many of the species that engage in clicking, such as arctiid moths, are also unpalatable which strongly suggests a function as a warning signal. If true, then it would be the type of selectively used aposematic signal described in Section 13.5.1, since the Lepidoptera only signal when the bat is close.

Evidence for a startle function rests on results presented by Miller (1991) that both of two bats tested modified their behaviour in a way consistent with abandoning an attack in laboratory experiments when they were first exposed to a recording of prey acoustic signalling. Both bats subsequently showed habituation to these clicks. Unfortunately, although intriguing, there is currently insufficient data to be confident that moths signal to startle bats in the wild.

### 13.5.5 Summary of empirical evidence

Although the work of Vaughan (1983), Schlenoff (1985), and Ingalls (1993) demonstrate that predators can behave in a way that suggests that startle signals should be possible, we have found no convincing evidence that real prey actually use such signals. This may simply reflect a lack of recent interest in this area, since the next section will outline that we consider startle signals to be at least logically plausible.

### 13.5.6 Why would predators be startled?

It seems possible that the startled predator is misidentifying the prey item as something that could be a threat to it, rather than as something that is merely an unattractive food item.

A useful analogy can be drawn with prey animals that can often be seen to take anti-predator action in circumstances where a human observer can identify no apparent threat. For example, studies of overwintering wading birds found that flocks would sometimes take flight in response to the approach of harmless birds such as curlews, as well as potential predators, such as peregrines (Cresswell et al., 2002). Indeed, such false alarm flights occurred more often than actual attacks. The flocking birds are highly energetically stressed, and such 'false alarm' flights are costly both in the considerable energetic expenses of taking flight and flying and in lost feeding time. Hence, one would expect such prey to face strong selection pressure to minimize the occurrence of mistakenly identifying a non-predator as a predator. The reason that false alarms are still commonplace is likely to be that prey also face a strong selection pressure to avoid mistakenly identifying a predator as a non-predator, and that these two selection pressures conflict, forcing the prey to evolve a cognitive mechanism that achieves the optimum trade-off between the two types of error.

Predators, similarly, may be presented with conflicting selection pressures acting on their response to unexpected stimuli. A startle response may represent the best compromise between the costs of less efficient prey capture (because of the delay in attacking allowing prey to escape) and the cost of failing to respond rapidly to unexpected and imminent danger. However, we would expect predators to evolve mechanisms that allow them to habituate to startle signals from harmless prey. That is, we would expect predators to react to startle signals not by completely fleeing the scene, but by retreating to as safer distance from which they may be able to assess the potential threat. Fleeing would deprive it of the ability to learn about the other organism (except in as much as the other organism did not successfully pursue it). Thus, the behaviour of a predator towards a startle signal should not be fixed, but rather should be responsive to that individual's experiences subsequent to previous exposures to similar signals. Such cognitive processes should be suitable to theoretical investigation, similar to those that have help to illuminate phenomena of aposematic signalling and mimicry.

## 13.6 Tonic immobility

Many animals when physically restrained adopt a relatively immobile state that can last after the physical constraint has been released. Sometimes such a behaviour involves tongue protrusion, and setting the eye's wide open, very reminiscent of a dead individual. This phenomenon has therefore been called *death feigning*, *playing dead*, *animal hypnosis*, and *thanotosis*, but the most common term is *tonic immobility* (TI). It is quite distinct from immobility used to reduce the risk of detection or tracking by a predator, since it acts later in the sequence of a predation event. TI is widely interpreted as a last-resort defence against predation after active physical resistance has proved unsuccessful. The lack of movement is suggested to inhibit further attack by the predator, and reduce the perceived need of the predator to continue to constrain the prey. This suggested adaptive function to TI is supported by many anecdotal works, but very limited purpose-designed experimentation. Here we briefly evaluate these works.

Arduino and Gould (1984) performed experiments where domestic chicks (*Gallus gallus*) were exposed to some stimulus for 15 s, then 15 s of being manually restrained whilst still being exposed to the stimulus. This induced TI, the chick was released and the duration of TI was then measured. They found in numerous paired comparisons that TI lasted longer in situations where it was thought that the chick would consider the chance of escape to be lower (e.g. when the stimulus was a hawk model facing the chick), compared to cases when this chance might be considered to be higher (when the hawk model was facing away from the chick). The authors conclude that TI is adaptive since these chicks demonstrate that they emerge from it more quickly when the chance of subsequent escape is perceived to be higher. This conclusion requires that the chick and the experimenters coincide in their interpretation of the different stimuli, which we consider to be by no means certain. But even if experimenters and chicks concur, these results do not demonstrate that entering TI provides a fitness benefit.

There have been several similar studies providing similar indirect evidence interpreted as supportive of an anti-predatory function to TI. Ewell et al. (1981) demonstrated in a laboratory that rabbits (*Oryctolagus cuniculus*) decreased the duration of TI as proximity to a human decreases or as proximity to their home cage increased. Gallup et al. (1971) found that chickens immobilized by manual restraints in the presence of a hawk model exhibited longer durations of TI than chickens immobilized without the presence of the model. Hennig et al. (1976) found that TI was of shorter duration in birds that had been immobilized near bushes than those immobilized in the absence of nearby protective cover. O'Brien and Dunlap (1975) reported that crabs immobilized on a sandy substrate (in which they could seek cover by burying themselves) exhibited shorter TI durations than individuals similarly immobilized on a hard surface. Whilst suggestive, these types of study fall short of being definitive. With some reluctance, we suspect that the adaptive significance of PI can only really be effectively probed in situations where real predators are involved.

Thompson et al. (1981) introduced a domestic cat (*Felis domesticus*) into an arena containing two Japanese quail (*Coturix coturix japonica*). One quail had been induced into a state of TI by constraint by a human, the other was unrestrained and mobile. Four different cats were used, each being used in four trials. In 14 of 16 trials, the first bird attacked by the cat was the mobile individual. In a further experiment a cat was introduced into an arena containing two mobile, unconstrained quail (again each of the four cats was used four times, each with two fresh quail). The cats was allowed to attack the quail, attacks by the cat were not always fatal, and TI was induced on 18 occasions. The authors report that,

> After TI was induced in a bird, the cat would leave it to stalk the remaining moving bird. A bird would be re-attacked if it came out of TI during a trial. In one case, a bird was re-attacked, killed and partially eaten after it came out of TI and moved while the second bird was still in TI.

Whilst none of these works are definitive, taken together they are strongly suggestive that TI does have a functional basis in reducing the probability of an attack ending in death. Given that more definitive work would almost certainly involve considerable animal suffering, we are not disposed to call for further exploration of this topic. However, one topic that we do feel requires further explanation is why TI causes the predator to reduce its attacking. Our suggestion is that the TI is effective against predators that commonly face a situation where they can subdue more than one prey item in quick succession. The predator faces a trade-off, since the more time it spends making certain that the first prey item is dead (or sufficiently injured that it cannot escape), the more likely it is that other nearby prey items will make good their escape. TI works by inducing the predator to switch prematurely from one prey item to the next, and the trade-off may make it maladaptive for the predator to invest time in making absolutely sure that the first item is dead. In order to explore this, we suggest that it would be useful for a systematic study of anecdotal accounts of the observation of TI to explore the percentage of occasions when the predator had the chance to attack another prey item soon after releasing the individual exhibiting TI. Our prediction would be that this percentage will be high. Additionally or alternatively, it would be worth exploring whether individuals of species exhibiting TI occur commonly in groups. TI can still be effective for single prey individuals, if the predator concerned has evolved to release prey quickly, because it often encounters prey in groups, against which quick release can aid in making multiple kills.

## 13.7 Distraction displays

It is well known that many species of ground-nesting birds exhibit conspicuous displays in response to approaching predators. These displays are generally interpreted as functioning to distract the predator, reducing the chance of the bird's offspring being discovered and predated. This area has been thoroughly reviewed by Gochfeld (1984), who commented that it was difficult to think of another field of enquiry where the ratio of anecdotal accounts to scientific study is as great as in distraction behaviour. This remains true today, although the lack of manipulative studies seems to us less problematic for this topic than for others discussed in this book. It seems difficult to interpret the wealth of information collected by Gochfeld and others using any other explanation that the conventional one of distracting the predator away from the animal's offspring. Gochfeld demonstrates that individuals use such behaviour flexibility: not using it when a nearby predator is moving on a course that will cause them to pass harmlessly by. Similarly, the behaviour is suppressed if the predator has approached so closely that the bird could not creep away from the nest undetected before initiating the display. Gochfeld lists several instances where the predators substantially modify their trajectory in order to move towards such a displaying bird. Displays are often apparently designed to make the bird look an easier target than usual: for example, by apparently feigning injury or mimicking a small rodent, in either case potentially suggesting that the adult bird is an easy target that cannot simply fly away if attacked. All these points are strongly supportive of the classical distraction interpretation.

As usual, we have to ask why predators allow themselves to be fooled in this way (if indeed it is a matter of fooling—the adult, if it can be caught, may make a better meal than eggs or chicks—Gochfeld (1984) cites several cases where the predator was able to successfully capture a bird exhibiting a distraction display). Of course, not all birds behaving as if they have a broken wing are faking it, and a predator that ignored such signals would forgo easy meals. However, there will be instances were a fox has a number of such breeding birds in its territory and encounters distraction displays much more commonly than genuinely injured birds. We might expect such a fox to become habituated to such displays and ignore them, or even use them to alert it to the close proximity of a nest. In support of this, Sonerud (1988) reports observations where foxes were not drawn to displaying grouse but did change their behaviour in a way that was interpreted as enhanced searching for a nest.

Distraction behaviour has also been reported in fish. Sticklebacks can show parental care, with males constructing a nest that they defend against conspecifics, which can show egg cannibalism.

There is a body of evidence (e.g. Whoriskey and FitzGerald, 1985; Ridgway and McPhail, 1987; Foster, 1988; Whoriskey, 1991) that suggests that when approached by a group of conspecifics, the male may use deception rather than aggression. This deception involves moving away from the nest and simulating feeding in the substrate, such apparent feeding behaviour attracting the nearby group away from the proximity of the nest. Whoriskey (1991) demonstrates that males with empty nests as well as those defending eggs use this behaviour. He thinks it unlikely that nesting materials were being defended, hence this aspect of behaviour requires further consideration: especially since Ridgway and McPhail found that males with empty nests did not show distraction behaviour.

## 13.8 Summary

All four mechanisms for deceiving predators discussed in this chapter seem plausible to us, but each has suffered from recent neglect and a consequent lack of data that meets modern standards of scientific rigour. It was surprising to us to find the flimsy empirical foundations on which our understanding of these widespread mechanisms currently relies. We hope that we have managed to identify new angles of approach to each of these topics that will help stimulate much needed further work.

CHAPTER 14

# General conclusions

The study of animal coloration and associated anti-predator adaptations has a long history. Although we have not emphasized historical development, we hope that a general theme evident throughout this book is that this field of research has been blessed from the earliest years with the insights of particularly gifted scientists. The writings of Wallace, Bates, Müller, Poulton, and Cott truly stand up to the test of time: These individuals deserve even better renown not just as great natural historians but as exceptional scientists too.

Despite the long history of many of the topics addressed in this book there was, surprisingly, no field of enquiry where we felt that all the pertinent questions have been addressed. In some fields of research we found that textbook examples of anti-predator adaptation stand on very limited empirical foundations: countershading and the startle function of eyespots are two notable examples. Over the last two or three decades academic interest in antipredator adaptations has been somewhat heterogenous in its application and considerable effort has been focussed on an understanding of warning signals and mimicry. We would be delighted if this book contributed to renewed interest in other areas that have been unjustly neglected.

In any text like this there is some danger of compartmentalization, and we would not want our chapter division in this book to encourage compartmentalization of thinking on the issues discussed. Indeed, we need to encourage the opposite. If there is an overriding theme to this book it is that the benefits of different anti-predatory adaptations have received much more consideration that the costs; but, of course, both are needed for a full understanding. In particular, we need to develop a comparative cost–benefit framework that will allow development of a general understanding of the distribution of anti-predatory sensory mechanisms throughout the natural world. In understanding why some hoverflies appear to evolve to be wasp mimics, we must be able to quantify not just the costs and benefits of this adaptation, but also those of alternatives such as crypsis. We must also seek to understand why this species has not evolved secondary defences of comparative efficacy to the wasps it resembles. These are challenging questions. But they are within our compass, and we hope that this book has suggested that investigation in this area is tremendously rewarding.

We have utilized mathematical modelling extensively throughout this book, and it is always important to remember that models are a simplification of reality. One important simplification to which we would like to draw attention is the assumption that the ecosystem under discussion consists of a single predatory species (sometimes a single predator!) and one or perhaps two or three prey species. This is a great simplification of reality, since most species exist in a complex food web. This complexity will have ramifications for the predictions of simplified models. For example, many species are both predatory themselves and are preyed upon by others. Hence, it may be that, in considering their relationship with their predators, we might expect certain species to have evolved aposematic patterning, but it may have remained cryptic because of the requirement to sneak up on its own prey undetected. A further complication to this story arises if weapons developed to aid this species in the subjugation of its own prey also affects its ability to defend itself against predators. We can expect further consequences of complex food webs. Many prey will face a suite of different predators, and there may often be compromises between the best defenses against each predator type. Similarly, many predators have broad diets, and we have already touched upon several cases when the availability of alternative prey types has a strong effect on the relation between a predator and a given prey species. We hope that the coming years will see the development of the ideas in this book into a framework that takes more explicit account of the complexity of ecosystems. Some of the ideas discussed here will be relatively unchanged by this development, but some will undoubtedly require considerable revision. In no sense does this book tie up the loose ends of our chosen topic, rather we hope it shows the way to expanses of yet-to–be-explored territory.

# Appendices

**Appendix A** A summary of mathematical and computer models that deal with Müllerian mimicry

| Study | Structure of model | Example conclusions | Comments |
|---|---|---|---|
| Müller (1879) | Predators in an area need to sample *n* items of a given morph before learning that it is distasteful. | Mimetic prey should benefit according to the reciprocal of their squared abundances. | A "*truly epoch-making paper*" Dixey (1908) |
| Blakiston & Alexander (1884) | As Müller (1879) | Mimetic prey should benefit only approximately according to the reciprocal of their squared abundances. | A question of how one measures the per capita success of a species (indeed a hypergeometric approach with reducing densities of prey as they are sampled may provide an even better approximation, especially for prey at low densities). |
| Marshall (1908) | A simple arithmetic case (not a detailed model) along the lines of Müller (1879) | If two unequally unpalatable species inhabit the same region then predation will lead the less numerous to resemble the more numerous but not vice-versa. | The first of many controversial theory papers on Müllerian mimicry. In fact, recent research has increasingly considered the evolutionary dynamic as one of advergence rather than convergence (Mallet 1999 [2001*a*]) |
| Fisher (1927) | A largely verbal response to Marshall's theory, emphasizing the importance of considering intermediate states. | A mutant of a common species which resembled a rare one, may not necessarily lose all of its advantage if it retains some resemblance to the common species. | Largely repeated in Fisher's classic (1930) text. Dixey (1908) had a similar argument, suggesting that an intermediate mutant may gain the benefits of both worlds (*cf* the multi-model approach of Edmunds 2000 applied to Batesian mimicry). |
| Holling (1965) | A detailed mechanistic model involving attack thresholds and learning and predicting domed functional responses for unpalatable prey (changes in predator attack rate with prey density) | Supports Müller's theory but suggests that the phenomenon is more complex than that envisaged by Müller, as a result of arguably more realistic predatory assumptions | Hunger (mediated by the absence of alternative prey) "*destroys the advantages of Müllerian mimicry for prey*" (page 39). These arguments have been rekindled 40 years later (Kokko et al. 2003; Sherratt et al. 2004b). |

| | | | |
|---|---|---|---|
| Huheey (1976) | An extension of Huheey (1964) in which attacking unpalatable species causes both species to be avoided for $n_i$ subsequent encounters. | When $n_i \neq n_j$ then the less unpalatable species will always act as a Batesian mimic (parasite) on the more unpalatable species, increasing the attack rate on the species. | In essence, a forgetting rather than a learning model. The model was quickly criticized by Sheppard & Turner (1977) and Benson (1977) who questioned its radical prediction (effectively, no Müllerian mimicry), pointing out that the theory was based on relative frequencies, and that time was not explicit. |
| Owen & Owen (1984) | A "recurrent sampling" model (temporary deterrence from attacking models/mimics after attacking an unpalatable prey item). Abundance-Regulated-Anamnesis (ARA) approach allows one to consider absolute rather than relative abundance (Huheey 1964, 1976). | Classical Batesian and Müllerian mimicry can arise, but in the case of unequal unpalatability of two distasteful models/mimics, the exact nature of the interaction depends on the abundances of the species involved. | The predator has only a short-term memory so that it always samples at an encounter immediately subsequent to a non-sampling encounter. This amnesia appears unrealistic, and results in a minimum attack rate of 50%. |
| Turner et al. (1984) | First of the published "Monte-Carlo" simulation approaches (long considered by JRG Turner & PM Sheppard) with predators stochastically encountering "Models", "Mimics" (and "Solo" and "Nasty" as controls). Probability of future attack of a phenotype ($P$) altered to $P = P * a$ (if "unpleasant") or $P = 1 - (P * b)$ (if "pleasant"), with a and b as pleasantness constants. The predator also forgets what it has learnt over time. | Predictions are consistent with traditional differences between Müllerian (mutualistic) and Batesian (parasitic) mimicry. | Asymptotic attack rates on models and mimics in the absence of forgetting are either 0 or 1. |
| Hadeler et al. (1982) | A population-dynamical model for prey (predator population held constant) with predators varying in their experience (learning and recruitment of inexperienced predators). | Classical Müllerian and Batesian mimicry (interpreted as an increase or decrease in equilibrium abundance caused by mimicry) can arise, but a distasteful mimic can be parasitic on a heterospecific dependent on learning and population parameters. | One of the few models to consider the population dynamical consequences of mimicry. |
| Speed (1993a) | A "Pavlovian" predator. Similar Monte-Carlo simulation to Turner et al. (1984) but with (arguably) a more realistic learning algorithm. Here: $\Delta P = (0.5 + \|\lambda_i - 0.5\|) * (\lambda_i - P)$ where $\lambda_i$ is the "asymptotic attack parameter" on prey type *i* when encountered alone in the absence of forgetting. | Weakly unpalatable prey may act as parasites on more highly unpalatable prey ("quasi-Batesian" mimicry). | The equilibrium value of P ($\Delta P = 0$) is $\lambda_i$. It is questionable whether learning behaviour with non-zero $\lambda$ values for unpalatable prey would be selectively advantageous to predators. |

| | | | |
|---|---|---|---|
| Gavrilets & Hastings (1998) | A haploid population genetic model involving 2 species each with two parallel morphs (A, a and B, b) and linear parameters characterizing both within-species and between-species interactions (Batesian and Müllerian situations) | The evolutionary dynamics are strongly affected by the relative strengths of within and between species interactions. | Non-equilibrium dynamics (a "chase") is a general feature of these linear models. As the authors note, more work should be conducted to explore the possibility that co-mimics in the natural world are in dynamic fluctuation. |
| MacDougall & Dawkins (1998) | Müllerian mimicry may reduce the information load on predators and thereby reduce the number of discrimination errors. | When discrimination errors are made then unpalatable prey may suffer a greater change in mortality than palatable prey. Thus, if mimicry reduces discrimination errors then it may be beneficial to all unpalatable prey whatever their unpalatability. | This paper addresses an important phenomenon that had previously been overlooked in the study of Müllerian mimicry (Ruxton 1998). However, no explicit relationship between mimicry and discrimination errors was described. |
| Speed (1999a) | A re-examination of unusual features of the Owen & Owen (1984) model, supported by stochastic simulation. | If prey forget not to attack less defended prey (mimic) more rapidly than more defended prey (model) then it may generate a humped relationship between mimic density and attack rate on encounter, even in the original model of Turner et al. (1984). | Predictions may be highly sensitive to relatively subtle changes in assumptions about the way predators behave. |
| Speed (2001) | For two species (A & B), numbers of distinct unpalatable morphs consumed before learning is complete is $n_A$ and $n_B$ respectively. Number attacked if they resemble one another ($n_{AB}$) is the weighted mean (according to the abundances of the two species). | Quasi-Batesian mimicry may arise at low density levels. | Although only numerical results are shown, this is an analytically tractable model in the same way as Box 1. |
| Speed & Turner (1999) | A comprehensive and systematic analysis of 29 different combinations of learning/forgetting algorithms. | Most of the 29 models included classical Batesian and Müllerian mimicry, but the vast majority also included "unconventional" forms of mimicry namely quasi-Batesian and quasi-Müllerian mimicry. | Qualitative predictions are dependent on both forgetting and learning assumptions. |
| Sasaki et al. (2002) | Reaction-diffusion model of the evolutionary dynamics of Müllerian mimicry in one and two dimensional habitats with positive frequency-dependent selection. | In heterogeneous habitats, areas of distinct co-mimics can be maintained due to the balance between frequency-dependent selection and biased gene flow. | Given the debate over polymorphisms in Müllerian mimicry, spatially-explicit models which can characterize the spatial scale at which polymorphisms occur will be very valuable. |

| | | | |
|---|---|---|---|
| Franks & Noble (2004) | An individual-based model in which palatable and unpalatable prey could evolve in appearance. Predators generalized to a degree and tended to avoid attacking phenotypes which were on average unprofitable. | The presence of Batesian mimics can influence the mean sizes of Müllerian mimicry rings that form. | One of the few models to allow phenotypes to evolve, and to consider mimicry evolution in complex communities. |
| Kokko et al. (2003) | A state dependent model (fit, starving, poisoned, dead) identifying optimal behaviours of predators on encountering models and mimics as well as alternative prey. | Both mutualistic and parasitic relationship between unpalatable model and unpalatable mimic are possible, mediated by the abundance of alternative prey. | If predators handle toxins relatively well but are likely to be limited by food intake, then "quasi-Batesian" (parasitic) mimicry becomes more likely between unpalatable species. The trade-off between food intake and toxin intake influences this relationship. |
| Sherratt et al. (2004b) | A state dependent model (different levels of energy and toxins) identifying optimal behaviours of predators on encountering models and mimics as well as alternative prey. The two toxic species can contain the same poison or different poisons. | When two unpalatable species contain high levels of different toxins then they may have a mutualistic relationship–predators do not know what sort of poison they will consume. | A case of convergent thinking with Kokko et al. (2003). This study also emphasizes the great importance of alternative palatable prey in mediating the nature of the mimetic relationship. |

**Appendix B** Quantitative models of the evolution and maintenance of Batesian mimicry[1]

| Study | Structure of model | Perfection of mimicry | Example conclusions | Comments |
|---|---|---|---|---|
| Huheey 1964 | Probability that an individual model/mimic will be eaten as a function of the proportions of models and mimics and a simple memory parameter (following attack of a model, mimics and models are avoided for a fixed number of encounters). | Perfect | Attack probability of a predator on encounter with a model or mimic rises with the proportion of mimics in the population. | Memory parameter is not time-based but encounter based, no costs or benefits associated with attacking models and mimics respectively, no alternative prey. |
| Holling 1965 | Attack threshold (hunger level) at which a predator attacks a model or mimic as a function of familiarity with prey, predator forgetting, prey attractiveness and palatability coefficients. | Perfect | Attack threshold increases with the proportion of models, increasing unpalatability of model, and decreasing the availability of alternative palatable prey. | A relatively mechanistic model, based on assumptions about how predators should behave rather than on any optimization criteria. |
| Emlen 1968 | A model based on information theory, with experience-mediated misclassification of the model and mimic. | Any level | Mimicry unlikely to evolve when the mimic is an important food item of the predator (when alternative prey are rare). | Some counter-intuitive predictions: when models outnumber mimics, the value of mimicry will sometimes wane as mimics become scarcer. |
| Matessi & Cori 1972 | A population genetics model with both sexes or just one sex mimetic, including both innate and learned avoidance behaviour of the predator | Perfect | Mimetic gene can become fixed in a population, persist in equilibrium or oscillate | Oscillations in mimetic gene frequency are only likely at the distributional limits of the unpalatable model |
| Williamson & Nelson 1972 | Graphical approach ("fitness set" analysis) to examine trade-off between avoiding predation through mimicry and gaining mates. | Any given level | Mimetic females and non-mimetic males will arise if predators feed more heavily on females, if females exhibit stronger visual mating preferences ,and/or if there is proportionately more selection on males to be sexually attractive. | Mate preference assumed fixed rather than frequency-dependent. |
| Estabrook & Jespersen 1974 | Markov chain analysis using conditional probabilities of encounter to control for prey distributions. The analysis estimated the optimal number of encounters (N) when models/mimics should be avoided that would maximize predator payoff. | Perfect | Sufficiently noxious models can protect even ubiquitous mimics. Reduces to the Huheey 1964 model as a special case (with randomly distributed model and mimic) | Based on "encounter-counting" after attacking a model, as Huheey (1964). Profitable alternative prey assumed to be available, but not considered explicitly. |

[1] We have omitted consideration of general population genetic models that implicitly cover Batesian mimicry as part of a wider phenomenon (such as frequency dependence, Clarke & O'Donald 1964) and focused on those papers that deal almost exclusively with Batesian mimicry (modeling papers that also deal with Mullerian mimicry are listed in Appendix A).

| | | | | |
|---|---|---|---|---|
| Oaten et al. 1975 | Signal detection model based on probability that an encountered model/mimic is a model and the costs and benefits of attacking a model and mimic respectively. | Any given level | Increasing model-mimic similarity often increases the attack rates of predators on models and mimics as a whole (although there may well be individual selection for mimics to increase their similarity). | To our knowledge the first paper to apply signal detection theory to Batesian mimicry. Not given the attention it deserves. |
| Charlesworth & Charlesworth 1975a | Population genetic simulation model with stage-structure and overlapping generations. Mimetic morphs differ in conspicuousness and similarity to model. Mimicry determined at a single locus. | Any given level | Even imperfect mimicry can be at a selective advantage and become fixed or form part of a polymorphism. | Also considers the spread of mimicry when the mimicry gene has particular dominance relations. |
| Charlesworth & Charlesworth 1975b | An extension of Charlesworth & Charlesworth 1975 with mimicry determined at two loci (the mimicry gene and a modifier). | Any given level | The only modifier mutations that can escape elimination are ones that occur at loci that are closely linked to the mimicry locus, thereby favouring mimetic "supergenes". When modifiers have effects on both mimetic and non-mimetic genotypes, imperfect mimicry may arise. | Modifiers may have several different effects, depending on whether the mimicry allele is present or absent. |
| Charlesworth & Charlesworth 1975c | An extension of Charlesworth & Charlesworth 1975a investigating the evolution of dominance in genes controlling mimetic patterns. | Any given level | Some types of mimicry genes are likely to evolve dominance, but recessivity could evolve under other conditions. | An example of a general phenomenon (evolution of dominance) being considered in the context of a specific system. |
| Bobisud & Portraz 1976 | A refinement of Estabrook & Jespersen 1974 that explicitly considered the number of mimics consumed between encounters with models, and assumes maximal avoidance behaviour when two models are attacked consecutively (multi-trial learning). | Perfect | Optimal predator behaviour (attack nothing, attack everything, avoid attacking models/mimics for N encounters) depends on conditional probabilities of encountering models and mimics. | Single-trial learners perform better in this case than multi-trial learners |
| Barrett 1976 | Population genetic model (Y-linked locus) with frequency-dependent benefit to mimetic form and frequency-independent (or frequency-dependent) mating preference for another form. | Any given level | Frequency-dependent selection can maintain the mimetic morph without any further balancing selection from mating. Non-mimetic form does not have to be preferentially mated with to maintain polymorphism. | An early model to understand the maintenance of non-mimetic forms in mimetic species. |

| | | | | |
|---|---|---|---|---|
| Arnold 1978 | Primarily a re-examination of the Estabrook & Jespersen 1974 model, investigating the relative success of modifiable and non-modifiable behaviour (unconditionally accept or reject models and mimics) when prey are distributed in different ways. | Perfect | Modifiable behaviour (varying N) produces higher payoff but only when prey items were clumped. For example, on encountering a clumped model a predator should skip the subsequent encounters, because they are likely to be models. | General method also extendable to predators searching for hidden prey. |
| Bobsuid 1978 | Predation rate on models declines over time, dependent on model density. Mimics reproduce at constant rate but are predated according to the rule above. | Perfect | Under a range of conditions mimics should delay their appearance until after the appearance of models. | Based on identifying a single appearance time that maximises the number of mimics alive at end of season. Did not consider a mimic strategy set, or how the model might respond. |
| Turner 1978 | A population genetical model to elucidate why some species show female-limited mimicry | Any given level | Sexual selection resists colour changes more strongly in males. Female-limited mimicry can arise when modifiers supressing the expression of mimicry in males becomes selected for. | An elegant extension of Barrett's (1976) work, which also draws the important distinction between number dependence and frequency dependence |
| Luedeman et al. 1981 | As Estabrook & Jespersen 1974, but alternative prey considered explicitly. Models and mimics can now have different distributions (e.g. model–clumped, mimic–random). | Perfect | Qualitative conclusions correspond to Estabrook & Jespersen 1974 although there are a richer set of properties. | Counting time based on exposures to all prey (not just models and mimics), which facilitates real-time comparisons. |
| Kannan 1983 | A generalized Markov chain model with Estabrook & Jespersen 1974 (single trial learning) and Bobisud & Potratz 1976 (multi-trial learning) as special cases. | Perfect | Predators would do better in most circumstances by adopting a multi-trial learning strategy | Perhaps one of the most detailed and authoritative investigations of the family of papers investigating mimicry using the Markov chain approach |
| Getty 1985 | A signal detection model | Any given level | Optimal attack rates on models/mimics neither 0 nor 1 (partial preferences). These partial preferences are dependent on the relative proportions of models and mimics. | An important paper which also notes the significance of alternative prey, showing that attack rates on models and mimics should decline as profitable alternative prey become more abundant. |
| Greenwood 1986 | A modification of the Staddon & Gendron 1983 signal detection model for detecting cryptic prey, which is functionally equivalent in form to Getty 1985 (see Getty 1987). | Any given level | As with Getty 1985, imperfect mimicry generates a gradual (sigmoidal) relationship between optimal attack probability and proportion of mimics, and is least abrupt when mimicry is imperfect. | Most of the analysis centres on two prey that mimic different models and are not mistaken for one another. |

| | | | | |
|---|---|---|---|---|
| Yamauchi 1993 | A population dynamic model of the consequences of Batesian mimicry, involving a modified version of the Huheey 1964 attack rule dependent on the frequency of the mimic. | Perfect | The presence of the mimic significantly modifies the dynamics of the model in this predator-prey system. | One of the few formulations to consider the population dynamics of models and mimics explicitly. |
| Holmgren & Enquist 1999 | An artificial neural network (multi-layered perceptron) used to model predators' perception of multi-dimensional signals produced by models and mimics. Behaviour of predator and morphologies of models and mimics could evolve subject to selection. | Any given level | The mimic generally evolved similarity to the model faster than the model could evolve away. | Perhaps the first application of artificial neural networks to an evolutionary simulation of mimicry. |
| Sherratt 2002b | A signal detection model assuming the classical receiver operation characteristic (ROC–Staddon & Gendron 1983). | Any given level | Little or no further selection to improve mimetic similarity beyond a certain level if the model is numerous or noxious. The mimic may evolve a "jack of all trades" phenotype with an intermediate appearance between two or more model species (Edmunds 2000). | As with the majority of signal detection models, no initial learning is assumed: the predator is assumed to react appropriately for a given set of costs and benefits and probability of getting it wrong. |
| Johnstone 2002 | A signal detection model that considers kin selection in the evolution of mimetic perfection. | Any given level | Kin selection may allow an optimal level mimetic imperfection since any relatives of a more perfect mimic would be subject to greater predation. | Makes the counter-intuitive prediction that imperfect mimicry will arise by kin selection when the model is weakly aversive and common compared to the model (opposite to that of Sherratt 2002). |
| Sherratt 2003 | A stochastic dynamic programming model with one state (energy levels) that formalises the trade-off between attacking defended prey (and risking injury) and avoiding starvation. | Perfect | Attack probability on encounter with models and mimics declines in a sigmoidal fashion with increasing availability of alternative prey. Predators become risk averse when there are abundant alternative prey even if the risk from attacking a model/mimic is small. | Another way of considering the effects of unpalatable models on predators is to consider the level of toxins they contain, with predators capable of metabolizing toxins at certain rates (Speed 1993; Sherratt et al. 2004b). |

# References


Abrams, P. A. 1991. Life-history and the relationship between food availability and foraging effort. *Ecology,* **72,** 1242–1252.

Acuña, J. L. 2001. Pelagic tunicates: why gelatinous? *American Naturalist,* **158,** 100–107.

Agrawal, A. F. and Karban, R. 1999. Adaptive status of localized and systematic defense responses. In: *The ecology and evolution of inducible defenses* (Ed. by Tollrian, R. and Harvell, C. D.), pp. 33–44. Princeton: Princeton University Press.

Alatalo, R. V. and Mappes, J. 1996. Tracking the evolution of warning signals. *Nature,* **382,** 708–710.

Alatalo, R. V. and Mappes, J. 2000. Initial evolution of warning coloration: comments on the novel world method. *Animal Behaviour,* **60,** F1–F2.

Alcock, J. 1970. Punishment levels and response of white-throated sparrows (*Zonotrichia albicollis*) to three kinds of artificial models and mimics. *Animal Behaviour,* **18,** 733–739.

Alcock, J. 1971. Interspecific differences in avian feeding behavior and the evolution of batesian mimicry. *Behaviour,* **40,** 1–9.

Allen, J. A. 1988*a*. Frequency-dependent selection by predators. *Philosophical Transactions of the Royal Society of London B,* **319,** 251–253.

Allen, J. A. 1988*b*. Reflexive selection is apostatic selection. *Oikos,* **51,** 251–253.

Allen, J. A. 1989*a*. Colour polymorphism, predation and frequency-dependent selection. *Genetics (Life Science Advances),* **8,** 27–43.

Allen, J. A. 1989*b*. Searching for search image. *Trends in Ecology & Evolution,* **4,** 361–361.

Allen, J. A. and Cooper, J. M. 1985. Crypsis and masquerade. *Journal of Biological Education,* **19,** 268–270.

Allen, J. A., Raymond, D. L. and Geburtig, M. A. 1988. Wild birds prefer the familiar morph when feeding on pastry-filled shells of the landsnail Cepaea hortensis (Muell.). *Biological Journal of the Linnean Society,* **33,** 395–401.

Alonso-Mejia, L. and Brower, L. P. 1994. From model to mimic: age-dependent unpalatability in monarch butterflies. *Experientia,* **50,** 176–181.

Alvarez, F. 1993. Alertness signalling in two rail species. *Animal Behaviour,* **46,** 1229–1231.

Amsler, C. D., McClintock, J. B. and Baker, B. J. 2001. Secondary metabolites as mediators of trophic interactions among antarctic marine organisms. *American Zoologist,* **41,** 17–26.

Andersen, K. K., Bernstein, D. T., Caret, R. L. and Romanczyk, L. J. 1982. Chemical-constituents of the defensive secretion of the striped skunk (*Mephitis mephitis*). *Tetrahedron,* **38,** 1965–1970.

Andres, J. A. and Cordero, A. 1999. The inheritance of female colour morphs in the damselfly *Ceriagrion tenellum* (Odonata, Coenagrionidae). *Heredity,* **82,** 328–335.

Anthony, P. D. 1981. Visual contrast thresholds in the cod Gadus morhua. *Journal of Fish Biology,* **19,** 87–103.

Archetti, M. 2000. The origin of autumn colours by coevolution. *Journal of Theoretical Biology,* **205,** 625–630.

Arduino, P. J. J. and Gould, J. L. 1984. Is tonic immobility adaptive? *Animal Behaviour,* **32,** 921–923.

Armbruster, J. W. and Page, L. M. 1996. Convergence of a cryptic saddle pattern in benthic freshwater fishes. *Environmental Biology of Fishes,* **45,** 249–257.

Arnold, S. J. 1978. Evolution of a special class of modifiable behaviors in relation to environmental pattern. *American Naturalist,* **112,** 415–427.

Augner, M. and Bernays, E. A. 1998. Plant defence signals and Batesian mimicry. *Evolutionary Ecology,* **12,** 667–679.

Avery, M. L. 1985. Application of mimicry theory to bird damage control. *Journal of Wildlife Management,* **49,** 1116–1121.

Azmeh, S., Owen, J., Sorensen, K., Grewcock, D. and Gilbert, F. 1998. Mimicry profiles are affected by human-induced habitat changes. *Proceedings of the Royal Society of London Series B—Biological Sciences,* **265,** 2285–2290.

Badcock, J. 1970. The vertical distribution of mesopelagic fishes collected on the SOND cruise. *Journal of the Marine Biological Association of the UK,* **50,** 1001–1044.

Baker, B. J., McClintock, J. B. and Amsler, C. D. 1999. The chemical ecology of antarctic marine organisms. *American Zoologist,* **39,** 4.

Baker, R. R. and Parker, G. A. 1979. The evolution of bird colouration. *Philosophical Transactions of the Royal Society of London Series B—Biological Sciences*, **287**, 63–130.

Balgooyen, T. G. 1997. Evasive mimicry involving a butterfly model and grasshopper mimic. *American Midland Naturalist*, **137**, 183–187.

Ball, P. 1999. *The self-made tapestry: Pattern formation in nature.* Oxford: Oxford University Press.

Barlow, G. W. 1972. The attitude of fish eye-lines in relation to body shape and to stripes and bars. *Copeia*, **1972**, 4–12.

Barnard, C. J. 1984. When cheats may prosper. In: *Producers and scroungers: strategies of exploitation and parasitism* (Ed. by Barnard, C. J.), pp. 6–33. London: Chapman & Hall.

Barrett, J. A. 1976. Maintenance of non-mimetic forms in a dimorphic Batesian mimic species. *Evolution*, **30**, 82–85.

Bates, H. W. 1862. Contributions to an insect fauna of the Amazon valley. Lepidoptera: Heliconidae. *Transactions of the Linnean Society of London*, **23**, 495–566.

Beccaloni, G. W. 1997. Vertical stratification of ithomiine butterfly (Nymphalidae : Ithomiinae) mimicry complexes: the relationship between adult flight height and larval host-plant height. *Biological Journal of the Linnean Society*, **62**, 313–341.

Becerro, M. A., Thacker, R. W., Turon, X., Uriz, M. J. and Paul, V. J. 2003. Biogeography of sponge chemical ecology: comparisons of tropical and temperate defenses. *Oecologia*, **135**, 91–101.

Beddard, F. E. 1892. *Animal coloration: an account of the principle facts and theories relating to the colours and markings of animals.* London: Swan Sonnenschein & Co.

Beldale, P., Koops, K. and Brakefield, P. M. 2002. Modularity, individuality, and evo-devo in butterfly wings. *Proceedings of the National Academy of Sciences of the United States of America*, **99**, 14262–14267.

Belt, T. 1874. *The naturalist in Nicaragua.* New York: E.P. Dutton.

Benard, M. F. and Fordyce, J. A. 2003. Are induced defenses costly? Consequences of predator-induced defenses in western toads, *Bufo boreas*. *Ecology*, **84**, 68–78.

Benson, W. W. 1971. Evidence for the evolution of unpalatability through kin selection in the Heliconiinae. *American Naturalist*, **105**.

Benson, W. W. 1972. Natural selection for Müllerian mimicry in *Heliconius erato* in Costa Rica. *Science*, **176**, 936.

Benson, W. W. 1977. On the supposed spectrum between Batesian and Müllerian mimicry. *Evolution*, **31**, 454–455.

Berenbaum, M. R. and Miliczky, E. 1984. Mantids and milweed bugs: efficacy of aposematic coloration against invertebrate predators. *American Midland Naturalist*, **111**, 64–68.

Bergstrom, C. T. and Lachmann, M. 2001. Alarm calls as costly signals of antipredatory vigilance: the watchful babbler game. *Animal Behaviour*, **61**, 535–543.

Bernays, E. and Graham, M. 1988. On the evolution of host specificity in phytophagous arthropods. *Ecology*, **69**, 886–892.

Bernays, E. A. and Wcislo, W. T. 1994. Sensory capabilities, information processing and resource specialization. *Quarterly Review of Biology*, **69**, 187–204.

Bigot, L. and Jouventin, P. 1974. Quelques expériences de cmesttibilité de Lépidoptères gabonais faites avec le mandrill, le cercocèbe à joues grises et le garde-boeufs. *Terre Vie*, **28**, 521–543.

Björkman, C. and Larsson, S. 1991. Pine sawfly defense and variation in host plant resin acids—a trade-off with growth. *Ecological Entomology*, **16**, 283–289.

Blakiston, T. and Alexander, T. 1884. Protection by mimicry—a problem in mathematical zoology. *Nature*, **29**, 405–406.

Bleiweiss, R. 1992. Reversed plumage ontogeny in a female hummingbird: implications for the evolution of iridescent colors and sexual dichromatism. *Biological Journal of the Linnean Society*, **47**, 183–195.

Blest, A. D. 1957. The function of eyespot patterns in Lepidoptera. *Behaviour*, **11**, 209–255.

Blum, M. S. 1981. *Chemical defenses of arthropods.* London: Academic Press.

Bobisud, L. E. 1978. Optimal time of appearance of mimics. *American Naturalist*, **112**, 962–965.

Bobisud, L. E. and Potratz, C. J. 1976. One-trial versus multi-trial learning for a predator encountering a model–mimic system. *American Naturalist*, **110**, 121–128.

Bohl, E. 1982. Food supply and prey selection in planktivorous cyprinidae. *Oecologia*, **53**, 134–138.

Bond, A. B. 1983. Visual search and selection of natural stimuli in the pigeon: the attention threshold hypothesis. *Journal of Experimental Psychology*, **9**, 292–386.

Bond, A. B. and Kamil, A. C. 1998. Apostatic selection by blue jays produces balanced polymorphism in virtual prey. *Nature*, **395**, 594–596.

Bond, A. B. and Kamil, A. C. 1999. Searching image in blue-jays: facilitation and interference in sequential priming. *Animal Learning and Behaviour*, **27**, 461–471.

Bond, A. B. and Kamil, A. C. 2002. Visual predators select for crypticity and polymorphism in virtual prey. *Nature*, **415**, 609–613.

Bouton, M. E. 1993. Context, time, and memory retrieval in the interference paradigms of pavlovian learning. *Psychological Bulletin*, **114**, 80–99.

Bouton, M. E. 1994. Conditioning, remembering, and forgetting. *Journal of Experimental Psychology—Animal Behavior Processes*, **20**, 219–231.

Bovey, P. 1941. Contribution a l'etude genetique et biogeographique de *Zygaena ephialtes* L. *Revue Suisse de Zoologie*, **48**, 1–90.

Bowdish, T. I. and Bultman, T. L. 1993. Visual cues used by mantids in learning aversion to aposematically colored prey. *American Midland Naturalist*, **129**, 215–222.

Bowers, M. D. 1980. Unpalatability as a defense strategy of *Euphydryas phaeton* (Lepidoptera: Nymphalidae). *Evolution*, **34**, 586–600.

Bowers, M. D. 1984. Iridoid glycosides and host-plant specificity in larvae of the buckeye butterfly, *Junonia coenia* (Nymphalidae). *Journal of Chemical Ecology*, **10**, 1567–1577.

Bowers, M. D. 1986. Population differences in larval host-plant use in the checkerspot butterfly, *Euphydryas chalcedona*. *Entomologia Experimentalis Et Applicata*, **40**, 61–69.

Bowers, M. D. 1988. Chemistry and coevolution: iridoid glycosides, plants and herbivorous insects. In: *Chemical mediation of coevolution* (Ed. by Spencer, K.). New York: Academic Press.

Bowers, M. D. 1990. Recycling plant natural products for insect defense. In: *Insect defenses: Apative mechanisms and strategies of prey and predators* (Ed. by Evans, D. L. and Schmid, J. O.), pp. 353–386. New York: State University of New York Press.

Bowers, M. D. 1992. The evolution of unpalatability and the cost of chemical defense in insects. In: *Insect chemical ecology* (Ed. by Roitberg, B. D. and Isman, M. B.). London: Chapman & Hall.

Bowers, M. D. 1993. Aposematic caterpillars: life-styles of the warningly colored and unpalatable. In: *Caterpillars: ecological and evolutionary constraints on foraging* (Ed. by Stamp, N. E. and Casey, T. M.), pp. 331–371. London: Chapman & Hall.

Bowers, M. D. 1999. Eating and being eaten: The role of host plant chemistry in multitrophic interactions. *American Zoologist*, **39**, 282.

Bowers, M. D., Brown, I. L. and Wheye, D. 1985. Bird predation as a selective agent in a butterfly population. *Evolution*, **39**, 93–103.

Bowers, M. D. and Collinge, S. K. 1992. Fate of iridoid glycosides in different life stages of the buckeye, *Junonia coenia* (Lepidoptera, Nymphalidae). *Journal of Chemical Ecology*, **18**, 817–831.

Bowers, M. D. and Stamp, N. E. 1992. Chemical variation within and between individuals of *Plantago lanceolata* (Plantaginaceae). *Journal of Chemical Ecology*, **18**, 985–995.

Bowers, M. D. and Williams, E. H. 1995. Variable chemical defence in the checkerspot butterfly *Euphydryas gilletii* (Lepidoptera: Nymphalidae). *Ecological Entomology*, **20**, 208–212.

Boyden, T. C. 1976. Butterfly palatability and mimicry—experiments with *Ameiva* lizards. *Evolution*, **30**, 73–81.

Brakefield, P. M. 1985. Polymorphic Müllerian mimicry and interactions with thermal melanism in ladybirds and a soldier beetle—a hypothesis. *Biological Journal of the Linnean Society*, **26**, 243–267.

Braude, S., Ciszek, D., Berg, N. E. and Shefferly, N. 2001. The ontogeny and distribution of countershading in colonies of the naked mole-rat (*Heterocephalus glaber*). *Journal of Zoology*, **253**, 351–357.

Bretagnolle, V. 1993. Adaptive significance of seabird coloration: the case of procellariiforms. *American Naturalist*, **142**, 141–173.

Brodie, E. D. 1981. Phenological relationships of model and mimic salamanders. *Evolution*, **35**, 988–994.

Brodie, E. D. 1993. Differential avoidance of coral snake banded patterns by free-ranging avian predators in Costa Rica. *Evolution*, **47**, 227–235.

Brodie, E. D. 1999. Predator-prey arms races. *Bioscience*, **49**, 557–568.

Brodie, E. D. and Agrawal, A. F. 2001. Maternal effects and the evolution of aposematic signals. *Proceedings of the National Academy of Sciences of the United States of America*, **98**, 7884–7887.

Brodie, E. D., Ducey, P. K. and Baness, E. A. 1991. Antipredator skin secretions of some tropical salamanders (Bolitoglossa) are toxic to snake predators. *Biotropica*, **23**, 58–62.

Brodie, E. D. and Formanowicz, D. R. 1991. Predator avoidance and antipredator mechanisms—distinct pathways to survival. *Ethology Ecology and Evolution*, **3**, 73–77.

Brodie, E. D. and Howard, R. R. 1973. Experimental study of Batesian mimicry in the salamanders *Plethodon jordani* and *Desmognathus ochrophaeus*. *American Midland Naturalist*, **90**, 38–46.

Brodie, E. D. and Janzen, F. J. 1995. Experimental studies of coral snake mimicry—generalized avoidance of ringed snake patterns by free-ranging avian predators. *Functional Ecology*, **9**, 186–190.

Brodie, E. D. I. and Brodie, E. D. J. 1999. Predator–prey arms races. *Bioscience*, **49**, 557–568.

Brodie, E. D. J. and Brodie, E. D. I. 1980. Differential avoidance of mimetic salamanders by free-ranging birds. *Science*, **208**, 181–182.

Brönmark, C. and Miner, J. G. 1992. Predator-induced phenotypical change in body morphology in Crucian carp. *Science*, **258**, 1348–1350.

Brönmark, C., Pettersson, L. B. and Nilsson, P. A. 1999. Predator-induced defense in crucial carp. In: *The ecology and evolution of inducible defences* (Ed. by Tollrian, R. and Harvell, C. D.), pp. 203–217. Princeton, New Jersey: Princeton.

Brower, A. V. Z. 1995. Locomotor mimicry in butterflies? A critical review of the evidence. *Philosophical Transactions of the Royal Society of London Series B—Biological Sciences*, **347,** 413–425.

Brower, A. V. Z. 1996. Parallel race formation and the evolution of mimicry in *Heliconius* butterflies: a phylogenetic hypothesis from mitochondrial DNA sequences. *Evolution*, **50,** 195–221.

Brower, J. V. 1958*a*. Experimental studies of mimicry in some North-American butterflies. 1. The monarch, *Danaus plexippus*, and Viceroy, *Limenitis archippus-Archippus*. *Evolution*, **12,** 32–47.

Brower, J. V. 1958*b*. Experimental studies of mimicry in some North-American butterflies. 2. *Battus philenor* and *Papilio troilus*, *P. polyxenes* and *P. glaucus*. *Evolution*, **12,** 123–136.

Brower, J. V. 1958*c*. Experimental studies of mimicry in some North-American butterflies. 3. *Danaus gilippus berenice* and *Limenitis archippus floridensis*. *Evolution*, **12,** 273–285.

Brower, J. V. 1960. Experimental studies of mimicry. IV. The reactions of starlings to different proportions of models and mimics. *American Naturalist*, **94,** 271–282.

Brower, L. P. 1969. Ecological chemistry. *Scientific American*, **220,** 22–29.

Brower, L. P. 1984. Chemical defence in butterflies. In: *The biology of butterflies* (Ed. by Vane-Wright, R. I. and Ackery, P.), pp. 109–134. Princeton: Princeton University Press.

Brower, L. P. 1988. Avian predation on the monarch butterfly and its implications for mimicry theory. *American Naturalist (Supplement)*, **131,** S4–S6.

Brower, L. P., Brower, J. V. and Collins, C. T. 1963. Experimental studies of mimicry. 7. Relative palatability and Müllerian mimicry among neotropical butterflies of the subfamily Heliconiinae. *Zoologica*, **48,** 65–84.

Brower, L. P., Brower, J. V. and Westcott, P. W. 1960. Experimental studies of mimicry. 5. The reactions of toads (*Bufo terrestris*) to bumblebees (*Bombus-americanorum*) and their robberfly mimics (*Mallophora bomboides*), with a discussion of aggressive mimicry. *American Naturalist*, **94,** 343–355.

Brower, L. P., Brower, J. V. Z. and Corvino, J. M. 1967. Plant poisons in a terrestrial food chain. *Proceedings of the National Academy of Sciences of the United States of America*, **57,** 893–898.

Brower, L. P. and Calvert, W. H. 1985. Foraging dynamics of bird predators on overwintering monarch butterflies in Mexico. *Evolution*, **39,** 852–868.

Brower, L. P., Cook, L. M. and Croze, H. J. 1967. Predator responses to artificial Batesian mimics released in a neotropical environment. *Evolution*, **21,** 11–23.

Brower, L. P., Edmunds, M. and Moffitt, C. M. 1975. Cardenolide content and palatability of *Danaus chrysippus* butterflies from West Africa. *J. Entomol.*, **49,** 183–196.

Brower, L. P. and Glazier, S. C. 1975. Localization of heart poisons in the monarch butterfly. *Science*, **188,** 19–25.

Brower, L. P., Hower, A. S., Croze, H. J., Brower, J. V. Z. and Stiles, F. G. 1964. Mimicry—differential advantage of color patterns in the natural environment. *Science*, **144,** 183–185.

Brower, L. P. and Moffitt, C. M. 1974. Palatability dynamics of cardenolides in the monarch butterfly. *Nature*, **249,** 280–283.

Brower, L. P., Pough, F. H. and Meck, H. R. 1970. Theoretical investigations of automimicry. I. Single trial learning. *Proceedings of the National Academy of Sciences of the United States of America*, **66,** 1059–1066.

Brower, L. P., Reyerson, W. N., Coppinger, L. L. and Glazier, S. C. 1968. Ecological chemistry and the palatability spectrum. *Science*, **161,** 1349–1351.

Brower, L. P., Seiber, J. N., Nelson, C. J., Lynch, S. P. and Holland, M. M. 1984. Plant-determined variation in the cardenolide content, thin layer chromatography profiles, and emetic potency of monarch butterflies, *Danaus plexippus* L. reared on milkweed plants in California. *Journal of Chemical Ecology*, **10,** 601–639.

Browman, H. I., Novales-Flamarique, I. and Hawryshyn, C.W. 1994. Ultraviolet photoreception contributes to prey search behaviour in two species of zooplanktivorous fish. *Journal of Experimental Biology*, **186,** 187–198.

Brown, G. E., Godin, J.-G. J. and Pedersen, J. 1999. Fin-flicking behaviour: a visual anti-predator alarm signal in characin fish, *Hemigrammus erthrozonus*. *Animal Behaviour*, **58,** 469–475.

Brown, J. H. and Kodric-Brown, A. 1979. Convergence, competition and mimicry in a temperate community of hummingbird-pollinated flowers. *Ecology*, **60,** 1022–1035.

Brown, K. S. 1981. The biology of *Heliconius* and related genera. *Annual Review of Entomology*, **26,** 427–456.

Brown, K. S. 1984. Adult-obtained pyrrolozidine alkaloids defend ithomiine butterflies against a spider predator. *Nature*, **309,** 707–709.

Brown, K. S. and Benson, W. W. 1974. Adaptive polymorphism associated with multiple Müllerian mimicry in *Heliconius numata* (Lepid.: Nymph). *Biotropica*, **6,** 205–228.

Brown, K. S., Trigo, J. R., Francini, R. B., de Morais, A. D. B. and Motta, P. C. 1991. Aposematic insects on toxic host plants: coevolution, colonisation and chemical emancipation. In: *Plant–Animal interactions: evolutionary ecology in tropical and temperate regions* (Ed. by Price, P. W., Lewinsohn, T. M., Wilson Fernandes, G. and Benson, W. W.), pp. 375–402. New York: John Wiley and Sons.

Brownell, C. L. 1985. Laboratory analysis of cannibalism by larvae of the cape anchovy (*Engraulis capenis*). *Transactions of the American Fisheries Society*, **114,** 512–518.

Bryan, P. J., Yoshida, W. Y., McClintock, J. B. and Baker, B. J. 1995. Ecological role for pteroenone, a novel antifeedant from the conspicuous antarctic pteropod *Clione antarctica* (Gymnosomata, Gastropoda). *Marine Biology*, **122,** 271–277.

Bullard, S. B. and Hay, M. E. 2002. Palatability of marine macro-holoplankton: nematocysts, nutritional quality, and chemistry as defenses against consumers. *Limnology and Oceanography*, **47,** 1456–1467.

Bullini, L., Sbordoni, V. and Ragazzini, P. 1969. Mimetismo mulleriano in popolazione italiane di *Zygaena ephialtes* (L.) (Lepidoptera, Zygaenidae). *Arch. zool. ital.*, **44,** 181–214.

Burd, M. 1994. Butterfly wing colour patterns and flying heights in the seasonally wet forest of Barro Colorado Island, Panama. *Journal of Tropical Ecology*, **10,** 601–610.

Cain, A. J. and Curry, J. D. 1963*a*. Area effects in Cepaea. *Philosophical Transaction of the Royal Society of London B*, **246,** 1–81.

Cain, A. J. and Curry, J. D. 1963*b*. The causes of area effects. *Heredity*, **18,** 467–471.

Cain, A. J. and Sheppard, P. M. 1950. Selection in the polymorphic land snail Cepaea nemoralis. *Heredity*, **4,** 275–294.

Cairns, D. K. 1986. Plumage colour in pursuit-diving seabirds: why do penguins wear tuxedos. *Bird Behaviour*, **6,** 58–65.

Caldwell, G. S. and Rubinoff, R. W. 1983. Avoidance of venemous sea snakes by naive herons and egrets. *Auk*, **100,** 195–198.

Caldwell, J. P. 1982. Disruptive selection: a tail color polymorphism in Acris tadpoles in response to differential predation. *Canadian Journal of Zoology*, **60,** 2818–2827.

Caldwell, J. P. 1986. A description of the tadpole of *Hyla smithii* with comments on tail coloration. *Copeia*, **1986,** 1004–1006.

Caley, M. J. and Schluter, D. 2003. Predators favour mimicry in a tropical reef fish. *Proceedings of the Royal Society of London Series B—Biological Sciences*, **270,** 667–672.

Camara, M. D. 1997. Physiological mechanisms underlying the costs of chemical defence in *Junonia coenia* Hubner (Nymphalidae): A gravimetric and quantitative genetic analysis. *Evolutionary Ecology*, **11,** 451–469.

Cardoso, M. Z. 1997. Testing chemical defence based on pyrrolizidine alkaloids. *Animal Behaviour*, **54,** 985–991.

Caro, T. M. 1986*a*. The functions of stotting: a review of the hypotheses. *Animal Behaviour*, **34,** 649–662.

Caro, T. M. 1986*b*. The functions of stotting in Thomson's gazelles: some tests of the predictions. *Animal Behaviour*, **34,** 663–684.

Caro, T. M. 1994. Ungulate anti-predator behaviour: preliminary and comparative data from African Bovids. *Behaviour*, **128,** 189–228.

Caro, T. M. 1995. Pursuit-deterrence revisited. *Trends in Ecology and Evolution*, **10,** 500–503.

Caro, T. M., Lombardo, L., Goldizen, A. W. and Kelly, M. 1995. Tail flagging and other anti-predatory signals in white-tailed deer: new data and synthesis. *Behavioural Ecology*, **6,** 442–450.

Carpenter, G. D. H. and Ford, E. B. 1933. *Mimicry*. London: Methuen and Co.

Carrascal, L. M., Diaz, J. A., Huertas, D. L. and Mozetich, I. 2001. Behavioral thermoregulation by three creepers: Trade-off between saving energy and reducing crypsis. *Ecology*, **82,** 1642–1654.

Case, J. F., Warner, J., Barnes, A. T. and Lowenstine, M. 1977. Bioluminescence of lantern fish (Myctophidae) in response to changes in light intensity. *Nature*, **265,** 179–181.

Castilla, A. M., Gosá, A., Galán, P. and Pérez-Mellado, V. 1999. Green tails in lizards of the genus *Podarcis*: do they influence the intensity of predation. *Herpetologica*, **55,** 530–537.

Chai, P. 1986. Field observations and feeding experiments on the responses of rufous-tailed Jacamars (*Galbula ruficauda*) to free-flying butterflies in a tropical rain-forest. *Biological Journal of the Linnean Society*, **29,** 161–189.

Chai, P. 1996. Butterfly visual characteristics and ontogeny of responses to butterflies by a specialized tropical bird. *Biological Journal of the Linnean Society*, **59,** 37–67.

Chai, P. and Srygley, R. B. 1990. Predation and the flight, morphology, and temperature of neotropical rainforest butterflies. *American Naturalist*, **135,** 748–765.

Chapman, L. J., Kaufman, L. and Chapman, C. A. 1994. Why swim upside down? A comparative study of two mochokid catfishes. *Copeia*, **1994,** 130–135.

Charlesworth, B. 1994. The genetics of adaptation: lessons from mimicry. *The American Naturalist*, **144,** 839–847.

Charlesworth, D. and Charlesworth, B. 1975*a*. Theoretical genetics of Batesian mimicry. I. Single-locus models. *Journal of Theoretical Biology*, **55,** 283–303.

Charlesworth, D. and Charlesworth, B. 1976*b*. Theoretical genetics of Batesian mimicry. II. Evolution of supergenes. *Journal of Theoretical Biology*, **55,** 305–324.

Charlesworth, D. and Charlesworth, B. 1976*c*. Theoretical genetics of Batesian mimicry. III. Evolution of dominance. *Journal of Theoretical Biology*, **55**, 325–337.

Chen, Z. L. and Zhu, D. Y. 1987. Aristolochia alkaloids. In: *The alkaloids: chemistry and pharmacology* (Ed. by Brossi, A.), pp. 29–65. San Diego: Academic Press.

Chiao, C.-C. and Hanlon, R. T. 2001. Cuttlefish camouflage: visual perception of size, contrast and number of white squares in artificial checkerboard substrata initiates disruptive coloration. *Journal of Experimental Biology*, **204**, 2119–2125.

Cimino, G. and Ghiselin, M. 1998. Chemical defense and evolution in the Sacoglossa (Mollusca: Gastropoda: Opisthobranchia). *Chemoecology*, **8**, 51–60c.

Cimino, G. and Ghiselin, M. T. 1999. Chemical defense and evolutionary trends in biosynthetic capacity among dorid nudibranchs. *Chemoecology*, **9**, 187–207.

Clark, B. R. and Faeth, S. H. 1997. The consequences of larval aggregation in the butterfly Chlosyne lacinia. *Ecological Entomology*, **22**, 408–415.

Clark, B. R. and Faeth, S. H. 1998. The evolution of egg clustering in butterflies: a test of the egg desiccation hypothesis. *Evolutionary Ecology*, **12**, 543–552.

Clark, C. W. and Mangel, M. 2000. *Dynamic state variable models in ecology; methods and applications*. Oxford: Oxford University Press.

Clarke, B. and O'Donald, P. 1964. Frequency-dependent selection. *Heredity*, **19**, 201.

Clarke, B. C. 1962. Natural selection in mixed populations of two polymorphic snails. *Heredity*, **17**, 319–345.

Clarke, C., Clarke, F. M. M., Collins, S. C., Gill, A. C. L. and Turner, J. R. G. 1985. Male-like females, mimicry and transvestism in butterflies (Lepidoptera: Papilionidae). *Systematic Entomology*, **10**, 257–283.

Clarke, C. and Sheppard, P. M. 1975. Genetics of mimetic butterfly *Hypolimnas bolina* (L). *Philosophical Transactions of the Royal Society of London Series B—Biological Sciences*, **272**, 229–265.

Clarke, C. A., Clarke, F. M. M., Gordon, I. J. and Marsh, N. A. 1989. Rule-breaking mimics—palatability of the butterflies *Hypolimnas bolina* and *Hypolimnas misippus*, a sister species pair. *Biological Journal of the Linnean Society*, **37**, 359–365.

Clarke, C. A. and Sheppard, P. M. 1959. The genetics of some mimetic forms of *Papilio dardanus*, Brown, and *Papilio glaucus*, Linn. *Journal of Genetics*, **56**, 237–259.

Clarke, C. A. and Sheppard, P. M. 1960*a*. The evolution of mimicry in the butterfly *Papilio dardanus*. *Heredity*, **14**, 163–173.

Clarke, C. A. and Sheppard, P. M. 1960*b*. Supergenes and mimicry. *Heredity*, **14**, 175–185.

Clarke, C. A. and Sheppard, P. M. 1971. Further studies on the genetics of the mimetic butterfly *Papilio memnon*. *Philosophical Transactions of the Royal Society*, **263**, 431–458.

Clough, M. and Summers, K. 2000. Phylogenetic systematics and biogeography of the poison frogs: evidence from mitochondrial DNA sequences. *Biological Journal of the Linnean Society*, **70**, 515–540.

Cock, M. J. W. 1978. The assessment of perferences. *Journal of Animal Ecology*, **47**, 805–816.

Cohen, J. A. 1985. Differences and similarities in cardenolide contents of Queen and Monarch butterflies in Florida and their ecological and evolutionary implications. *Journal of Chemical Ecology*, **11**, 85–103.

Confer, J. L., Howick, G. L., Corzette, M. H., Kramer, S. L., Fitzgibbon, S. and Landesberg, R. 1978. Visual predation by planktivores. *Oikos*, **31**, 27–37.

Congdon, J. D., Vitt, L. J. and King, W. W. 1974. Geckos: adaptive significance and energetics of tail autotomy. *Science*, **184**, 1379–1380.

Conner, W. E., Boada, R., Schroeder, F. C., Gonzalez, A., Meinwald, J. and Eisner, T. 2000. Chemical defense: bestowal of a nuptial alkaloidal garment by a male moth on its mate. *Proceedings of the National Academy of Sciences of the United States of America*, **97**, 14406–14411.

Conover, M. R. 1994. Stimuli eliciting distress calls in adult passerines and response of predators to their broadcast. *Beahviour*, **131**, 19–37.

Cook, L. M. 1998. A two-stage model for *Cepaea* polymorphism. *Philosophical Transaction of the Royal Society of London B*, **353**, 1577–1593.

Cook, L. M. 2000. Changing views on melanic moths. *Biological Journal of the Linnean Society*, **69**, 431–441.

Cook, L. M., Brower, L. P. and Alcock, J. 1969. An attempt to verify mimetic advantage in a neotropical environment. *Evolution*, **23**, 339–345.

Cook, S. E., Vernon, J. G., Bateson, M. and Guilford, T. 1994. Mate choice in the polymorphic African swallowtail butterfly, *Papilio dardanus*: male-like females may avoid sexual harassment. *Animal Behaviour*, **47**, 389–397.

Cooper, J. M. and Allen, J. A. 1994. Selection by wild birds on artificial dimorphic prey on varied backgrounds. *Biological Journal of the Linnean Society*, **51**, 433–446.

Cooper, W. E., Jr. 1998*a*. Conditions favouring anticipatory and reactive displays deflecting predatory attacks. *Behavioral Ecology*, **9**, 598–604.

Cooper, W. E., Jr. 1998*b*. Reactive and anticipatory display to deflect predatory attack to an autotomous lizard tail. *Canadian Journal of Zoology*, **76**, 1507–1510.

Cooper, W. E., Jr. and Vitt, L. J. 1985. Blue tails and autotomy: enhancement of predation avoidance in juvenile skinks. *Zeitschrift fur Tierpsychologie*, **70,** 265–276.

Cooper, W. E., Jr. and Vitt, L. J. 1991. Influence of detectability and ability to escape on natural selection of conspicuous autotomous defenses. *Canadaian Journal of Zoology*, **69,** 757–764.

Cooper, W. E. J. 2000. Pursuit deterrence in lizards. *Saudi Journal of Biological Science*, **7,** 15–29.

Cooper, W. E. J. 2001. Multiple roles of tail display by the curly-tailed lizard Leicephalus carinatus: pursuit deterrent and deflective roles of a social signal. *Ethology*, **107,** 1137–1149.

Coppinger, R. P. 1970. The effect of experience and novelty on avian feeding behaviour with reference to the evolution of warning coloration in butterflies. II. Reactions of naive birds to novel insects. *American Naturalist*, **104,** 323–335.

Corbet, P. S. 1999. *Dragonflies: Behaviour and Ecology of Odonata.*: Harley Books, Essex.

Cordero, C. 2001. A different look at the false head of butterflies. *Ecological Entomology*, **26,** 106–108.

Cott, H. B. 1932. Reply to McAtee. *Nature*, **130,** 962.

Cott, H. B. 1940. *Adaptive coloration in animals.* London: Methuen.

Cowan, P. J. 1972. The contrast and coloration of seabirds: an experimental approach. *Ibis*, **114,** 390–393.

Coyne, J. A. 1998. Not black and white. *Nature*, **396,** 35–36.

Craig, J. L. 1982. On the evidence for a 'pursuit deterrent' function of alarm signals in swamphens. *American Naturalist*, **119,** 753–755.

Cresswell, W. 1994. Song as a pursuit-deterrent signal, and its occurrence relative to other anti-predator behaviours of skylark (*Alauda arvensis*) on attack by merlins (*Falco columbarius*). *Behavioural Ecology and Sociobiology*, **23,** 217–223.

Cresswell, W., Hilton, G. M. and Ruxton, G. D. 2002. Evidence for a rule governing the avoidance of superfluous escape flights. *Proceedings of the Royal Society of London B*, **267,** 733–737.

Cummings, M. E., Rosenthal, G. G. and Ryan, M. J. 2003. A private ultraviolet channel in visual communication. *Proceedings of the Royal Society of London B*, **270,** 897–904.

Cundy, J. M. and Allen, J. A. 1988. Two models for exploring the anti-predator function of eyespots. *Journal of Biological Education*, **22,** 207–210.

Curio, E. 1976. *The ethology of predation.* Berlin: Springer.

Cuthill, I. C. and Bennett, A. T. D. 1993. Mimicry and the eye of the beholder. *Proceedings of the Royal Society of London Series B—Biological Sciences*, **253,** 203–204.

Dafni, A. 1984. Mimicry and deception in pollination. *Annual Review of Ecology and Systematics*, **15,** 259–278.

Dafni, A. and Ivri, Y. 1981. Floral mimicry between *Orchis israelitica* Baumann and Dafni (Orchidaceae) and *Bellevalia flexuousa* Boiss. (Liliaceae). *Oecologia*, **49,** 229–232.

Dale, G. and Pappantoniou, A. P. 1986. Eye-picking behavior of the cutlips minnow, Exoglossum maxillingua: applications to studies of eyepot mimicrys. *Annals of the New York Academy of Sciences*, **463,** 177–178.

Dall, S. R. X. and Cuthill, I. C. 1997. The information costs of generalism. *Oikos*, **80,** 197–202.

Daly, J. W., Myers, C. W. and Whittaker, N. 1987. Further classification of skin alkaloids from neotropical poison frogs (Dendrobatidae) with general survey of toxic/noxious substances in the Amphibia. *Toxicon*, **25,** 1023–1095.

Daly, J. W., Padgett, W. L., Saunders, R. L. and Cover, J. F. 1997. Absence of tetrodotoxins in a captive-raised riparian frog, *Atelopus varius*. *Toxicon*, **35,** 705–709.

Darwin, C. 1887. *The life and letters of Charles Darwin: including an autobiographical chapter, edited by his son Francis Darwin.* London: Murray.

Darwin, C. 1903. *More letters of Charles Darwin, a record of his work in a series of hitherto unpublished letters* (Ed. by Francis Darwin and Seward A.C.). New York: D. Appleton and Co.

Davey, G. 1989. *Ecological Learning Theory.* London: Routledge.

Davies, N. B., Kilner, R.M., and Noble, D.G. 1998. Nestling cuckoos, *Cuculus canorus*, exploit hosts with begging calls that mimic a brood. *Proceedings of the Royal Society of London Series B—Biological Sciences*, **265,** 673–678.

Dawkins, M. 1971. Perceptual changes in chicks: another look at the 'search image' concept. *Animal Behaviour*, **19,** 566–574.

De Cock, R. and Matthysen, E. 2001. Aposematism and bioluminescence: Experimental evidence from glow-worm Larvae(Coleoptera : Lampyridae). *Evolutionary Ecology*, **13,** 619–639. [1999/2001]

de Jong, P. W. and Brakefield, P. M. 1998. Climate and change in clines for melanism in the two-spot ladybird, *Adalia bipunctata* (Coleoptera: Coccinellidae). *Proceedings of the Royal Society of London Series B—Biological Sciences*, **265,** 39–43.

De Ruiter, L. 1952. Some experiments on the camouflage of stick caterpillars. *Behaviour*, **4,** 222–232.

De Ruiter, L. 1956. Countershading in caterpillars: an analysis of its adaptive significance. *Archives Neerlandaises de Zoologie*, **11,** 285–341.

De Ruiter, L. 1959. Some remarks on problems of the ecology and evolution of mimicry. *Archives Neerlandaises de Zoologie*, **13 Suppl.**, 351–368.

Deml, R. and Dettner, K. 2003. Comparative morphology and secretion chemistry of the scoli caterpillars of *Hyalophora cecropia*. *Naturwissenschaften*, **90,** 460–463.

Denno, R. F. and Benrey, B. 1997. Aggregation facilitates larval growth in the neotropical nymphalid butterfly Chlosyne janais. *Ecological Entomology*, **22**, 133–141.

Denny, M. W. 1993. *Air and water: the biology and physics of life's media*. Princeton: Princeton University Press.

Denton, E. J. 1970. On the organisation of reflecting surfaces in some marine animals. *Proceedings of the Royal Society of London B*, **258**, 285–313.

Denton, E. J. 1971. Reflectors in fishes. *Scientific American*, **224**, 64–72.

Denton, E. J., Gilpin-Brown, J. B. and Wright, P. G. 1972. The angular distribution of light produced by some mesopelagic fish in relation to their camouflage. *Proceedings of the Royal Society of London B*, **182**, 145–158.

DeVries, P. J. 1987. *The butterflies of Costa Rica and their natural history*. Princeton, NJ: Princeton University Press.

DeVries, P. J., Lande, R. and Murray, D. 1999. Associations of co-mimetic ithomiine butterflies on small spatial and temporal scales in a neotropical rainforest. *Biological Journal of the Linnean Society*, **67**, 73–85.

D'Heursel, A. and Haddad, C. F. B. 1999. Unpalatability of *Hyla semilineata* tadpoles (Anura) to captive and free-ranging vertebrate predators. *Ethology, Ecology and Evolution*, **11**, 339–348.

Dial, B. E. 1986. Tail displays in two species of iguanid lizards: a test of the predator signal hypothesis. *American Naturalist*, **127**, 103–111.

Dill, L. M. 1975. Calculated risk-taking by predators as a factor in Batesian mimicry. *Canadian Journal of Zoology—Revue Canadienne De Zoologie*, **53**, 1614–1621.

Dittrich, W., Gilbert, F., Green, P., McGregor, P. and Grewcock, D. 1993. Imperfect mimicry—a pigeons perspective. *Proceedings of the Royal Society of London Series B—Biological Sciences*, **251**, 195–200.

Dixey, F. A. 1909. On Müllerian mimicry and diaposematism. A reply to Mr GAK Marshall. *Transactions of the Entomological Society of London*, **XXIII**, 559–583.

Dixey, F. A. 1919. The geographical factor in mimicry. In: *Presidential Address, Section D, Report of the British Association for the Advancement of Science*, pp. 199–207.

Dixon, C. A., Erickson, J. M., Kellett, D. N. and Rothschild, M. 1978. Some adaptations between *Danaus plexippus* and its food plant, with notes on *Danaus chrysippus* and *Euploea core* (Insecta: Lepidoptera). *J. Zool.*, **185**, 437–467.

Dobler, S. 2001. Evolutionary aspects of defense by recycled plant compounds in herbivorous insects. *Basic and Applied Ecology*, **2**, 15–26.

Dobler, S., Mardulyn, P., Pasteels, J. M. and Rowell-Rahier, M. 1996. Host-plant switches and the evolution of chemical defense and life history in the leaf beetle genus Oreina. *Evolution*, **50**, 2373–2386.

Dobler, S. and Rowell-Rahier, M. 1994. Response of a leaf beetle to two food plants, only one of which provides a sequestrable defensive chemical. *Oecologia*, **97**, 271–277.

Doesburg, P. H. V. 1968. A revision of the New World species of *Dysdercus* Guerin Meneville (Heteroptera, Pyrrhocoridae). *Zool. Verh.*, **97**, 1–215.

Dolinger, P. M., Ehrlich, P. R., Fitch, W. L. and Breedlove, D. E. 1973. Alkaloid and predation patterns in Colorado lupine populations. *Oecologia*, **13**, 191–204.

Duffey, S. S. 1980. Sequestration of plant natural products by insects. *Annual Review of Entomology*, **25**, 447–477.

Duffy, J. E. and Hay, M. E. 1994. Herbivore resistance to seaweed chemical defense—the roles of mobility and predation risk. *Ecology*, **75**, 1304–1319.

Dukas, R. and Ellner, S. 1993. Information processing and prey detection. *Ecology*, **74**, 1337–1346.

Dukas, R. and Kamil, A. C. 2001. Limited attention: the constraint underlying search image. *Behavioural Ecology*, **12**, 192–199.

Dumbacher, J. P. 1999. Evolution of toxicity in pitohuis: I. Effects of homobatrachotoxin on chewing lice (order Phthiraptera). *Auk*, **116**, 957–963.

Dumbacher, J. P., Beehler, B. M., Spande, T. F. and Garraffo, H. M. 1992. Homobatrachotoxin in the genus *Pitohui*—chemical defense in birds. *Science*, **258**, 799–801.

Dumbacher, J. P. and Fleischer, R. C. 2001. Phylogenetic evidence for color pattern convergence in toxic *pitohuis*: Müllerian mimicry in birds? *Proceedings of the Royal Society of London Series B—Biological Sciences*, **268**, 1971–1976.

Dumbacher, J. P., Spande, T. F. and Daly, J. W. 2000. Batrachotoxin alkaloids from passerine birds: A second toxic bird genus (*Ifrita kowaldi*) from New Guinea. *Proceedings of the National Academy of Sciences of the United States of America*, **97**, 12970–12975.

Duncan, C. J. and Sheppard, P. M. 1965. Sensory discrimination and its role in the evolution of Batesian mimicry. *Behaviour*, **24**, 269–282.

Dunham, D. W. and Tierney, A. J. 1983. The communicative cost of crypsis in a hermit crab *Pagurus marshi*. *Animal Behaviour*, **31**, 783–&.

Dyer, L. A. and Bowers, M. D. 1996. The importance of sequestered iridoid glycosides as a defense against an ant predator. *Journal of Chemical Ecology*, **22**, 1527–1539.

Eberhard, W. G. 1977. Aggressive chemical mimicry by a bolas spider. *Science*, **198**, 1173–1175.

Edgar, J. A., Cockrum, P. A. and Frahn, J. L. 1976. Pyrrolizidine alkaloids in *Danaus plexippus* and *Danaus chrysippus*. *Experientia*, **32**, 1535–1537.

Edmunds, J. and Edmunds, M. 1974. Polymorphic mimicry and natural selection—a reappraisal. *Evolution*, **28**, 402–407.

Edmunds, M. 1974. *Defence in Animals: A survey of anti-predator defences.* Harlow, Essex: Longman.

Edmunds, M. 1981. On defining 'mimicry'. *Biological Journal of the Linnean Society*, **16**, 9–11.

Edmunds, M. 1991. Does warning coloration occur in nudibranchs? *Malacologia*, **32**, 241–255.

Edmunds, M. 2000. Why are there good and poor mimics? *Biological Journal of the Linnean Society*, **70**, 459–466.

Edmunds, M. and Dewhirst, R. A. 1994. The survival value of countershading with wild birds as predators. *Biological Journal of the Linnean Society*, **51**, 447–452.

Eggenberger, F., Daloze, D., Pasteels, J. M. and Rowell-Rahier, M. 1992. Identification and seasonal quantification of defensive secretion components of *Oreina gloriosa* (Coleoptera: Chrysomelidae). *Experientia*, **48**, 1173–1182.

Eggenberger, F. and Rowell-Rahier, M. 1991. Chemical defense and genetic-variation—interpopulational study of *Oreina gloriosa* (Coleoptera, Chrysomelidae). *Naturwissenschaften*, **78**, 317–320.

Eggenberger, F. and Rowell-Rahier, M. 1992. Genetic component of variation in chemical defense of *Oreina gloriosa* (Coleoptera, Chrysomelidae). *Journal of Chemical Ecology*, **18**, 1375–1404.

Eggenberger, F. and Rowell-Rahier, M. 1993. Physiological sources of variation in chemical defense of *Oreina gloriosa* (Coleoptera, Chrysomelidae). *Journal of Chemical Ecology*, **19**, 395–410.

Eisner, T., Meinwald, J., Monro, A. and Ghent, R. 1961. Defence mechanisms of arthropods-I. The composition and function of the spray of the whipscorpian, *Mastigoproctus giganteus* (Lucas) (Arachnida, Pedipalpida). *Journal of Insect Physiology*, 6, 272–298.

Eisner, H. E., Alsop, D. W. and Eisner, T. 1967. Defense mechanisms of arthropods. XX. Quantitative assessment of hydrogen cyanide production in two species of millipedes. *Psyche*, **74**, 107–117.

Eisner, T. 1965. Defensive spray in a phasmid insect. *Science*, **148**, 966–968.

Eisner, T. and Aneshansley, D. J. 1999. Spray aiming in the bombardier beetle: Photographic evidence. *Proceedings of the National Academy of Sciences of the United States of America*, **96**, 9705–9709.

Eisner, T., Goetz, M. A., Hill, D. E., Smedley, S. R. and Meinwald, J. 1997. Firefly 'femmes fatales' acquire defensive steroids (lucibufagins) from their firefly prey. *Proceedings of the National Academy of Sciences of the United States of America*, **94**, 9723–9728.

Eisner, T. and Grant, R. P. 1981. Toxicity, odor aversion and 'olfactory aposematism'. *Science*, **213**, 476.

Eisner, T., Hicks, K., Eisner, M. and Robson, D. S. 1978. Wolf-in-sheeps-clothing strategy of a predaceous insect larva. *Science*, **199**, 790–794.

Eisner, T., Meinwald, J., Monro, A. and Ghent, R. 1961. Defence mechanisms of arthropods—I. The composition and function of the spray of the whipscorpian, *Mastigoproctus giganteus* (Lucas) (Arachnida, Pedipalpida). *Journal of Insect Physiology*, **6**, 272–298.

El-Sayed, S. Z., Van Dijken, G. L. and Gonzalez-Rodas, G. 1996. Effects of ultraviolet radiation on marine ecosystems. *International Journal of Environmental Studies*, **51**, 199–216.

Emlen, J. M. 1968. Batesian mimicry—a preliminary theoretical investigation of quantitative aspects. *American Naturalist*, **102**, 235–241.

Endler, J. A. 1978. A predator's view of animal color patterns. *Evolutionary Biology*, **11**, 319–364.

Endler, J. A. 1981. An overview of the relationships between mimicry and crypsis. *Biological Journal of the Linnean Society*, **16**, 25–31.

Endler, J. A. 1984. Progressive background matching in moths, and a quantitative measure of crypsis. *Biological Journal of the Linnean Society*, **22**, 187–231.

Endler, J. A. 1986. Defense against predators. In: *Predator-prey relationships* (Ed. by Feder, M. E. and Lauder, G. V.). Chicago: University of Chicago Press.

Endler, J. A. 1988. Frequency-dependent predation, crypsis and aposematic coloration. *Philosophical Transactions of the Royal Society of London Series B—Biological Sciences*, **319**, 505–523.

Endler, J. A. 1990. On the measurement and classification of color in studies of animal color patterns. *Biological Journal of the Linnean Society*, **41**, 315–352.

Endler, J. A. 1991. Interactions between predators and prey. In: *Behavioural ecology: an evolutionary approach* (Ed. by Krebs, J. A. and Davies, N. B.), pp. 169–196. Oxford: Blackwell Scientific.

Endler, J. A. and Mappes, J. 2004. Predator mixes and the conspicuosness of aposematic signals. *American Naturalist*, **163**, 532–547.

Engen, S., Järvi, T. and Wiklund, C. 1986. The evolution of aposematic coloration by individual selection—a life-span survival model. *Oikos*, **46**, 397–403.

Erickson, J. M. 1973. The utilisation of various *Asclepias* species by larvae of the monarch butterfly *Danaus plexippus*. *Psyche*, **80**, 230–244.

Estabrook, G. F. and Jespersen, D. C. 1974. Strategy for a predator encountering a model–mimic system. *The American Naturalist*, **108**, 443–457.

Evans, H. E. and Eberhard, M. J. W. 1970. *The wasps*. Ann Arbor: University of Michigan Press.

Ewell, A. H., Cullen, J. M. and Woodruff, M. L. 1981. Tonic immobility as a predator-defense in the rabbit (*Oryctolagus cuniculus*). *Behavioral and Neural Biology*, **31**, 483–389.

Feltmate, B. W. and Williams, D. D. 1989. A test of crypsis and predation avoidance in the stonefly Paragnetina media (Plecoptera: Perlidae). *Animal Behaviour*, **37**, 992–999.

Ferguson, G. P., Messenger, J. B. and Budelmann, B. U. 1994. Gravity and light influence the countershading reflexes of the cuttlefish, Sepia officinalis. *Journal of Experimental Biology*, **191**, 247–256.

Fink, L. S. and Brower, L. P. 1981. Birds can overcome the cardenolide defence of monarch butterflies in Mexico. *Nature*, **291**, 67–70.

Fisher, R. A. 1927. On some objections to mimicry theory: statistical and genetic. *Transactions of the Entomological Society of London*, **1909**, 269–278.

Fisher, R. A. 1930. *The genetical theory of natural selection*. Oxford: Clarendon Press.

Fitze, P. S. and Richner, H. 2002. Differential effects of a parasite on ornamental structures based on melanins and carotenoids. *Behavioural Ecology*, **13**, 401–407.

Fitzgibbon, C. D. and Fanshawe, J. H. 1988. Stotting in Thomson's Gazelles: an honest signal of condition. *Behavioural Ecology and Sociobiology*, **23**, 69–74.

Forbes, M. R., Schalk, G., Miller, J. G. and Richardson, J. M. L. 1997. Male–female morph interactions in the damselfly *Nehalennia irene* (Hagen). *Canadian Journal of Zoology*, **75**, 253–260.

Ford, E. B. 1936. The genetics of *Papilio dardanus* Brown (Lep.). *Transactions of the Royal Entomological Society of London*, **85**, 435–466.

Ford, E. B. 1940. Polymorphism and taxonomy. In: *The new systematics* (Ed. by Huxley, J.), pp. 495–513. Oxford: Clarendon Press.

Ford, E. B. 1957. *Buterflies*. London: Collins.

Forkman, B. and Enquist, M. 2000. A method for simulating signal evolution using real animals. *Ethology*, **106**, 887–897.

Forsyth, A. and Alcock, J. 1990. Female mimicry and resource defense polygyny by males of a tropical rove beetle, *Leistotrophus versicolor* (Coleoptera: Staphylilidae). *Behavioral Ecology and Sociobiology*, **26**, 325–330.

Foster, S. A. 1988. Diversionary displays of paternal stickleback. *Behavioral Ecology and Sociobiology*, **22**, 335–340.

Frank, T. M. and Widder, E. A. 2002. Effects of a decrease in downwelling irradiance on the daytime vertical distribution patterns of zooplankton and micronekton. *Marine Biology*, **118**, 279–284.

Franks, D. W. and Noble, J. 2004. Batesian mimics influence mimicry ring evolution. *Proceedings of the Royal Society of London Series B*, **271**, 191–196.

Fryer, J. C. F. 1913. An investigation by pedigree breeding into the polymorphism of *Papilio polytes*. *Philosophical Transactions of the Royal Society*, **204**, 227–254.

Fricke, H. W. 1970. Ein mimetisches Kolloektiv—Beobachtungen an Fischschwarmen, die Seeigel nchahmen. *Mar. Biol.*, **5**, 307–314.

Gagliardo, A. and Guilford, T. 1993. Why do warning-colored prey live gregariously. *Proceedings of the Royal Society of London Series B—Biological Sciences*, **251**, 69–74.

Gallup, G. G. J., Nash, R. F., Donegan, N. H. and McClure, M. K. 1971. The immobility response. A predator-induced reaction in chickens. *Psychological Record*, **21**, 513–519.

Gamberale, G. and Tullberg, B. S. 1996*a*. Evidence for a more effective signal in aggregated aposematic prey. *Animal Behaviour*, **52**, 597–601.

Gamberale, G. and Tullberg, B. S. 1996*b*. Evidence for a peak-shift in predator generalization among aposematic prey. *Proceedings of the Royal Society of London Series B—Biological Sciences*, **263**, 1329–1334.

Gamberale, G. and Tullberg, B. S. 1998. Aposematism and gregariousness: the combined effect of group size and coloration on signal repellence. *Proceedings of the Royal Society of London Series B—Biological Sciences*, **265**, 889–894.

Gamberale-Stille, G. 2000. Decision time and prey gregariousness influence attack probability in naive and experienced predators. *Animal Behaviour*, **60**, 95–99.

Gamberale-Stille, G. 2001. Benefit by contrast: an experiment with live aposematic prey. *Behavioral Ecology*, **12**, 768–772.

Gamberale-Stille, G. and Guilford, T. 2003. Contrast versus colour in aposematic signals. *Animal Behaviour*, **65**, 1021–1026.

Gamberale-Stille, G. and Tullberg, B. S. 1999. Experienced chicks show biased avoidance of stronger signals: an experiment with natural colour variation in live aposematic prey. *Evolutionary Ecology*, **13**, 579–589.

Gamberale-Stille, G. and Tullberg, B. S. 2001. Fruit or aposematic insect? Context-dependent colour preferences in domestic chicks. *Proceedings of the Royal Society of London Series B—Biological Sciences*, **268**, 2525–2529.

Garcia, J. and Koelling, R. A. 1966. Relation of cue to consequence in avoidance learning. *Psychonomic Science*, **4**, 123–124.

Gavagnin, M., Mollo, E., Castelluccio, F., Ghiselin, M. T., Calado, G. and Cimino, G. 2001*a*. Can molluscs biosynthesize typical sponge metabolites? The case of the nudibranch Doriopsilla areolata. *Tetrahedron*, **57**, 8913–8916.

Gavagnin, M., Mollo, E., Calado, G., Fahey, S., Ghiselin, M., Ortea, J. and Cimino, G. 2001*b*. Chemical studies of porostome nudibranchs: comparative and ecological aspects. *Chemoecology*, **11**, 131–136.

Gavish, L. and Gavish, B. 1981. Patterns than conceal a bird's eye. *Zwitschrift fuer Tierpsychologie*, **56,** 193–204.

Gavrilets, S. and Hastings, A. 1998. Coevolutionary chase in two-species systems with applications to mimicry. *Journal of Theoretical Biology*, **191,** 415–427.

Gelperin, A. 1968. Feeding behaviour of praying mantis—a learned modification. *Nature*, **219,** 399–400.

Gendron, R. P. and Staddon, J. E. R. 1983. Searching for cryptic prey: the effect of search rate. *American Naturalist*, **121,** 172–186.

Gentry, G. L. and Dyer, L. A. 2002. On the conditional, nature of neotropical caterpillar defenses against their natural enemies. *Ecology*, **83,** 3108–3119.

Getty, T. 1985. Discriminability and the sigmoid functional response—how optimal foragers could stabilize model–mimic complexes. *American Naturalist*, **125,** 239–256.

Getty, T. 1987. Crypsis, mimicry and switching: the basic similarity of superficially different analyses. *American Naturalist*, **130,** 793–795.

Getty, T. 2002. The discriminating babbler meets the optimal diet hawk. *Animal Behaviour*, **63,** 397–402.

Gibson, D. O. 1974. Batesian mimicry without distastefulness? *Nature*, **250,** 77–79.

Gibson, D. O. 1980. The role of escape in mimicry and polymorphism. I. The response of captive birds to artifical prey. *Biological Journal of the Linnean Society*, **14,** 201–214.

Gibson, D. O. 1984. How is automimicry maintained? In: *The Biology of Butterflies* (Ed. by Vane-Wright, R. I. and Ackery, P. R.). London: Academic Press.

Giguère, L. A. and Northcote, T. G. 1987. Ingested prey increase risks of visual predation in transparent Chaborus larvae. *Oecologia*, **73,** 48–52.

Gilbert, F. 2004. Imperfect mimicry. *Royal Entomological Society Symposium* (in press).

Gilbert, J. J. 1994. Jumping behaviour in the Oligotrich Ciliates *Strobilidium velox* and *Halteria grandinella*, and its significance as a defence against rotifer predators. *Microbial Ecology*, **27,** 189–200.

Gilbert, J. J. 1999. Kariomone-induced morphological defenses in rotifers. In: *The ecology and evolution of inducible defenses* (Ed. by Tollrian, R. and Harvell, C.D.), pp. 127–141. Princeton, NJ: Princeton University Press.

Gillette, R., Huang, R.-C., Hatcher, N. and Moroz, L. 2000. Cost–benefit analysis potential in feeding behavior of a predator snail by integration of hunger, taste, and pain. *Proceedings of the National Academy of Sciences of the United States of America*, **97,** 3585–3590.

Gil-Turnes, M. S., Hay, M. E. and Fenical, W. 1989. Symbiotic marine-bacteria chemically defend crustacean embryos from a pathogenic fungus. *Science*, **246,** 116–118.

Giske, J., Aksnes, D. L., Kaartvedt, S., Lie, U., Nordeide, T., Gro Vea Salvanes, A., Wakili, S. M. and Aadnesen, A. 1990. Vertical distribution and trophic interactions of zooplankton and fish in Masfjorden Norway. *Sarsia*, **75,** 65–81.

Gittleman, J. L. and Harvey, P. H. 1980. Why are distasteful prey not cryptic? *Nature*, **28,** 897–899.

Gittleman, J. L., Harvey, P. H. and Greenwood, P. J. 1980. The evolution of conspicuous colouration: some experiments in bad taste. *Animal Behaviour*, **28,** 897–899.

Glanville, P. W. and Allen, J. A. 1997. Protective polymorphism in populations of computer-simulated moth-like prey. *Oikos*, **80,** 565–571.

Gochfeld, M. 1984. Antipredator behaviour: aggressive and distraction displays of shorebirds. In: *Shorebirds: breeding behaviour and populations. Behaviour of marine mammals* (Ed. by Burger, J. and Olla, B.), pp. 289–377. New York: Plenum Press.

Godin, J.-G. J. and Davis, S. A. 1995*a*. Who dares, benefits: predator approach behaviour in the guppy (*Poecilia reticulata*) deters predator pursuit. *Proceedings of the Royal Society of London B*, **259,** 193–200.

Godin, J.-G. J. and Davis, S. A. 1995*b*. Boldness and predation deterrence: a reply to Milinski and Boltshauser. *Proceedings of the Royal Society of London B*, **262,** 107–112.

Golding, Y. C. and Edmunds, M. 2000. Behavioural mimicry of honeybees (*Apis mellifera*) by droneflies (Diptera : Syrphidae : *Eristalis* spp.). *Proceedings of the Royal Society of London Series B—Biological Sciences*, **267,** 903–909.

Goldschmidt, R. B. 1945. Mimetic polymorphism, a controversial chapter of Darwinism. *Quarterly Review of Biology*, **20,** 147–164.

Goodale, M. A. and Sneddon, I. 1977. Effect of distastefulness of model on predation of artificial Batesian mimics. *Animal Behaviour*, **25,** 660–665.

Götmark, F. 1987. White underparts in gulls function as hunting camouflage. *Animal Behaviour*, **35,** 1786–1792.

Götmark, F. and Hohlfält, A. 1995. Bright male plumage and predation risk in passerine birds: are males easier to detect than females? *Oikos*, **74,** 475–484.

Gould, S. J. 1991. *Bully for Brontosaurus*. London: Penguin.

Grant, B. S. 1999. Fine tuning the peppered moth paradigm. *Evolution*, **53,** 980–984.

Grant, B. S. and Clarke, C. A. 2000. Industrial melanism. In: *Nature encyclopedia of life sciences*. London: Nature Publishing Group.

Greene, H. W. and McDiarmid, R. W. 1981. Coral snake mimicry: does it occur? *Science*, **213,** 1207–1212.

Greenwood, J. J. D. 1984. The functional basis for frequency-dependent food selection. *Biological Journal of the Linnean Society*, **23,** 177–199.

Greenwood, J. J. D. 1986. Crypsis, mimicry, and switching by optimal foragers. *American Naturalist*, **128,** 294–300.

Greenwood, J. J. D., Cotton, P. A. and Wilson, D. M. 1989. Frequency-dependent delection on aposematic prey—some experiments. *Biological Journal of the Linnean Society*, **36,** 213–226.

Greenwood, J. J. D. and Elton, R. A. 1979. Analysing experiments on frequency dependent selection by predators. *Journal of Animal Ecology*, **48,** 721–737.

Greenwood, J. J. D., Wood, E. M. and Batchelor, S. 1981. Apostatic selection of distasteful prey. *Heredity*, **47,** 27–34.

Grill, C. P. and Moore, A. J. 1998. Effects of a larval antipredator response and larval diet on adult phenotype in an aposematic ladybird beetle. *Oecologia*, **114,** 274–282.

Guilford, T. 1985. Is kin selection involved in the evolution of warning coloration. *Oikos*, **45,** 31–36.

Guilford, T. 1986. How do warning colors work? Conspicuousness may reduce recognition errors in experienced predators. *Animal Behaviour*, **34,** 286–288.

Guilford, T. 1988. The evolution of conspicuous coloration. *American Naturalist (Supplement)*, **131,** S7-S21.

Guilford, T. 1990*a*. The evolution of aposematism. In: *Insect defenses: Apative mechanisms and strategies of prey and predators* (Ed. by Schmidt, J. O. and Evans, D. L.), pp. 23–61. New York: State University of New York Press.

Guilford, T. 1990*b*. Evolutionary pathways to aposematism. *Acta Oecologica—International Journal of Ecology*, **11,** 835–841.

Guilford, T. 1992. Predator psychology and the evolution of prey coloration. In: *Natural enermies: the population biology of predators, parasites and diseases* (Ed. by Crawley, M. J.). Oxford: Blackwell.

Guilford, T. 1994. Go-slow signaling and the problem of automimicry. *Journal of Theoretical Biology*, **170,** 311–316.

Guilford, T. and Cuthill, I. 1989. Aposematism and bioluminescence. *Animal Behaviour*, **37,** 339–341.

Guilford, T. and Cuthill, I. 1991. The evolution of aposematism in marine gastropods. *Evolution*, **45,** 449–451.

Guilford, T. and Dawkins, M. S. 1987. Search images not proven: a reappraisal of recent evidence. *Animal Behaviour*, **35,** 1838–1845.

Guilford, T. and Dawkins, M. S. 1989*b*. Search image versus search rate—2 different ways to enhance prey capture. *Animal Behaviour*, **37,** 163–165.

Guilford, T. and Dawkins, M. S. 1989*a*. Search image versus search rate—a reply. *Animal Behaviour*, **37,** 160–162.

Guilford, T. and Dawkins, M. S. 1991. Receiver psychology and the evolution of animal signals. *Animal Behaviour*, **42,** 1–14.

Guilford, T. and Dawkins, M. S. 1993. Are warning colors handicaps. *Evolution*, **47,** 400–416.

Guilford, T., Nicol, C., Rothschild, M. and Moore, B. P. 1987. The biological roles of pyrazines—evidence for a warning odor function. *Biological Journal of the Linnean Society*, **31,** 113–128.

Gullan, P. J. and Cranston, P. S. 1994. *The insects: an outline of entomology*. London: Chapman & Hall.

Guthrie, R. D. and Petocz, R. G. 1970. Weapon automimicry among animals. *American Naturalist*, **104,** 585–588.

Hadeler, K. P., Demottoni, P. and Tesei, A. 1982. Mimetic gain in Batesian and Müllerian mimicry. *Oecologia*, **53,** 84–92.

Hafernik, J., and Saul-Gershenz, L. 2000. Beetle larvae cooperate to mimic bees. *Nature*, **405,** 35–36.

Hagen, S. B., Folstad, I. and Jakobsen, J. W. 2003. Autumn colouration and herbivore resistance in mountain birch (*Betula pubescens*). *Ecology Letters*, **6,** 807–811.

Hagman, M. and Forsman, A. 2003. Correlated evolution of conspicuous coloration and body size in poison frogs (Dendrobatidae). *Evolution*, **57,** 2904–2910.

Hamilton, W. D. and Brown, S. P. 2001. Autumn tree colours as a handicap signal. *Proceedings of the Royal Society of London Series B—Biological Sciences*, **268,** 1489–1493.

Hamner, W. M. 1995. Predation, cover, and convergent evolution in epipelagic oceans. *Marine and Freshwater Behaviour and Physiology*, **26,** 71–89.

Hancox, A. P. and Allen, J. A. 1991. A simulation of evasive mimicry in the wild. *Journal of Zoology*, **223,** 9–13.

Hanlon, R. T., Forsythe, J. W. and Joneschild, D. 1999. Crypsis, conspicuousness, mimicry and polyphenism as antipredator defences of foraging octopuses on Indo-Pacific coral reefs, with a method of quantifying crypsis from video tapes. *Biological Journal of the Linnean Society*, **66,** 1–22.

Hanson, H. M. 1959. Effects of discrimination training on stimulus generalization. *Journal of Experimental Psychology*, **58.**

Härlin, C. and Härlin, M. 2003. Towards a historization of aposematism. *Evolutionary Ecology*, **17,** 197–212.

Harper, G. and Whittaker, J. B. 1976. Role of natural enemies in color polymorphism of *Philaenus spumarius* (L.). *Journal of Animal Ecology*, **45,** 91–104.

Harper, R. D. and Case, J. F. 1999. Disruptive counter-illumination and its antipredatory value in the plainfish midshipman Porichthys notatus. *Marine Biology*, **134,** 529–540.

Harvell, C. D. and Tollrian, R. 1999. Why inducible defences? In: *The ecology and evolution of inducible defenses* (Ed. by Tollrian, R. and Harvell, C. D.). Princeton, NJ: Princeton University Press.

Harvey , P. H., Birley, N. and Blackstock, T. H. 1975. The effect of experience on selective behaviour of song thrushes feeding on artificial populations of Cepaea (Held.). *Genetica*, **45**, 211–216.

Harvey, P. H., Bull, J. J., Pemberton, M. and Paxton, R. J. 1982. The evolution of aposematic coloration in distasteful prey—a family model. *American Naturalist*, **119,** 710–719.

Hassell, M. P. 1978. The dynamics of arthropod predation. Princeton, NJ: Princeton University Press.

Hasson, O. 1991. Pursuit-deterent signals: communication between prey and predator. *Trends in Ecology and Evolution*, **6,** 325–329.

Hasson, O., Hibbard, R. and Ceballos, G. 1989. The pursuit deterrent function of tail wagging in the zebra-tailed lizard. *Canadian Journal of Zoology*, **67,** 1203–1209.

Hay, M. E. 1996. Marine chemical ecology: What's known and what's next? *Journal of Experimental Marine Biology and Ecology*, **200,** 103–134.

Hay, M. E. and Fenical, W. 1988. Marine plant–herbivore interactions—the ecology of chemical defense. *Annual Review of Ecology and Systematics*, **19,** 111–145.

Haynes, K. F., Gemeno, C., Yeargan, K. V., Millar, J. G. and Johnson, K. M. 2002. Aggressive chemical mimicry of moth pheromones by a bolas spider: how does this specialist predator attract more than one species of prey? *Chemoecology*, **12,** 99–105.

Hebert, P. 1974. Spittlebug morph mimics avian excrement. *Nature*, **250,** 352.

Hecht, M. K. and Marien, D. 1956. The coral snake mimic problem: a re-interpretation. *Journal of Morphology*, **98,** 335–356.

Heikertinger, F. 1919. Zur Losung des trutzfarbungsproblems der fall *Pyrrhocoris apterus* und das prinzip der ungerohntfarbung. *Wien. Entomol. Ztg.*, **37,** 179–196.

Heiling, A. M., Herberstein, M. E. and Chittka, L. 2003. Crab-spiders manipulate flower signals. *Nature*, **421,** 334.

Hennig, C. W., Dunlap, W. P. and Gallup, G. G. J. 1976. Effect of defensive distance and oppurtunity to escape on tonic immobility in Anolis carolinensis. *Psychological Record*, **26,** 313–320.

Herring, P. J. 1994. Reflective systems in aquatic animals. *Comparative Biochemistry and Physiology*, **109A,** 513–546.

Herring, P. 2002. The biology of the deep ocean. Oxford: Oxford University Press.

Hespenheide, H. A. 1975. Reversed sex-limited mimicry in a beetle. *Evolution*, **29,** 780–783.

Hessen, D. O. 1985. Selective zooplankton predation by pre-adult roach (*Rutilus rutilus*): the size selective versus the visibility-selective hypothesis. *Hydrobiology*, **124,** 73–79.

Hetz, M. and Slobodchikoff, C. N. 1988. Predation pressure on an imperfect Batesian mimicry complex in the presence of alternative prey. *Oecologia*, **76,** 570–573.

Hetz, M. and Slobodchikoff, C. N. 1990. Reproduction and the energy-cost of defense in a Batesian mimicry complex. *Oecologia*, **84,** 69–73.

Hileman, K. S., Brodie, E. D. and Formanowicz, D. R. 1995. Avoidance of unpalatable prey by predaceous diving beetle larvae—the role of hunger level and experience (Coleoptera, Dytiscidae). *Journal of Insect Behaviour*, **8,** 241–249.

Högstedt, G. 1983. Adaptation unto death: function of fear screams. *American Naturalist*, **121,** 562–570.

Holley, A. J. F. 1993. Do brown hares signal to foxes? *Ethology*, **94,** 21–30.

Holling, C. S. 1965. The functional response of predators to prey density, and its role in mimicry and population regulation. *Memoirs of the Entomological Society of Canada*, **45,** 1–60.

Holloway, G. J., de Jong, P. W., Brakefield, P. M. and de Vos, H. 1991. Chemical defence in ladybird beetles (Coccinellidae). I. Distribution of coccinelline and individual variation in defence in 7-spot ladybirds (*Coccinella septempunctata*). *Chemoecology*, **2,** 7–14.

Holm, E. and Kirsten, J. F. 1979. Pre-adaptation and speed mimicry among Namib Desert scarabaeids with orange elytra. *Journal of Arid Environments*, **2,** 263–271.

Holmgren, N. M. A. and Enquist, M. 1999. Dynamics of mimicry evolution. *Biological Journal of the Linnean Society*, **66,** 145–158.

Holopainen, J. K. and Peltonen, P. 2002. Bright autumn colours of deciduous trees attract aphids: nutrient retranslocation hypothesis. *Oikos*, **99,** 184–188.

Honig, W. K. 1993. The stimulus revisited: my how you've grown? In: *Animal cognition: a tribute to Donald A. Riley* (Ed. by Zentall, T. R.), pp. 19–33: L. Erlbaum.

Hooper, J. 2002. *Of moths and men: intrigue, tragedy and the peppered moth*. London: Fourth Estate.

Hosking, E. 1970. *An Eye for a Bird*. London: Hutchinson.

Howarth, B. and Edmunds, M. 2000. The phenology of Syrphidae (Diptera): are they Batesian mimics of Hymenoptera. *Biological Journal of the Linnean Society*, **71,** 437–457.

Howse, P. E. and Allen, J. A. 1994. Satyric mimicry—the evolution of apparent imperfection. *Proceedings of the Royal Society of London Series B—Biological Sciences*, **257,** 111–114.

Huey, R. B. and Pianka, E. R. 1977. Natural selection for juvenile lizards mimicking noxious beetles. *Science*, **195,** 201–203.

Huheey, J. E. 1964. Studies of warning coloration and mimicry. IV. A mathematical model of model–mimic frequencies. *Ecology*, **45,** 185–188.

Huheey, J. E. 1976. Studies in warning coloration and mimicry. VII. Evolutionary consequences of a Batesian–Müllerian spectrum: a model for Müllerian mimicry. *Evolution*, **30,** 86–93.

Huheey, J. E. 1980. Studies in warning coloration and mimicry. VIII. Further evidence for a frequency-dependent model of predation. *Journal of Herpetology*, **14,** 223–230.

Huheey, J. E. 1988. Mathematical models of mimicry. *American Naturalist*, **31,** S22–S41.

Humphries, D. A. and Driver, P. M. 1970. Protean defence by prey animals. *Oecologia*, **5,** 285–302.

Hunter, A. F. 2000. Gregariousness and repellent defences in the survival of phytophagous insects. *Oikos*, **91,** 213–224.

Ingalls, V. 1993. Startle and habituation responses of blue jays (*cyanocitta cristata*) in a laboratory simulation of anti-predator defense of catocala moth (*Lepidoptera noctuidae*). *Behaviour*, **1993,** 77–96.

Iyengar, E. V. and Harvell, C. D. 2002. Specificity of cues inducing defensive spines in the bryozoan Membranipora membranacea. *Marine Ecology-Progress Series*, **225,** 205–218.

Jackson, J. F. and Drummond, B. A. 1974. Batesian ant-mimicry complex from Mountain Pine Ridge of British Honduras, with an example of transformational mimicry. *American Midland Naturalist*, **91,** 248–251.

Janssen, J. 1981. Searching for zooplankton just outside Snell's window. *Limnology and Oceanography*, **26,** 1168–1171.

Jansson, L. and Enquist, M. 2003. Receiver bias for colourful signals. *Animal Behaviour*, **66,** 965–971.

Järvi, T., Sillèn-Tullberg, B. and Wiklund, C. 1981. The cost of being aposematic—an experimental-study of predation on larvae of *Papilio-machaon* by the great tit, *Parus major*. *Oikos*, **36,** 267–272.

Jeffords, M. R., Sternburg, J. G. and Waldbauer, G. P. 1979. Batesian mimicry—field demonstration of the survival value of pipevine swallowtail and monarch color patterns. *Evolution*, **33,** 275–286.

Jeffries, M. 1988. Individual vulnerability to predation: the effect of alternative prey types. *Freshwater Biology*, **19,** 49–56.

Jetz, W., Rowe, C. and Guilford, T. 2001. Non-warning odors trigger innate color aversions—as long as they are novel. *Behavioral Ecology*, **12,** 134–139.

Jiggins, C. D. and McMillan, W. O. 1997. The genetic basis of an adaptive radiation: warning colour in two *Heliconius* species. *Proceedings of the Royal Society of London Series B—Biological Sciences*, **264,** 1167–1175.

Johannesson, K. and Ekendahl, A. 2002. Selective predation favouring cryptic individuals of marine snails (Littorina). *Biological Journal of the Linnean Society*, **76,** 137–144.

Johansson, F. 2002. Reaction norms and production costs of predator-induced morphological defences in a larval dragonfly (Odonata: *Leucorrhinia dubia*). *Canadian Journal of Zoology—Revue Canadienne De Zoologie*, **80,** 944–950.

Johnsen, S. 2001. Hidden in plain sight: the ecology and physiology of organism transparency. *Biological Bulletin*, **201,** 301–318.

Johnsen, S. and Widder, E. A. 1998. Transparency and visibility of gelatinous zooplankton from the Northwestern Atlantic and the Gulf of Mexico. *Biological Bulletin*, **195,** 337–348.

Johnsen, S. and Widder, E. A. 2001. Ultraviolet absorption in transparent zooplankton and its implications for depth distribution and visual predation. *Marine Biology*, **138,** 717–730.

Johnson, S. D. 1994. Evidence for Batesian mimicry in a butterfly-pollinated orchid. *Biological Journal of the Linnean Society*, **53,** 91–104.

Johnson, S. D. 2000. Batesian mimicry in the non-rewarding orchid *Disa pulchra*, and its consequences for pollinator behaviour. *Biological Journal of the Linnean Society*, **71,** 119–132.

Johnstone, R. A. 2002. The evolution of inaccurate mimics. *Nature*, **418,** 524–526.

Johnstone, R. A. and Grafen, A. 1993. Dishonesty and the handicap principle. *Animal Behaviour*, **46,** 759–764.

Jones, B. R. 1986. Responses of domestic chicks to novel food as a function of sex, strain and previous experience. *Behavioural Processes*, **12,** 261–271.

Jones, D. A., Parsons, J. and Rothschild, M. 1962. Release of Hydocyanic acid from crushed tissues of all stages in the life-cycle of species of the Zygaeninae (Lepidoptera). *Nature*, **193,** 52–53.

Jones, F. M. 1932. Insect colouration and the relative acceptability of insects to birds. *Transactions of the Royal Entomological Society of London*, **80,** 345–386.

Jones, J. S., Leith, B. H. and Rawlings, P. 1977. Polymorphism in Cepaea: a problem with too many solutions. *Annual Reviews in Ecology and Systematics*, **8,** 109–143.

Jormalainen, V., Merilaita, S. and Tuomo, J. 1995. Differential predation on sexes affects colour polymorphism of the isopod Idotea baltica (Pallas). *Biological Journal of the Linnean Society*, **55,** 45–68.

Joron, M. 2003. Mimicry. In: *Encyclopedia of Insects* (Ed. by Cardé, R. T. and Resh, V. H.), pp. 417–426. New York: Academic Press.

Joron, M. and Mallet, J. L. B. 1998. Diversity in mimicry: paradox or paradigm? *Trends in Ecology and Evolution*, **13,** 461–466.

Joron, M., Wynne, I. R., Lamas, G. and Mallet, J. 2001. Variable selection and the coexistence of multiple mimetic forms of the butterfly *Heliconius numata*. *Evolutionary Ecology*, **13,** 721–754.

Kannan, D. 1983. A Markov-chain analysis of predator strategy in a model–mimic system. *Bulletin of Mathematical Biology*, **45,** 347–400.

Kapan, D. D. 2001. Three-butterfly system provides a field test of Müllerian mimicry. *Nature*, **409,** 338–340.

Karplus, I. and Algom, D. 1981. Visual cues for predator face recognition by reef fish. *Zeitschrift fur Tierpsychologie*, **55,** 343–364.

Kauffman, S. A. 1993. *The origins of order: self-organisation and selection in evolution*. Oxford: Oxford University Press.

Kauppinen, J. and Mappes, J. 2003. Why are wasps so intimidating: field experiments on hunting dragonflies (Odonata: *Aeshna grandis*). *Animal Behaviour*, **66,** 505–511.

Kearsley, M. J. C. and Whitham, T. G. 1992. Guns and butter—a no cost defense against predation for *Chrysomela-confluens*. *Oecologia*, **92,** 556–562.

Kelley, R. B., Seiber, J. N., Jones, A. D., Segall, A. D. and Brower, L. P. 1987. Pyrrolizidine alkaloids in overwintering monarch butterflies (*Danaus plexippus*) from Mexico. *Experientia*, **43,** 943–946.

Kelly, D. J. and Marples, N. M. 2004. The effects of novel odour and colour cues on food acceptance by the zebra finch (*Taenipygia gvttata*). *Animal Behaviour* (in press).

Kerfoot, W. C. 1982. A question of taste—crypsis and warning coloration in fresh-water zooplankton communities. *Ecology*, **63,** 538–554.

Kettlewell, H. B. D. 1955. Selection experiments on industrial melanism in the Lepidoptera. *Heredity*, **9,** 323–342.

Kettlewell, H. B. D. 1956. Further selection experiments on industrial melanism in the lepidoptera. *Heredity*, **10,** 214–224.

Kettlewell, H. B. D. 1973. *The evolution of melanism*. Oxford: Clarendon Press.

Kiltie, R. A. 1988. Countershading: universally deceptive or deceptively universal. *Trends in Ecology and Evolution*, **3,** 21–33.

Kiltie, R. A. 1989. Testing Thayer's countershading hypothesis: an image processing approach. *Animal Behaviour*, **38,** 542–552.

Kingsolver, J. G. and Srygley, R. B. 2000. Experimental analyses of body size, flight and survival in pierid butterflies. *Evolutionary Ecology Research*, **2,** 593–612.

Kirby, W. and Spence, W. 1823. *An Introduction to Entomology, Volume 2*. Longman, Hurst, Rees, Orme and Brown.

Klump, G. M., Kretzschmar, E. and Curio, E. 1986. The hearing of an avian predator and its prey. *Behavioural Ecology and Sociobiology*, **18,** 317–323.

Klump, G. M. and Shalter, M. D. 1984. Accoustic behaviour of birds and mammals in the predator context. *Z. tierpsychol.*, **66,** 189–226.

Knill, R. and Allen, J. A. 1995. Does polymorphism protect? An experiment with human 'predators'. *Ethology*, **99,** 127–138.

Kobayashi, J. and Ishibashi, M. 1993. Bioactive metabolites of symbiotic marine microorganisms. *Chemical Reviews*, **93,** 1753–1769.

Kokko, H., Mappes, J. and Lindström, L. 2003. Alternative prey can change model–mimic dynamics between parasitism and mutualism. *Ecology Letters*, **6,** 1068–1076.

Korner, H. K. 1982. Countershading by physiological colour change in the fish louse Anilocra physodes L. (Crustacea: Isopoda). *Oecologia*, **55,** 248–250.

Kraemer, P. J. 1984. Forgetting of visual discriminations by pigeons. *Journal of Experimental Psychology: Animal Behavior Processes*, **10,** 530–542.

Kraemer, P. J. and Golding, J. M. 1997. Adaptive forgetting in animals. *Psychonomic Bulletin and Review*, **4,** 480–491.

Krebs, R. A. and West, D. A. 1988. Female mate preference and the evolution of female-limited Batesian mimicry. *Evolution*, **42,** 1101–1104.

Kuhlmann, H. W., Kusch, J. and Heckman, K. 1999. Predator-induced defenses in ciliated protozoa. In: *The ecology and evolution of inducible defenses* (Ed. by Tollrian, R. and Harvell, C.D.), pp. 142–159. Princeton, NJ: Princeton University Press.

Kusch, J. 1995. Adaptation of inducible defense in *Euplotes daidaleos* (Ciliophora) to predation risks by various predators. *Microbial Ecology*, **30,** 79–88.

Kusch, J. and Kuhlmann, H. W. 1994. Cost of stenostomum-induced morphological defense in the Ciliate, *Euplotes octocarinatus*. *Archiv Fur Hydrobiologie*, **130,** 257–267.

Land, M. F. 2000. On the functions of double eyes in mid-water animals. *Philosophical Transactions of the Royal Society of London B*, **355,** 1147–1150.

Land, M. F. and Nilsson, D.-E. 2002. *Animal eyes*. Oxford: Oxford University Press.

Langley, C. M., Riley, D. A., A. C., B. and Goel, N. 1996. Visual search for natural grains by pigeons (*Columba livia*): search images and selective attention. *Journal of Experimental Psychology*, **22**, 139–151.

Lardner, B. 1998. Plasticity or fixed adaptive traits? Strategies for predation avoidance in *Rana arvalis* tadpoles. *Oecologia*, **117**, 119–126.

Lariviere, S. and Messier, F. 1996. Aposematic behaviour in the striped skunk, *Mephitis mephitis*. *Ethology*, **102**, 986–992.

Latz, M. I. and Case, J. F. 1982. Light organ and eyeshalk compensation to body tilt in the luminescent shrimp, *Sergestes similis*. *Journal of Experimental Biology*, **98**, 83–104.

Layberry, R. A., Hall, P. W. and Lafontaine, J. D. 1998. *The butterflies of Canada*. Toronto: University of Toronto press.

Leal, M. 1999. Honest signalling during prey-predator interactions in the lizard Anolis cristatellus. *Animal Behaviour*, **58**, 521–526.

Leal, M. and Rodriguez-Robles, J. A. 1995. Anti-predator responses of *Anolis cristatellus* (Sauria: Polychrotidae). *Copeia*, **1995**, 155–161.

Leal, M. and Rodriguez-Robles, J. A. 1997. Signalling displays during predator–prey interactions in a Puerto Rican anole Anolis cristatellus. *Animal Behaviour*, **54**, 1147–1154.

Lederhouse, R. C. and Scriber, J. M. 1996. Intrasexual selection constrains the evolution of the dorsal color pattern of male black swallowtail butterflies, *Papilio polyxenes*. *Evolution*, **50**, 717–722.

Leimar, O., Enquist, M. and Sillén-Tullberg, B. 1986. Evolutionary stability of aposematic coloration and prey unprofitability—a theoretical-analysis. *American Naturalist*, **128**, 469–490.

Leimar, O. and Tuomi, J. 1998. Synergistic selection and graded traits. *Evolutionary Ecology*, **12**, 59–71.

Leonard, G. H., Bertness, M. D. and Yund, P. O. 1999. Crab predation, waterborne cues, and inducible defenses in the blue mussel, *Mytilus edulis*. *Ecology*, **80**, 1–14.

Lev-Yadun, S. 2001. Aposematic (warning) coloration associated with thorns in higher plants. *Journal of Theoretical Biology*, **210**, 385-U1.

Lev-Yadun, S. 2003. Weapon (thorn) automimicry and mimicry of aposematic colorful thorns in plants. *Journal of Theoretical Biology*, **224**, 183–188.

Lev-Yadun, S. and Inbar, M. 2002. Defensive ant, aphid and caterpillar mimicry in plants? *Biological Journal of the Linnean Society*, **77**, 393–398.

Li, G., Roze, U. and Locke, D. C. 1997. Warning odor of the North America porcupine (*Erethizon dorsatum*). *Journal of Chemical Ecology*, **23**, 2737–2754.

Lincoln, D. E. and Langenheim, J. H. 1978. Effect of light and temperature on monoterpenoid yield and composition in *Satureja douglassi*. *Biochemical Systematics and Ecology*, **6**, 21–32.

Lincoln, D. E. and Langenheim, J. H. 1979. Variation of *Satureja douglassi* monoterpenoids in relation to light intensity and herbivory. *Biochemical Systematics and Ecology*, **7**, 289–298.

Lincoln, D. E. and Mooney, H. A. 1984. Herbivory on *Diplacus aurantiacus* shrubs in sun and shade. *Oecologia*, **64**, 173–177.

Lindell, L. E. and Forsman, A. 1996. Sexual dichromatism in snakes: support for the flicker-fusion hypothesis. *Canadian Journal of Zoology*, **74**, 2254–2256.

Lindquist, N. 2002. Chemical defense of early life stages of benthic marine invertebrates. *Journal of Chemical Ecology*, **28**, 1987–2000.

Lindquist, N. and Hay, M. E. 1996. Palatability and chemical defense of marine invertebrate larvae. *Ecological Monographs*, **66**, 431–450.

Lindquist, N., Hay, M. E. and Fenical, W. 1992. Defense of ascidians and their conspicuous larvae—adult vs larval chemical defenses. *Ecological Monographs*, **62**, 547–568.

Lindroth, C. H. 1971. Disappearance as a protective factor. *Entomologica Scandinavica*, **2**, 41–48.

Lindström, L. 2001. Experimental approaches to studying the initial evolution of conspicuous aposematic signalling. *Evolutionary Ecology*, **13**, 605–618.

Lindström, L., Alatalo, R. V., Lyytinen, A. and Mappes, J. 2001*a*. Strong antiapostatic selection against novel rare aposematic prey. *Proceedings of the National Academy of Sciences of the United States of America*, **98**, 9181–9184.

Lindström, L., Alatalo, R. V., Lyytinen, A. and Mappes, J. 2001*b*. Predator experience on cryptic prey affects the survival of conspicuous aposematic prey. *Proceedings of the Royal Society of London Series B—Biological Sciences*, **268**, 357–361.

Lindström, L., Rowe, C. and Guilford, T. 2001*c*. Pyrazine odour makes visually conspicuous prey aversive. *Proceedings of the Royal Society of London Series B—Biological Sciences*, **268**, 159–162.

Lindström, L., Alatalo, R. V. and Mappes, J. 1997. Imperfect Batesian mimicry—the effects of the frequency and the distastefulness of the model. *Proceedings of the Royal Society of London Series B—Biological Sciences*, **264**, 149–153.

Lindström, L., Alatalo, R. V. and Mappes, J. 1999*a*. Reactions of hand-reared and wild-caught predators toward warningly colored, gregarious, and conspicuous prey. *Behavioral Ecology*, **10**, 317–322.

Lindström, L., Alatalo, R. V., Mappes, J., Riipi, M. and Vertainen, L. 1999*b*. Can aposematic signals evolve by gradual change? *Nature*, **397**, 249–251.

Linsley, E. G., Eisner, T. and Klots, A. B. 1961. Mimetic assemblages of sibling species of lycid beetles. *Evolution*, **15**, 15–29.

Lloyd, J. E. 1965. Aggressive mimicry in *Photuris*—firefly femmes fatales. *Science*, **149**, 653–654.

Lloyd, J. E. 1975. Aggressive mimicry in *Photuris* fireflies—signal repertoires by femmes fatales. *Science*, **187**, 452–453.

Lloyd Morgan, C. 1896. *Habit and instinct*. London: Arnold.

Losey, G. S. 1972. Predation protection in the poison-fang blenny, *Meiacanthus atrodorsalis*, and its mimics, *Escenius bicolor* and *Runula laudadus* (Blenniidae). *Pac. Sci.*, **26**, 129–139.

Losey, G. S., Cronin, T. H., Goldsmith, T. H., Hyde, D., Marshall, N. J. and McFarland, W. N. 1999. The UV visual world of fishes: a review. *Journal of Fish Biology*, **54**, 921–943.

Losey, J. E., Ives, A. R., Harmon, J., Ballantyne, F. and Brown, C. 1997. Polymorphism maintained by opposite patterns of parasitism and predation. *Nature*, **388**, 269–272.

Louda, S. M. and Rodman, J. E. 1983*a*. Ecological patterns in the glucosinolate content of a native mustard, *Cardamine cordifolia*, in the Rocky Mountains. *Journal of Chemical Ecology*, **9**, 397–421.

Louda, S. M. and Rodman, J. E. 1983*b*. Concentrations of glucosinolates in relation to habitat and insect herbivory for the native crucifer *Cardamine cordifolia*. *Biochemical Systematics and Ecology*, **11**, 199–207.

Luedeman, J. K., McMorris, F. R. and Warner, D. D. 1981. Predators encountering a model–mimic system with alternative prey. *American Naturalist*, **117**, 1040–1048.

Lythgoe, J. N. 1979. *The ecology of vision*. London: Clarendon Press.

Lyytinen, A., Brakefield, P. M., Lindström, L. and Mappes, J. 2004. Does predation maintain eyespot plasticity in *Bicyclus anynana*? *Proceedings of the Royal Society of London B*, **271**, 279–284.

Lyytinen, A., Brakefield, P. M. and Mappes, J. 2003. Significance of butterfly eyespots as an anti-predator device in ground-based and aerial attacks. *Oikos*, **100**, 373–379.

MacDougall, A. and Dawkins, M. S. 1998. Predator discrimination error and the benefits of Müllerian mimicry. *Animal Behaviour*, **55**, 1281–1288.

Mackintosh, J. A. 2001. The antimicrobial properties of melanoctes, melanosomes and melanin and the evolution of black skin. *Journal of Theoretical Biology*, **212**, 128.

Majerus, M. E. N. 1988. *Melanism: evolution in action*. Oxford: Oxford University Press.

Majerus, M. E. N., Burton, C. F. A. and Stalker, J. 2000. A bird's eye view of the peppered moth. *Journal of Evolutionary Biology*, **13**, 155–159.

Malcolm, S. B. 1990. Mimicry—status of a classical evolutionary paradigm. *Trends in Ecology and Evolution*, **5**, 57–62.

Malcolm, S. B. and Brower, L. P. 1989. Evolutionary and ecological implications of cardenolide sequestration in the monarch butterfly. *Experientia*, **45**, 284–294.

Mallet, J. 2001*a*. Causes and consequences of a lack of coevolution in Müllerian mimicry. *Evolutionary Ecology*, **13**, 777–806.

Mallet, J. 2001*b*. Mimicry: an interface between psychology and evolution. *Proceedings of the National Academy of Sciences of the United States of America*, **98**, 8928–8930.

Mallet, J., Barton, N., Lamas, G., Santisteban, J., Muedas, M. and Eeley, H. 1990. Estimates of selection and gene flow from measures of cline width and linkage disequilibrium in *Heliconius* hybrid zones. *Genetics*, **124**, 921–936.

Mallet, J. and Barton, N. H. 1989. Strong natural selection in a warning color hybrid zone. *Evolution*, **43**, 421–431.

Mallet, J. and Gilbert, L. E. 1995. Why are there so many mimicry rings—correlations between habitat, behavior and mimicry in *Heliconius* butterflies. *Biological Journal of the Linnean Society*, **55**, 159–180.

Mallet, J., Jiggins, C. D. and McMillan, W. O. 1996. Evolution: mimicry meets the mitochondrion. *Current Biology*, **6**, 937–940.

Mallet, J. and Joron, M. 1999. Evolution of diversity in warning color and mimicry: polymorphisms, shifting balance, and speciation. *Annual Review of Ecology and Systematics*, **30**, 201–233.

Mallet, J., McMilan, W. O. and Jiggins, C. D. 1998. Mimicry and warning color at the boundary between races and species. In: *Endless forms: species and speciation* (Ed. by Howard, D. J. and Berlocher, S. H.), pp. 390–403. New York: Oxford University Press.

Mallet, J. and Singer, M. C. 1987. Individual selection, kin selection, and the shifting balance in the evolution of warning colors—the evidence from butterflies. *Biological Journal of the Linnean Society*, **32**, 337–350.

Mallet, J. L. B. and Turner, J. R. G. 1997. Biotic drift or the shifting balance—did forest islands drive the diversity of warningly coloured butterflies? In: *Evolution on Islands* (Ed. by Grant, P. R. and Clarke, B.), pp. 262–280. Oxford: Oxford University Press.

Mangel, M. and Clark, C. W. 1988. *Dynamic Modeling in Behavioral Ecology*. Princeton, NJ: Princeton University Press.

Mappes, J. and Alatalo, R. V. 1997*a*. Effects of novelty and gregariousness in survival of aposematic prey. *Behavioral Ecology*, **8**, 174–177.

Mappes, J. and Alatalo, R. V. 1997*b*. Batesian mimicry and signal accuracy. *Evolution*, **51,** 2050–2053.

Marden, J. H. and Chai, P. 1991. Aerial predation and butterfly design—how palatability, mimicry, and the need for evasive flight constrain mass allocation. *American Naturalist*, **138,** 15–36.

Marples, N. M., Brakefield, P. M. and Cowie, R. J. 1989. Differences between the 7-spot and 2-spot ladybird beetles (Coccinellidae) in their toxic effects on a bird predator. *Ecological Entomology*, **14,** 79–84.

Marples, N. M. and Kelly, D. J. 2001. Neophobia and dietary conservatism: two distinct processes? *Evolutionary Ecology*, **13,** 641–653.

Marples, N. M. and Roper, T. J. 1996. Effects of novel colour and smell on the response of naive chicks towards food and water. *Animal Behaviour*, **51,** 1417–1424.

Marples, N. M. and Roper, T. J. 1997. Response of domestic chicks to methyl anthranilate odour. *Animal Behaviour*, **53,** 1263–1270.

Marples, N. M., Roper, T. J. and Harper, D. G. C. 1998. Responses of wild birds to novel prey: evidence of dietary conservatism. *Oikos*, **83,** 161–165.

Marples, N. M., Vanveelen, W. and Brakefield, P. M. 1994. The relative importance of color, taste and smell in the protection of an aposematic insect *Coccinella septempunctata*. *Animal Behaviour*, **48,** 967–974.

Marsh, N. A., Clarke, C. A., Rothschild, M. and Kellett, D. N. 1977. *Hypolimnas bolina* L, a mimic of Danaid butterflies, and its model *Euploea core* (Cram) store cardioactive substances. *Nature*, **268,** 726–728.

Marshall, G. A. K. and Poulton, E. B. 1902. Five years' observations and experiments (1896–1901) on the bionomics of South African insects, chiefly directed to the investigation of mimicry and warning colours. *Transactions of the Entomological Society of London* **1902,** 287–697.

Marshall, G. A. K. 1908. On diaposematism, with reference to some limitations of the Müllerian hypothesis of mimicry. *Transactions of the Entomological Society of London*, **1908,** 93–142.

Marshall, N. B. 1971. *Explorations of the life of fishes*. Cambridge, MA: Harvard University Press.

Marshall, N. J. 2000. Communication and camouflage with the same 'bright' colours in reef fishes. *Philosophical Transactions of the Royal Society of London Series B—Biological Sciences*, **355,** 1243–1248.

Mastrota, F. N. and Mench, J. A. 1994. Avoidance of dyed food by the northern bobwhite. *Applied Animal Behaviour Science*, **42,** 109–119.

Mastrota, F. N. and Mench, J. A. 1995. Color avoidance in Northern bobwhites—effects of age, sex and previous experience. *Animal Behaviour*, **50,** 519–526.

Matessi, C. and Cori, R. 1972. Models of population genetics of Batesian mimicry. *Theoretical Population Biology*, **3,** 41-67.

Matthews, E. G. 1977. Signal-based frequency-dependent defense strategies and the evolution of mimicry. *American Naturalist*, **111,** 213–222.

Maynard Smith, J. 2000. *Evolutionary genetics*. Oxford: Oxford University Press.

Maynard Smith, J. and Harper, D. 2003. *Animal signals*. Oxford: Oxford University Press.

McAtee, W. L. 1932*a*. Effectiveness in nature of the so-called protective adaptations in the animal kingdom, chiefly as illustrated by the food habits of nearctic birds. *Smithsonian Miscellaneous Collections*, **85,** 1–201.

McAtee, W. L. 1932*b*. 'Protective' adaptations of animals. *Nature*, **130,** 961–962.

McClintock, J. B. and Baker, B. J. 1997*a*. A review of the chemical ecology of Antarctic marine invertebrates. *American Zoologist*, **37,** 329–342.

McClintock, J. B. and Baker, B. J. 1997*b*. Palatability and chemical defense of eggs, embryos and larvae of shallow-water Antarctic marine invertebrates. *Marine Ecology-Progress Series*, **154,** 121–131.

McClintock, J. B., Baker, B. J. and Steinberg, D. K. 2001. The chemical ecology of invertebrate meroplankton and holoplankton. In: *Marine chemical ecology* (Ed. by McClintock, J. B. and Baker, B. J.), pp. 196–225. New York: CRC Press.

McClintock, J. B. and Janssen, J. 1990. Pteropod abduction as a chemical defense in a Pelagic Antarctic amphipod. *Nature*, **346,** 462–464.

McCollum, S. A. and Leimberger, J. D. 1997. Predator-induced morphological changes in an amphibian: predation by dragonflies affects tadpole shape and color. *Oecologia*, **109,** 615–621.

McCollum, S. A. and VanBuskirk, J. 1996. Costs and benefits of a predator-induced polyphenism in the gray treefrog Hyla chrysoscelis. *Evolution*, **50,** 583–593.

McCosker, J. E. 1977. Fright posture of plesiopid fish *Calloplesiops altivelis*: example of Batesian mimicry. *Science*, **197,** 400–401.

McFall-Ngai, M. and Morin, J. G. 1991. Camouflage by disruptive illumination in leiognathids, a family of shallow-water bioluminescent fishes. *Journal of Experimental Biology*, **156,** 119–137.

McFall-Ngai, M. J. 1990. Crypsis in the pelagic environment. *American Zoologist*, **30,** 175–188.

McIver, J. D. and Stonedahl, G. 1993. Myrmecomorphy—morphological and behavioral mimicry of ants. *Annual Review of Entomology*, **38,** 351–379.

McLain, D. K. and Shure, D. J. 1985. Host plant toxins and unpalatability of *Neacoryphus bicrusis* (Hemiptera: Lygaeidae). *Ecological Entomology*, **10,** 291–298.

McPhail, J. D. 1977. A possible function of the caudal spots in characid fishes. *Canadian Journal of Zoology*, **55,** 1063–1066.

Meadows, D. W. 1993. Morphological variation in eyespots of the foureye butterflyfish (*Chaetodon capistratus*): implications for eyespot function. *Copeia*, **1993,** 235–240.

Mebs, D. 2001. Toxicity in animals. Trends in evolution? *Toxicon*, **39,** 87–96.

Mebs, D. and Kornalik, F. 1984. Intraspecific variation in content of a basic toxin in eastern diamondback rattlesnake (*Crotalus adamanteus*) venum. *Toxicon*, **22,** 831–833.

Merilaita, S. and Kaitala, V. 2002. Community structure and the evolution of aposematic coloration. *Ecology Letters*, **5,** 495–501.

Merilaita, S., Lyytinen, A. and Mappes, J. 2001. Selection for cryptic coloration in visually heterogeneous habitat. *Proceedings of the Royal Society of London B*, **268,** 1925–1929.

Merilaita, S., Tuomi, J. and Jormalainen, V. 1999. Optimization of cryptic coloration in heterogeneous habitats. *Biological Journal of the Linnean Society*, **67,** 151–161.

Merlaita, S. 1998. Crypsis through disruptive coloration in an isopod. *Proceedings of the Royal Society of London B*, **265,** 1059–1064.

Merrill, D. N. and Elgar, M. A. 2000. Red legs and golden gasters: Batesian mimicry in Australian ants. *Naturwissenschaften*, **87,** 212–215.

Messenger, J. B. 1997. Consequences of colour-blindness for cuttlefish camouflage. *Journal of Physiology*, **504P,** 23P.

Milinski, M. and Boltshauser, P. 1995. Boldness and predator deterrence: a critique of Godin and Davis. *Proceedings of the Royal Society of London B*, **262,** 103–105.

Milinski, M., Luthi, J. H., Eggler, R. and Parker, G. A. 1997. Cooperation under predation risk: experiments on costs and benefits. *Proceedings of the Royal Society of London B*, **264,** 831–837.

Miller, L. A. 1991. Arctiid moth clicks can degrade the accuracy of range difference discrimination in ecolocating big brown bats, *Eptesicus fusucs*. *Comparative Physiology A*, **168,** 571–579.

Miller, L. A. and Surlykke, A. 2001. How some insects detect and avoid being eaten by bats: tactics and countertactics of prey and predator. *Bioscience*, **51,** 570–581.

Miller, M. N. and Fincke, O. M. 1999. Cues for mate recognition and the effect of prior experience on mate recognition in *Enallagma* damselflies. *Journal of Insect Behaviour*, **12,** 801–814.

Milne, A. A. 1926 (1994 Edition). *Winnie the Pooh: the complete collection of stories and poems*. Meuthen Children's Books Limited.

Mobley, C. D. 1994. *Light and water: radiative transfer in natural waters*. London: Academic Press.

Moment, G. B. 1962. Reflexive selection: a possible answer to an old puzzle. *Science*, **136,** 262–263.

Moranz, R. and Brower, L. P. 1998. Geographic and temporal variation of cardenolide-based chemical defenses of queen butterfly (*Danaus gilippus*) in northern Florida. *Journal of Chemical Ecology*, **24,** 905–932.

Morgan, S. G. and Christy, J. H. 1996. Survival of marine larvae under countervailing selection pressures of photodamage and predation. *Limnology and Oceanography*, **21,** 498–504.

Morton, T. C. and Vencl, F. V. 1998. Larval beetles form a defense from recycled host-plant chemicals discharged as fecal wastes. *Journal of Chemical Ecology*, **24,** 765–785.

Mostler, G. 1935. Beobachtungen zur frage der wespenmimikry. *Zeitschrift für Morphologie und Ökologie der Tiere*, **29,** 381–454.

Muhtasib, H. and Evans, D. L. 1987. Linamarin and histamine in the defense of adult *Zygaena*- Filipendulae. *Journal of Chemical Ecology*, **13,** 133–142.

Müller, C., Agerbirk, N., Olsen, C. E., Boeve, J. L., Schaffner, U. and Brakefield, P. M. 2001. Sequestration of host plant glucosinolates in the defensive hemolymph of the sawfly *Athalia rosae*. *Journal of Chemical Ecology*, **27,** 2505–2516.

Müller, C., Boeve, J. L. and Brakefield, P. 2002. Host plant derived feeding deterrence towards ants in the turnip sawfly *Athalia rosae*. *Entomologia Experimentalis et Applicata*, **104,** 153–157.

Müller, C., Zwaan, B. J., de Vos, H. and Brakefield, P. M. 2003. Chemical defence in a sawfly: genetic components of variation in relevant life-history traits. *Heredity*, **90,** 468–475.

Müller, F. 1877. *Kosmos*, **December,** vi–vii.

Müller, F. 1878. Über die vortheile der mimicry bei schmetterlingen. *Zoologischer Anzeiger*, **1,** 54–55.

Müller, F. 1879. Ituna and Thyridia: a remarkable case of mimicry in butterflies. *Proceedings of the Entomological Society*, **1879,** xx–xxiv.

Murdoch, W. W. 1969. Switching in general predators: experiments on predator specificity on switching. *Ecological Monographs*, **39,** 335–354.

Nagaishi, H., Nishi, H., Fujii, R. and Oshima, N. 1989. Correlation between body colour and behaviour in the upside-down catfish, Synodontis nigriventis. *Comparative Biochemsitry and Physiology*, **92A,** 323–326.

Nahrstedt, A. 1988. Cyanogenesis and the role of cyanogenic compounds in insects. In: *Cyanide compounds in biology* (Ed. by Evered, D. and Harnett, S.). Chichester: John Wiley/Ciba Foundation.

Nappi, A. J. and Vass, E. 1993. Melanogenesis and the generation of cytotoxic molecules during insect vellular immune reactions. *Pigment Cell Research*, **6,** 117–126.

Neff, B. D., Fu, P. and Gross, M. R. 2003. Sperm investment and alternative mating tactics in bluegill sunfish (*Lepomis macrochirus*). *Behavioral Ecology*, **14,** 634–641.

Neudecker, S. 1989. Eye camouflage and false eyespots: chaetodontid responses to predators. *Environmental Biology of Fishes*, **25,** 143–157.

Neudorf, D. L. and Sealy, S. G. 2002. Distress calls of birds in neotropical cloud forest. *Biotropica*, **34,** 118–126.

Nicholson, A. J. 1927. Presidential Address. A new theory of mimicry in insects. *Australian Zoologist*, **5,** 10–24.

Nijhout, H. F. 1991. *The Development and Evolution of Butterfly Wing Patterns*. Washington: Smithsonian Institute Press.

Nilsson, M. and Forsman, A. 2003. Evolution of conspicuous colouration, body size and gregariousness: a comparative analysis of lepidopteran larvae. *Evolutionary Ecology*, **17,** 51–66.

Nilsson, P. A., Brönmark, C. and Pettersson, L. B. 1995. Benefits of a predator-induced morphology in Crucian carp. *Oecologia*, **104,** 291–296.

Nishida, R. 1994/1995. Sequestration of plant secondary compounds by butterflies and moths. *Chemoecology*, **5/6,** 127–38.

Nishida, R. 2002. Sequestration of defensive substances from plants by Lepidoptera. *Annual Review of Entomology*, **47,** 57–92.

Nonacs, P. 1985. Foraging in a dynamic mimicry complex. *American Naturalist*, **126,** 165–180.

Norman, M. D., Finn, J. and Tregenza, T. 2001. Dynamic mimicry in an Indo-Malayan octopus. *Proceedings of the Royal Society of London Series B—Biological Sciences*, **268,** 1755–1758.

Norris, K. S. and Lowe, C. H. 1964. An analysis of background matching in amphibians and reptiles. *Ecology*, **45,** 565–580.

Novales-Flamarique, I. and Browman, H. I. 2001. Foraging and prey search of small juvenile brown trout (Oncorhynchus mykiss) under polarised light. *Journal of Experimental Biology*, **204,** 2415–2422.

Nur, U. 1970. Evolutionary rates of models and mimics in Batesian mimicry. *American Naturalist*, **104,** 477–486.

Oaten, A., Pearce, C. E. M. and Smyth, M. E. B. 1975. Batesian mimicry and signal-detection theory. *Bulletin of Mathematical Biology*, **37,** 367–387.

O'Brien, T. J. and Dunlap, W. P. 1975. Tonic immobility in the blue crab (Callinectes sapidus, Rathbun): its relation to threat of predation. *Journal of Comparative and Physiological Psychology*, **89,** 86–94.

O'Brien, W. J., Kettle, D. and Riessen, H. 1979. Helmets and invisible armour: structures reducing predation from tactile and visual planktivores. *Ecology*, **60,** 287–294.

O'Donald, P. and Barrett, J. A. 1973. Evolution of dominance in polymorphic Batesian mimicry. *Theoretical Population Biology*, **4,** 173–192.

O'Donald, P. and Pilecki, C. 1970. Polymorphic mimicry and natural selection. *Evolution*, **24,** 395–401.

O'Donald, P. and Pilecki, C. 1974. Polymorphic mimicry and natural selection—reply. *Evolution*, **28,** 484–485.

O'Donnell, S. 1996. Dragonflies (*Gynacantha nervosa* Rambur) aviod wasps (*Polybia aequatorialis* Zavattari and *Mischocyttarus* sp.) as prey. *Journal of Insect Behaviour*, **9,** 159–162.

O'Donnell, S. 1999. Dual mimicry in the dimorphic eusocial wasp *Mischocyttarus mastigophorus* Richards (Hymenoptera : Vespidae). *Biological Journal of the Linnean Society*, **66,** 501–514.

Ohsaki, N. 1995. Preferential predation of female butterflies and the evolution of Batesian mimicry. *Nature*, **378,** 173–175.

Ortolani, A. 1999. Sopts, stripes, tail tips and dark eyes: predicting the function of carnivore colour patterns using the comparative method. *Biological Journal of the Linnean Society*, **67,** 433–476.

Owen, A. R. G. and Owen, R. E. 1978. Mathematical paradigms for Batesian and Müllerian mimicry. *American Zoologist*, **18,** 584.

Owen, D. F., Smith, D. A. S., Gordon, I. J. and Owiny, A. M. 1994. Polymorphic Müllerian mimicry in a group of African butterflies—a reassessment of the relationship between *Danaus-Chrysippus*, *Acraea-Encedon* and *Acraea-Encedana* (Lepidoptera, Nymphalidae). *Journal of Zoology*, **232,** 93–108.

Owen, D. F. and Whiteley, D. 1986. Reflexive selection: Moment's hypothesis resurrected. *Oikos*, **47,** 117–120.

Owen, D. F. and Whiteley, D. 1989. Evidence that reflexive polymorphisms are maintained by visual selection by predators. *Oikos*, **55,** 130–133.

Owen, R. E. and Owen, A. R. G. 1984. Mathematical paradigms for mimicry—recurrent sampling. *Journal of Theoretical Biology*, **109,** 217–247.

Owings, D. H., Rowe, M. P. and Rundus, A. S. 2002. The rattling sound of rattlesnakes (*Crotalus viridis*) as a communicative resource for ground squirrels (*Spermophilus beecheyi*) and burrowing owls (*Athene cunicularia*). *Journal of Comparative Psychology*, **116,** 197–205.

Oxford, G. S. and Gillespie, R. G. 1998. Evolution and ecology of spider coloration. *Annual Review of Entomology*, **43,** 619–643.

Papageorgis, C. 1975. Mimicry in neotropical butterflies. *American Scientist*, **63,** 522–532.

Parrish, M. D. and Fowler, H. G. 1983. Contrasting foraging related behaviors in 2 sympatric wasps (*Vespula-Maculifrons* and *Vespula-Germanica*). *Ecological Entomology*, **8,** 185–190.

Pasteels, J. M. 1982. Is Kairomone a valid and useful term. *Journal of Chemical Ecology*, **8,** 1079–1081.

Pasteels, J. M. 1993. The value of defensive compounds as taxonomic characters in the classification of leaf beetles. *Biochemical Systematics and Ecology*, **21,** 135–142.

Pasteels, J. M., Daloze, D. and Rowell-Rahier, M. 1986. Chemical defense in chrysomelid eggs and neonate larvae. *Physiological Entomology*, **11,** 29–37.

Pasteels, J. M., Dobler, S., Rowell-Rahier, M., Ehmke, A. and Hartmann, T. 1995. Distribution of autogenous and host-derived chemical defenses in *Oreina* leaf beetles (Coleoptera: Chrysomelidae). *Journal of Chemical Ecology*, **21,** 1163–1179.

Pasteels, J. M., Duffey, S. and Rowell-Rahier, M. 1990. Toxins in Chrysomelid beetles—possible evolutionary sequence from denovo synthesis to derivation from food-plant chemicals. *Journal of Chemical Ecology*, **16,** 211–222.

Pasteels, J. M. and Gregoire, J. C. 1984. Selective predation on chemically defended chrysomelid larvae—a conditioning process. *Journal of Chemical Ecology*, **10,** 1693–1700.

Pasteels, J. M., Gregoire, J. C. and Rowell-Rahier, M. 1983. The chemical ecology of defense in arthropods. *Annual Review of Entomology*, **28,** 263–289.

Pasteels, J. M., Rowell-Rahier, M., Braekman, J. C. and Daloze, D. 1984. Chemical defenses in leaf beetles and their larvae—the ecological, evolutionary and taxonomic significance. *Biochemical Systematics and Ecology*, **12,** 395–406.

Pasteels, J. M., Rowell-Rahier, M., Braekman, J. C., Daloze, D. and Duffey, S. 1989. Evolution of exocrine chemical defense in leaf beetles (Coleoptera, Chrysomelidae). *Experientia*, **45,** 295–300.

Pasteur, G. 1982. A classificatory review of mimicry systems. *Annual Review of Ecology and Systematics*, **13,** 169–199.

Paul, N. D. and Gwynn-Jones, D. 2003. Ecological roles of solar UV radiation: towards an integrated approach. *Trends in Ecology and Evolution*, **18,** 48–55.

Pawlik, J. R. 1993. Marine invertebrate chemical defenses. *Chemical Reviews*, **93,** 1911–1922.

Pearce, J. M. and Bouton, M. E. 2001. Theories of associative learning in animals. *Annual Review of Psychology*, **52,** 111–139.

Peck, D. C. 2000. Reflex bleeding in froghoppers (Homoptera : Cercopidae): Variation in behavior and taxonomic distribution. *Annals of the Entomological Society of America*, **93,** 1186–1194.

Peckham, E. G. 1889. Protective resemblances of spiders. *Occasional Papers of the Natural History Society Wiscussion*, **1,** 61–113.

Penney, B. K. 2002. Lowered nutritional quality supplements nudibranch chemical defense. *Oecologia*, **132,** 411–418.

Peterson, G. S., Johnson, L. B., Axler, R. P. and Diamond, S. A. 2002. Assessment of the risk of solar ultraviolet radiation to amphibians II. In situ characterisation of exposure in amphibian habitats. *Environmental Science and Technology*, **36,** 2859–2865.

Pettersson, L. B. and Brönmark, C. 1997. Density-dependent costs of an inducible morphological defense in Crucian carp. *Ecology*, **78,** 1805–1815.

Pfennig, D. W., Harcombe, W. R. and Pfennig, K. S. 2001. Frequency-dependent batesian mimicry : Predators avoid look- alikes of venomous snakes only when the real thing is around. *Nature*, **410,** 323–323.

Phillips, G. C. 1962. Survival value of the white coloration of gulls and other seabirds. University of Oxford.

Pielowski, Z. 1959. Studies on the relationship: predator (goshawk)—prey (pigeon). *Bulletin of the Academy of Polish Science*, **7,** 401–403.

Pietrewicz, A. T. and Kamil, A. C. 1979. Search image formation in the blue jay (*Cyanocitta cristata*). *Nature*, **204,** 1332–1333.

Pilecki, C. and O'Donald, P. 1971. Effects of predation on artificial mimetic polymorphisms with perfect and imperfect mimics at varying frequencies. *Evolution*, **25,** 365-370.

Pinheiro, C. E. G. 1996. Palatability and escaping ability in neotropical butterflies: tests with wild kingbirds (*Tyrannus melancholicus*, Tyrannidae). *Biological Journal of the Linnean Society*, **59,** 351–365.

Pinheiro, C. E. G. 2003. Does Müllerian mimicry work in nature? Experiments with butterflies and birds (Tyrannidae). *Biotropica*, **35,** 356–364.

Pires, O. R., Sebben, A., Schwartz, E. F., Largura, S. W. R., Bloch, C., Morales, R. A. V. and Schwartz, C. A. 2002. Occurrence of tetrodotoxin and its analogues in the Brazilian frog *Brachycephalus ephippium* (Anura: Brachycephalidae). *Toxicon*, **40,** 761–766.

Plaisted, K. C. and Mackintosh, N. J. M. 1995. Visual search for cryptic stimuli in pigeons: implications for the search image and search rate hypotheses. *Animal Behaviour*, **50**, 1219–1232.

Platt, A. P. and Brower, L. P. 1968. Mimetic versus disruptive coloration in intergrading populations of *Limenitis arthemis* and *astyanax* butterflies. *Evolution*, **22**, 699–718.

Platt, A. P., Coppinger, R. P. and Brower, L. P. 1971. Demonstration of selective advantage of mimetic *Limenitis* butterflies presented to caged avian predators. *Evolution*, **25**, 692–701.

Plough, F. H. 1978. Multiple cryptic effects of cross-banded and ringed patterns of snakes. *Copeia*, **4**, 834–836.

Plowright, R. C. and Owen, R. E. 1980. The evolutionary significance of bumble bee color patterns—a mimetic interpretation. *Evolution*, **34**, 622–637.

Pough, F. H., Brower, L. P., Meck, H. R. and Kessell, S. R. 1973. Theoretical investigations of automimicry: multiple trial learning and the palatability spectrum. *Proceedings of the National Academy of Sciences of the United States of America*, **70**, 2261–2265.

Poulton, E. B. 1887. The experimental proof of the protective value of colour and markings in insects in reference to their vertebrate enemies. *Transactions of the Entomological Society of London*, **March**, 14–274.

Poulton, E. B. 1888. Notes in 1887 upon Lepidopterus Larvae & c. *Transactions of the entomological Society of London*, **1888**, 595–596.

Poulton, E. B. 1890. *The colours of animals: their meaning and use especially considered in the case of insects*. London: Kegan Paul, Trench, Trubner and Co. Ltd.

Poulton, E. B. 1902. The meaning of the white undersides of animals. *Nature*, 596.

Poulton, E. B. 1904. The mimicry of Aculeata by the Asilidae and *Volucella*, and its probable significance. *Transaction of the Entemological Society of London*, **1904**, 661–665.

Poulton, E. B. 1909*a*. The value of colour in the struggle for life. In: *Darwin and modern science; essays in commemoration of the centenary of the birth of Charles Darwin and of the fiftieth anniversary of the publication of the Origin of Species* (Ed. by Seward, A. C.), pp. 207–227. Cambridge: Cambridge University Press.

Poulton, E. B. 1909*b*. Mimicry in the butterflies of North America. *Annals of the Entomological Society of America*, **2**, 203–242.

Poulton, E. B. 1914. Mimicry in North American butterflies: a reply. *Proceedings of the National Academy of Sciences of Philadelphia*, **66**, 161–195.

Powell, R. A. 1982. Evolution of black-tipped tails in weasels: predator confusion. *American Naturalist*, **119**, 126–131.

Preston-Mafham, R. and Preston-Mafham, K. 1993. *The encyclopedia of land invertebrate behaviour*. London: Blandford Press.

Pulliam, H. R. 1975. Diet optimisation with nutrient constraints. *American Naturalist*, **109**, 765–768.

Punnett, R. C. 1915. *Mimicry in Butterflies*. London: Cambridge University Press.

Purser, B. 2003. *Jungle bugs: masters of camouflage and mimicry*. Firefly Books.

Ramachandran, V. S. 1988. Perception of shape from shading. *Nature*, **331**, 163–166.

Reby, D., Cargnelutti, B. and Hewison, A. J. M. 1999. Contexts and possible functions of barking in roe deer. *Animal Behaviour*, **57**, 1121–1128.

Reid, D. G. 1987. Natural selection for apostasy and crypsis acting on the shell colour polymorphism of a mangrove snail, *Littoraria filosa* (Sowerby) (Gastropoda: Littorinidae). *Biological Journal of the Linnean Society*, **30**, 1–24.

Reid, P. J. and Shettleworth, S. J. 1992. Detection of cryptic prey: search image and search rate. *Journal of Experimental Psychology*, **18**, 273–286.

Rettenmeyer, C. W. 1970. Insect mimicry. *Annual Review of Entomology*, **15**, 43–74.

Rhoades, D. F. 1979. Evolution of plant chemical defense against herbivores. In: *Herbivores: their interactions with secondary plant metabolites* (Ed. by Rosenthal, G. A. and Janzen, D. H.), pp. 4–55. New York: Academic Press.

Ridgway, M. S. and McPhail, J. D. 1987. Raiding shoal size and the distraction display in male sticklebacks (Gasterosteus). *Canadaian Journal of Zoology*, **66**, 201–205.

Rigby, M. C. and Jokela, J. 2000. Predator avoidance and immune defence: costs and trade-offs in snails. *Proceedings of the Royal Society of London Series B—Biological Sciences*, **267**, 171–176.

Riipi, M., Alatalo, R. V., Lindström, L. and Mappes, J. 2001. Multiple benefits of gregariousness cover detectability costs in aposematic aggregations. *Nature*, **413**, 512–514.

Ritland, D. B. 1991. Palatability of aposematic Queen butterflies (*Danaus gilippus*) feeding on *Sarcostemma clausum* (Asclepiadaceae) in Florida. *Journal of Chemical Ecology*, **17**, 1593–1610.

Ritland, D. B. 1994. Variation in palatability of queen butterflies (*Danaus gilippus*) and implications regarding mimicry. *Ecology*, **75**, 732–746.

Ritland, D. B. and Brower, L. P. 1991. The viceroy butterfly is not a Batesian mimic. *Nature*, **350**, 497–498.

Robertson, H. M. 1985. Female dimorphism and mating behaviour in a damselfly, *Ischnura ramburi*: females mimicking males. *Animal Behaviour*, **33**, 805–809.

Robbins, R. K. 1980. The lycaenid 'false head' hypothesis: historical review and quantitative analysis. *Journal of the Lepidopterists' Society*, **34,** 194–208.

Robbins, R. K. 1981. The 'false head' hypothesis: predation and wing pattern variation of lycaenid butterflies. *American Naturalist*, **118,** 770–775.

Robinson, M. H. 1969. Defenses against visually hunting predators. *Evolutionary Biology*, **3,** 225–259.

Robinson, M. H. 1981. A stick is a stick and not worth eating—on the definition of mimicry. *Biological Journal of the Linnean Society*, **16,** 15–20.

Rocco, V., Barriga, J. P., Zagarese, H. and Lozada, M. 2002. How does ultraviolet radiation contribute to the feeding performance of rainbow trout, *Oncorhyynchus mykiss*, juveniles under natural illumination? *Environmental Biology of Fishes*, **63,** 223–228.

Roper, T. J. 1990. Responses of domestic chicks to artificially colored insect prey—effects of previous experience and background color. *Animal Behaviour*, **39,** 466–473.

Roper, T. J. 1993. Effects of novelty on taste-avoidance learning in chicks. *Behaviour*, **125,** 265–281.

Roper, T. J. 1994. Conspicuousness of prey retards reversal of learned avoidance. *Oikos*, **69,** 115–118.

Roper, T. J. and Cook, S. E. 1989. Responses of chicks to brightly colored insect prey. *Behaviour*, **110,** 276–293.

Roper, T. J. and Marples, N. M. 1997*a*. Colour preferences of domestic chicks in relation to food and water presentation. *Applied Animal Behaviour Science*, **54,** 207–213.

Roper, T. J. and Marples, N. M. 1997*b*. Odour and colour as cues for taste-avoidance learning in domestic chicks. *Animal Behaviour*, **53,** 1241–1250.

Roper, T. J. and Redston, S. 1987. Conspicuousness of distasteful prey affects the strength and durability of one-trial avoidance-learning. *Animal Behaviour*, **35,** 739–747.

Roper, T. J. and Wistow, R. 1986. Aposematic coloration and avoidance-learning in chicks. *Quarterly Journal of Experimental Psychology Section B—Comparative and Physiological Psychology*, **38,** 141–149.

Rothschild, M. 1963. Is the buff ermine (*Spilosoma lutea* (Huf.)) a mimic of the white ermine (*Spilosoma lubricipeda* (L.))? *Proceedings of the Royal Entomological Society of London, Series A*, **38,** 159–164.

Rothschild, M. 1964. An extension of Dr Lincoln Brower's theory on bird predation and food specificity, together with some observations on bird memory in relation to aposematic colour patterns. *Entomologist*, **97,** 73–78.

Rothschild, M. 1971. Speculations about mimicry with Henry Ford. In: *Ecological genetics and evolution* (Ed. by Creed, E. R.). Oxford: Blackwell.

Rothschild, M. 1972. Secondary plant substances and warning coloration in insects. In: *Insect–plant relationships* (Ed. by van Emden, H. F.), pp. 59–83. Oxford: Blackwell.

Rothschild, M. 1984. Aide-memoire mimicry. *Ecological Entomology*, **9,** 311–319.

Rothschild, M. and Marsh, N. 1978. Some peculiar aspects of danaid/plant relationships. *Entomologia Experimentalis et Applicata*, **24,** 437–450.

Rothschild, M., Valadon, G. and Mummery, R. 1977. Carotenoids of the pupae of the Large White butterfly (*Pieris brassicae*) and the small white butterfly (*Pieris rapae*). *Journal of the Zoological Society of London*, **181,** 323–39.

Rowe, C. 1999. Receiver psychology and the evolution of multicomponent signals. *Animal Behaviour*, **58,** 921–931.

Rowe, C. 2002. Sound improves visual discrimination learning in avian predators. *Proceedings of the Royal Society of London Series B—Biological Sciences*, **269,** 1353–1357.

Rowe, C. and Guilford, T. 1996. Hidden colour aversions in domestic clicks triggered by pyrazine odours of insect warning displays. *Nature*, **383,** 520–522.

Rowe, C. and Guilford, T. 1999. Novelty effects in a multimodal warning signal. *Animal Behaviour*, **57,** 341–346.

Rowe, C. and Guilford, T. 2001. The evolution of multimodal warning displays. *Evolutionary Ecology*, **13,** 655–671. [1999/2001]

Rowe, C., Lindstrom, L. and Lyytinen, A. 2004. The importance of pattern similarity between Müllerian mimics in predator avoidance learning. *Proceedings of the Royal Society of London Series B*, **271,** 407- 413.

Rowe, M. P., Coss, R. G. and Owings, D. H. 1986. Rattlesnake rattles and burrowing owl hisses—a case of acoustic Batesian mimicry. *Ethology*, **72,** 53–71.

Rowell-Rahier, M. and Pasteels, J. M. 1986. Economics of chemical defense in Chrysomelinae. *Journal of Chemical Ecology*, **12,** 1189–1203.

Rowell-Rahier, M., Pasteels, J. M., Alonso-Mejia, A. and Brower, L. P. 1995. Relative ipalatability of leaf beetles with either biosynthesized or sequestered chemical defense. *Animal Behaviour*, **49,** 709–714.

Roy, B. A. and Widmer, A. 1999. Floral mimicry: a fascinating yet poorly understood phenomenon. *Trends in Plant Science*, **4,** 325–330.

Rupp, L. 1989. The central European species of the genus *Volucella* (Diptera, Syrphidae) as commensals and parasitoids in the nests of bees and social wasps: studies on host-finding, larval biology and mimicry. In: Inaugural Dissertation, Albert-Ludwigs University, Freiburg-im-Breisgau.

Ruxton, G. D., Speed, M. and Kelly, D. J. 2004. What, if anything, is the adaptive function of countershading? *Animal Behaviour*.

Ryan, D. A., Bawden, K. M., Bermingham, K. T. and Elgar, M. A. 1996. Scanning and tail-flicking in the Australian dusky moorhen (*Gallinula tenebrosa*). *The Auk*, **113,** 499–501.

Saetre, G. P., and Slagsvold, T. 1996. The significance of female mimicry in male contests. *The American Naturalist*, **147,** 981–995.

Sargent, T. D. 1995. On the relative acceptabilities of local butterflies and moths to local birds. *Journal of the Lepidopterists' Society*, **49,** 148–162.

Sargent, T. D., Millar, C. D. and Lambert, D. M. 1998. The 'classical' explanation of industrial melanism. *Evolutionary Biology*, **30,** 299–322.

Sasaki, A., Kawaguchi, I. and Yoshimori, A. 2002. Spatial mosaic and interfacial dynamics in a Müllerian mimicry system. *Theoretical Population Biology*, **61,** 49–71.

Sbordoni, V., Bullini, L., Scarpelli, G., Forestiero, S. and Rampini, M. 1979. Mimicry in the burnet moth *Zygaena ephialtes* : population studies and evidence of a Batesian–Müllerian situation. *Ecological Entomology*, **4,** 83.

Schemske, D. W. 1981. Floral convergence and pollinator sharing in two bee-pollinated tropical herbs. *Ecology*, **62,** 946–954.

Schlenoff, D. H. 1984. Novelty—a basis for generalization in prey selection. *Animal Behaviour*, **32,** 919–921.

Schlenoff, D. H. 1985. The startle reponses of blue jays to Catocala (Lepidoptera: Noctuidae) prey models. *Animal Behaviour*, **33,** 1057–1067.

Schmidt, J. O. 1990. Hymenopteran venoms: striving toward the ultimate defense against vertebrates. In: *Insect defenses: adaptive mechanisms and strategies of prey and predators* (Ed. by Evans, D. L. and Schmid, J. O.), pp. 387–420. New York: State of New York University Press.

Schmidt, R. S. 1958. Behavioural evidence on the evolution of Batesian mimicry. *Animal Behaviour*, **6,** 129–138.

Schuler, W. and Hesse, E. 1985. On the function of warning coloration—a black and yellow pattern inhibits prey-attack by naive domestic chicks. *Behavioral Ecology and Sociobiology*, **16,** 249–255.

Schuler, W. and Roper, T. J. 1992. Responses to warning coloration in avian predators. *Advances in the Study of Behavior*, **21,** 111–146.

Servedio, M. R. 2000. The effects of predator learning, forgetting, and recognition errors on the evolution of warning coloration. *Evolution*, **54,** 751–763.

Sexton, O. J., Hoger, C. and Ortleb, E. 1966. *Anolis carolinensis*: effects of feeding on reactions to aposematic prey. *Science*, **153,** 1140.

Shashar, N., Hagan, R., Boal, J. G. and Hanlon, R. T. 2000. Cuttlefish use polarization sensitivity in predation on silvery fish. *Vision Research*, **40,** 71–75.

Shashar, N., Hanlon, R. T. and Petz, A. deM. 1998. Polarization vision helps detect transparent prey. *Nature*, **393,** 222–223.

Sheppard, P. M. 1959. The evolution of mimicry: a problem in ecology and genetics. *Cold Spring Harbor Symposia in Quantitative Biology*, **24,** 131–140.

Sheppard, P. M. 1975. *Natural selection and heredity*. London: Longman.

Sheppard, P. M. and Turner, J. R. G. 1977. The existence of Müllerian mimicry. *Evolution*, **31,** 452–453.

Sheppard, P. M., Turner, J. R. G., Brown, K. S., Benson, W. W. and Singer, M. C. 1985. Genetics and the evolution of Muellerian mimicry in *Heliconius* butterflies. *Philosophical Transactions of the Royal Society of London Series B—Biological Sciences*, **308,** 433-610.

Sherratt, T. N. 2001. The evolution of female-limited polymorphisms in damselflies: a signal detection model. *Ecology Letters*, **4,** 22–29.

Sherratt, T. N. 2002*a*. The coevolution of warning signals. *Proceedings of the Royal Society of London Series B—Biological Sciences*, **269,** 741–746.

Sherratt, T. N. 2002*b*. The evolution of imperfect mimicry. *Behavioral Ecology*, **13,** 821–826.

Sherratt, T. N. 2003. State-dependent risk-taking in systems with defended prey. *Oikos*, **103,** 93–100.

Sherratt, T. N. and Beatty, C. D. 2003. The evolution of warning signals as reliable indicators of prey defense. *American Naturalist*, **162,** 377–389.

Sherratt, T. N. and Harvey, I. F. 1993. Frequency-dependent food selection by arthropods: a review. *Biological Journal of the Linnean Society*, **48,** 167–186.

Sherratt, T. N. and MacDougall, A. D. 1995. Some population consequences of variation in preference among individual predators. *Biological Journal of the Linnean Society*, **55,** 93–107.

Sherratt, T. N., Rashed, A. and Beatty, C. D. 2004*a*. The evolution of locomotory behavior in profitable and unprofitable simulated prey. *Oecologia*, **138,** 143–150.

Sherratt, T. N., Speed, M. P. and Ruxton, G. D. 2004*b*. Natural selection on unpalatable species imposed by state-dependent foraging behaviour. *Journal of Theoretical Biology*, **228,** 217–226.

Shettleworth, S. J. 1972. The role of novelty in learned avoidance of unpalatable 'prey' by domestic chicks (*Gallus gallus*). *Animal Behaviour*, **20,** 29–35.

Shine, R., Philips, B., Waye, H., LeMaster, M. and Mason, R. T. 2001. Benefits of female mimicry in snakes. *Nature*, **414,** 267.

Sih, A. 1992. Prey uncertainty and the balancing of antipredator and feeding needs. *American Naturalist*, **139,** 1052–1069.

Silberglied, R. E., Aiello, A. and Windsor, D. M. 1980. Disruptive coloration in butterflies: lack of support in *Anartia fatima*. *Science*, **209,** 617–619.

Sillén-Tullberg, B. 1985*a*. Higher survival of an aposematic than of a cryptic form of a distasteful bug. *Oecologia*, **67,** 411–415.

Sillén-Tullberg, B. 1985*b*. The significance of coloration per se, independent of background, for predator avoidance of aposematic prey. *Animal Behaviour*, **33,** 1382–1384.

Sillén-Tullberg, B. 1988. Evolution of gregariousness in aposematic butterfly larvae—a Phylogenetic analysis. *Evolution*, **42,** 293–305.

Sillén-Tullberg, B. 1990. Do predators avoid groups of aposematic prey—an experimental test. *Animal Behaviour*, **40,** 856–860.

Sillén-Tullberg , B. 1993. The effect of biased inclusion of taxa on the correlation between discrete characters in phylogenetic trees. *Evolution*, **47,** 1182–1191.

Sillén-Tullberg, B. and Bryant, E. H. 1983. The evolution of aposematic coloration in distasteful prey—an individual selection model. *Evolution*, **37,** 993–1000.

Sillén-Tullberg, B. and Leimar, O. 1988. The Evolution of gregariousness in distasteful insects as a defense against oredators. *American Naturalist*, **132,** 723–734.

Sillén-Tullberg, B., Wiklund, C. and Jarvi, T. 1982. Aposematic coloration in adults and larvae of *Lygaeus equestris* and its bearing on Müllerian mimicry—an experimental study on predation on living bugs by the Great Tit *Parus major*. *Oikos*, **39,** 131–136.

Simmons, R. B. and Weller, S. J. 2002. What kind of signals do mimetic tiger moths send? A phylogenetic test of wasp mimicry systems (Lepidoptera : Arctiidae : Euchromiini). *Proceedings of the Royal Society of London Series B—Biological Sciences*, **269,** 983–990.

Simons, P. 1976. A specific visual response in dragonflies. *Odonatologica*, **5,** 285.

Smith, D. A. S. 1973. Negative non-random mating in the polymorphic butterfly *Danaus chrysippus* in Tanzania. *Nature*, **242,** 131–132.

Smith, D. A. S. 1978. The effect of cardiac glycoside storage on growth rate and adult size in the butterfly Danaus chrysippus (L.). *Experientia*, **34,** 845–846.

Smith, R. J. F. 1997. Avoiding and deterring predators. In: *Behavioural Ecology of Teleost Fishes* (Ed. by Godin, J.-G. J.), pp. 163–190. Oxford: Oxford University Press.

Smith, R. L. and Smith, T. M. 2001. *Ecology and field biology, 6th edn*. New York: Benjamin Cummings.

Smith, S. M. 1973*a*. A study of prey-attack behaviour in young Loggerhead Shrikes, *Lius ludovicianus* L. *Behaviour*, **44,** 113–141.

Smith, S. M. 1973*b*. Factors directing prey-attack by the young of three passerine species. *The Living Bird*, **12,** 55–67.

Smith, S. M. 1975. Innate recognition of coral snake pattern by a possible avian predator. *Science*, **187,** 759–760.

Smith, S. M. 1976. Predatory behaviour of young turquoise-brown motmots, *Eumota supercilosa*. *Behaviour*, **56,** 309–320.

Smith, S. M. 1977. Coral-snake pattern recognition and stimulus generalisation by naive great kiskadees (Aves: Tyrannidae). *Nature*, **265,** 535–536.

Smith, S. M. 1979. Responses of naive temperate birds to warning coloration. *American Midland Naturalist*, **103,** 340–352.

Sonerud, G. A. 1988. To distract display or not: grouse hens and foxes. *Oikos*, **51,** 233–237.

Spear, N. E. 1978. *The processing of memories: forgetting and retention*. New Jersey: L. Erlbaum.

Speed, M. P. 1993*a*. Muellerian mimicry and the psychology of predation. *Animal Behaviour*, **45,** 571–580.

Speed, M. P. 1993*b*. When is mimicry good for predators. *Animal Behaviour*, **46,** 1246–1248.

Speed, M. P. 1999. Robot predators in virtual ecologies: the importance of memory in mimicry studies. *Animal Behaviour*, **57,** 203–213.

Speed, M. P. 2000. Warning signals, receiver psychology and predator memory. *Animal Behaviour*, **60,** 269–278.

Speed, M. P. 2001*a*. Can receiver psychology explain the evolution of aposematism? *Animal Behaviour*, **61,** 205–216.

Speed, M. P. 2001*b*. Batesian, quasi-Batesian or Müllerian mimicry? Theory and data in mimicry research. *Evolutionary Ecology*, **13,** 755–776.

Speed, M. P. and Ruxton, G. D. 2002. Animal behaviour: Evolution of suicidal signals. *Nature*, **416,** 375–375.

Speed, M. P. and Ruxton, G. D. Unpublished. The costs of crypsis and the evolution of secondary defences and aposematism. Unpublished Manuscript.

Speed, M. P. 2003. Theoretical developments in the understanding of warning signals. *Comments on Theoretical Biology*, **8.**

Speed, M. P., Alderson, N. J., Hardman, C. and Ruxton, G. D. 2000. Testing Müllerian mimicry: an experiment with wild birds. *Proceedings of the Royal Society of London Series B—Biological Sciences*, **267,** 725–731.

Speed, M. P. and Turner, J. R. G. 1999. Learning and memory in mimicry: II. Do we understand the mimicry spectrum? *Biological Journal of the Linnean Society*, **67,** 281–312.

Spradberry, R. 1973. *Wasps: an account of the biology and natural history of solitary and social wasps*. New York: Academic Press.

Srygley, R. B. 1994. Locomotor mimicry in butterflies—the associations of positions of centers of mass among groups of mimetic, unprofitable prey. *Philosophical*

*Transactions of the Royal Society of London Series B—Biological Sciences*, **343,** 145–155.

Srygley, R. B. 2001. Incorporating motion into investigations of mimicry. *Evolutionary Ecology*, **13,** 691–708.

Srygley, R. B. 2004. The aerodynamic costs of warning signals in palatable, mimetic butterflies and their distasteful models. *Proceedings of the Royal Society of London Series B—Biological Sciences*, **271,** 589–594.

Srygley, R. B. and Chai, P. 1990*a*. Flight morphology of neotropical butterflies—palatability and distribution of mass to the thorax and abdomen. *Oecologia*, **84,** 491–499.

Srygley, R. B. and Chai, P. 1990*b*. Predation and the elevation of thoracic temperature in brightly colored neotropical butterflies. *American Naturalist*, **135,** 766–787.

Srygley, R. B. and Dudley, R. 1993. Correlations of the position of center of body-mass with butterfly escape tactics. *Journal of Experimental Biology*, **174,** 155–166.

Srygley, R. B. and Kingsolver, J. G. 1998. Red-wing blackbird reproductive behaviour and the palatability, flight performance, and morphology of temperate pierid butterflies (*Colias*, *Pieris*, and *Pontia*). *Biological Journal of the Linnean Society*, **64,** 41–55.

Stabell, O. B., Ogbebo, F. and Primicerio, R. 2003. Inducible defences in *Daphnia* depend on latent alarm signals from conspecific prey activated in predators. *Chemical Senses*, **28,** 141–153.

Stachowicz, J. J. 2001. Chemical ecology of mobile benthic invertebrates: predators and prey, allies and competitors. In: *Marine Chemical Ecology* (Ed. by McClintock, J. B. and Baker, B. J.), pp. 157–194. New York: CRC Press.

Staddon, J. E. R. and Gendron, R. P. 1983. Optimal detection of cryptic prey may lead to predator switching. *The American Naturalist*, **122,** 843–848.

Stamp, N. E. and Wilkens, R. T. 1993. On the cryptic side of life: being unapparent to enemies and the consequences for foraging and growth in caterpillars. In: *Caterpillars: ecological and evolutionary constraints on foraging* (Ed. by Stamp, N. E. and Casey, T. M.), pp. 283–330. New York: Chapman and Hall.

Starrett, A. 1993. Adaptive resemblance—a unifying concept for mimicry and crypsis. *Biological Journal of the Linnean Society*, **48,** 299–317.

Stauffer, J. A. J., Hale, E. A. and Seltzer, R. 1999. Hunting strategies of a Lake Malawi Cichlid with reverse countershading. *Copeia*, **1999,** 1108–1111.

Stemberger, R. S. 1988. Reproductive costs and hydrodynamic benefits of chemically-induced defenses in Keratella-Testudo. *Limnology and Oceanography*, **33,** 593–606.

Sternburg, J. G., Waldbauer, G. P. and Jeffords, M. R. 1977. Batesian mimicry—selective advantage of color pattern. *Science*, **195,** 681–683.

Stiles, E. W. 1979. Evolution of color pattern and pubescence characteristics in male bumblebees: automimicry vs. thermoregulation. *Evolution*, **33,** 941–957.

Stowe, M. K., Tumlinson, J. H. and Heath, R. R. 1987. Chemical mimicry—bolas spiders emit components of moth prey species sex-pheromones. *Science*, **236,** 964–967.

Summers, K. and Clough, M. E. 2001. The evolution of coloration and toxicity in the poison frog family (Dendrobatidae). *Proceedings of the National Academy of Sciences of the United States of America*, **98,** 6227–6232.

Sundberg, P. 1987. A possible mechanism for the evolution of aposematic coloration in solitary nemerteans (Phylum-Nemertea). *Oikos*, **48,** 289–296.

Sweet, S. S. 1985. Geographical variation, convergent crypsis and mimicry in gopher snakes (*Pituphis meanoleucus*) and Western Rattlesnakes (Crotalus viridis). *Journal of Herpetology*, **19,** 55–67.

Sword, G. A. 1999. Density-dependent warning coloration. *Nature*, **397,** 217–217.

Sword, G. A. 2002. A role for phenotypic plasticity in the evolution of aposematism. *Proceedings of the Royal Society of London Series B—Biological Sciences*, **269,** 1639–1644.

Sword, G. A., Simpson, S. J., El Hadi, O. T. M. and Wilps, H. 2000. Density-dependent aposematism in the desert locust. *Proceedings of the Royal Society of London Series B—Biological Sciences*, **267,** 63–68.

Swynnerton, C. F. M. 1915. A brief preliminary statement of a few of the results of five years' special testing of the theories of mimicry. *Proceedings of the Entomological Society of London*, **I,** xxxii–xliv.

Symula, R., Schulte, R. and Summers, K. 2001. Molecular phylogenetic evidence for a mimetic radiation in Peruvian poison frogs supports a Müllerian mimicry hypothesis. *Proceedings of the Royal Society of London Series B—Biological Sciences*, **268,** 2415–2421.

Termonia, A., Hsiao, T. H., Pasteels, J. M. and Milinkovitch, M. C. 2001. Feeding specialization and host-derived chemical defense in Chrysomeline leaf beetles did not lead to an evolutionary dead end. *Proceedings of the National Academy of Sciences of the United States of America*, **98,** 3909–3914.

Termonia, A., Pasteels, J. M., Windsor, D. M. and Milinkovitch, M. C. 2002. Dual chemical sequestration: a key mechanism in transitions among ecological specialization. *Proceedings of the Royal Society of London Series B—Biological Sciences*, **269,** 1–6.

Thayer, A. H. 1896. The law which underlies protective coloration. *Auk*, **13,** 124–129.

Theodoratus, D. H. and Bowers, M. D. 1999. Effects of sequestered iridoid glycosides on prey choice of the prairie wolf spider *Lycosa carolinensis*. *Journal of Chemical Ecology*, **25,** 283–95.

Thomas, R. J., Cuthill, I. and Marples, N. M. 2004. Dietary conservatism can explain the evolution of aposematism: an experiment with wild birds. *Oikos* (in press).

Thomas, R. J., Marples, N. M., Cuthill, I. C., Takahashi, M. and Gibson, E. A. 2003. Dietary conservatism may facilitate the initial evolution of aposematism. *Oikos*, **101**, 458–466.

Thompson, J. J. W., Armitage, S. A. O. and Siva-Jothy, M. T. 2002. Cuticular colour change after imaginal eclosion is time-constrained: blacker beetles darken faster. *Physiological Entomology*, **27**, 136–141.

Thompson, R. K. R., Foltin, R. W., Sweey, R. J. B. A., Graves, C. A. and Lowitz, C. E. 1981. Tonic immobility in Japanese quail can reduce the probabiltiy of sustained attack by cats. *Animal Learning and Behavior*, **9**, 145–149.

Thompson, V. 1973. Spittlebug polymorphic for warning coloration. *Nature*, **242**, 126–128.

Thompson, V. 1984. Polymorphism under apostatic and aposematic selection. *Heredity*, **53**, 677–686.

Thornhill, R. 1979. Adaptive female-mimicking behavior in a scorpionfly. *Science*, **295**, 412–414.

Till-Bottraud, I. and Gouyon, P.-H. 1992. Intra- versus interplant Batesian mimicry? A model on cyanogenesis and herbivory in clonal plants. *American Naturalist*, **139**, 509–520.

Tilson, R. L. and Norton, R. L. 1981. Alarm duetting and pursuit deterrence in an African antelope. *American Naturalist*, **118**, 455–462.

Tinbergen, L. 1960. The natural control of insects in pinewoods. 1. Factors influencing the intensity of predation by song birds. *Archives Neerlandaises de Zoologie*, **13**, 265–343.

Tollrian, R. and Dodson, S. I. 1999. Inducible defences in cladocera: constraints, costs and multi-predator environments. In: *The ecology and evolution of inducible defenses* (Ed. by Tollrian, R. and Harvell, C.D.), pp. 177–202. Princeton, NJ: Princeton University Press.

Tollrian, R. and Harvell, C. D. 1999. The evolution of inducible defenses: current ideas. In: *The ecology and evolution of inducible defenses* (Ed. by Tollrian, R. and Harvell, C.D.), pp. 306–322. Princeton, NJ: Princeton University Press.

Tonner, M., Novotny, V., Leps, J. and Komarek, S. 1993. False head wing pattern of the Burmese Junglequeen butterfly and the deception of avian predators. *Biotropica*, **25**, 474–478.

Trigo, J. R. 2000. The chemistry of antipredator defense by secondary compounds in neotropical Lepidoptera: facts, perspectives and caveats. *Journal of the Brazilian Chemical Society*, **11**, 551–561.

Trussell, G. C. and Nicklin, M. O. 2002. Cue sensitivity, inducible defense, and trade-offs in a marine snail. *Ecology*, **83**, 1635–1647.

Tsuda, A., Saito, H. and Hirose, T. 1998. Effect of gut content on the vulnerability of copepods to visual predation. *Limnology and Oceanography*, **43**, 1944–1947.

Tucker, G. M. 1991. Apostatic selection by song thrushes (*Turdus philomelos*) feeding on the snail *Cepaea hortensis*. *Biological Journal of the Linnean Society*, **43**, 149–156.

Tucker, G. M. and Allen, J. A. 1993. The behavioural basis of apostatic selection by humans searching for computer-generated cryptic 'prey'. *Animal Behaviour*, **46**, 713–719.

Tullberg, B. S., Gamberale-Stille, G. and Solbreck, C. 2000. Effects of food plant and group size on predator defence: differences between two co-occurring aposematic Lygaeinae bugs. *Ecological Entomology*, **25**, 220–225.

Tullberg, B. S. and Hunter, A. F. 1996. Evolution of larval gregariousness in relation to repellent defences and warning coloration in tree-feeding Macrolepidoptera: A phylogenetic analysis based on independent contrasts. *Biological Journal of the Linnean Society*, **57**, 253–276.

Tullberg, B. S., Leimar, O. and Gamberale-Stille, G. 2000. Did aggregation favour the initial evolution of warning coloration? A novel world revisited. *Animal Behaviour*, **59**, 281–287.

Tullrot, A. and Sundberg, P. 1991. The conspicuous nudibranch *Polycera quadrilineata*—aposematic coloration and individual selection. *Animal Behaviour*, **41**, 175–176.

Turner, E. R. A. 1961. Survival value of different methods of camouflage as shown in a model population. *Proceedings of the Zoological Society of London*, **136**, 273–284.

Turner, G. F. and Pitcher, T. J. 1986. Attack abatment: a model for group protection by combined avoidance and dilution. *American Naturalist*, **128**, 228–240.

Turner, J. R. G. 1971. Studies of Müllerian mimicry and its evolution in burnet moths and heliconid butterflies. In: *Ecological Genetics and Evolution* (Ed. by Creed, E. R.). Oxford: Blackwell.

Turner, J. R. G. 1975. A tale of two butterflies. *Natural History*, **84**.

Turner, J. R. G. 1977. Butterfly mimicry: the genetical evolution of an adaptation. *Evolutionary Biology*, **10**, 163–206.

Turner, J. R. G. 1978. Why male butterflies are non-mimetic : natural-selection, sexual selection, group selection, modification and sieving. *Biological Journal of the Linnean Society*, **10**, 385–432.

Turner, J. R. G. 1981. Adaptation and evolution in Heliconius—a defense of neodarwinism. *Annual Review of Ecology and Systematics*, **12**, 99–121.

Turner, J. R. G. 1984. The palatability spectrum and its consequences. In: *The Biology of Butterflies* (Ed. by Vane-Wright, R. I. and Ackery, P.). Princeton: Princeton University Press.

Turner, J. R. G. 1987. The evolutionary dynamics of batesian and muellerian mimicry: similarities and differences. *Ecological Entomology*, **12,** 81–95.

Turner, J. R. G. 2000. Mimicry. In: *Nature encyclopedia of life sciences*. London: Nature Publishing Group.

Turner, J. R. G., Kearney, E. P. and Exton, L. S. 1984. Mimicry and the Monte-Carlo predator—the palatability spectrum and the origins of mimicry. *Biological Journal of the Linnean Society*, **23,** 247–268.

Turner, J. R. G. and Speed, M. P. 2001. How weird can mimicry get? *Evolutionary Ecology*, **13,** 807–827.

Tutt, J. W. 1896. *British moths*. London: George Routledge.

Utne-Palm, A. C. 1999. The effect of prey mobility, prey contrast, turbidity and spectral composition on the reaction distance of Gobiusculus flavescens to its planktonic prey. *Journal of Fish Biology*, **54,** 1244–1258.

Van Buskirk, J. 2000. The costs of an inducible defense in anuran larvae. *Ecology*, **81,** 2813–2821.

Van Buskirk, J. and Relyea, R. A. 1998. Selection for phenotypic plasticity in *Rana sylvatica* tadpoles. *Biological Journal of the Linnean Society*, **65,** 301–328.

Van Gossum, H., Stoks, R. and De Bruyn, L. 2001. Reversible frequency-dependent switches in male mate choice. *Proceedings of the Royal Society B*, **268,** 1–3.

Van Someren, V. G. L. 1922. Notes on certain colour patterns in Lyaenidae. *Journal of the East Africa and Uganda Natural History Society*, **17,** 18–21.

Van Someren, V. G. L. and Jackson, T. H. E. 1959. Some comments on the adaptive resemblance amongst African Lepidoptera (Rhopalocera). *Journal of the Lepidopterist's Society*, **13,** 121–150.

Vane-Wright, R. I. 1980. On the definition of mimicry. *Biological Journal of the Linnean Society*, **13,** 1–6.

Vane-Wright, R. I., Raheem, D. C., Cieslak, A. and Vogler, A. P. 1999. Evolution of the mimetic African swallowtail butterfly *Papilio dardanus*: molecular data confirm relationships with *P. phorcas* and *P. constantinus*. *Biological Journal of the Linnean Society*, **66,** 215–229.

Vaughan, F. A. 1983. Startle responses of blue jays to visual stimuli presented during feeding. *Animal Behaviour*, **31,** 385–396.

Vega-Redondo, F. and Hasson, O. 1993. A game-theoretic model of predator-prey signalling. *Journal of theoretical Biology*, **162,** 309–319.

Vences, M., Kosuch, J., Lotters, S., Widmer, A., Jungfer, K. H., Kohler, J. and Veith, M. 2000. Phylogeny and lassification of poison frogs (Amphibia : Dendrobatidae), based on mitochondrial 16S and 12S ribosomal RNA gene sequences. *Molecular Phylogenetics and Evolution*, **15,** 34–40.

Vencl, F. V., Morton, T. C., Mumma, R. O. and Schultz, J. C. 1999. Shield defense of a larval tortoise beetle. *Journal of Chemical Ecology*, **25,** 549–566.

Vinyard, G. L. and O'Brien, W. J. 1975. Dorsal light response as an index of prey perference in bluegill (Lepomis macrochirus). *Journal of the Fisheries Research Board of Canada*, **32,** 1860–1863.

Vitt, L. J. 1992. Lizard mimics millipede. *Research Exploration*, **8,** 76–95.

Vulinec, K. 1990. Collective security: aggregations by insects as a defence. In: *Insect defenses: Apative mechanisms and strategies of prey and predators* (Ed. by Evans, D. L. and Schmid, J. O.), pp. 251–288. New York: State University of New York Press.

Waldbauer, G. P. 1970. Mimicry of hymenopteran antennae by Syrphidae. *Psyche*, **77,** 45–49.

Waldbauer, G. P. 1988. Aposematism and Batesian mimicry—measuring mimetic advantage in natural habitats. *Evolutionary Biology*, **22,** 227–259.

Waldbauer, G. P. 1996. *Insects Through the Seasons*. Cambridge, MA: Harvard University Press.

Waldbauer, G. P. and LaBerge, W. E. 1985. Phenological relationships of wasps, bumblebees, their mimics and insectivorous birds in Northern Michigan. *Ecological Entomology*, **10,** 99–110.

Waldbauer, G. P. and Sheldon, J. K. 1971. Phenological relationships of some aculeate Hymenoptera, their dipteran mimics, and insectivorous birds. *Evolution*, **25,** 371–382.

Waldbauer, G. P. and Sternburg, J. G. 1975. Saturniid moths as mimics—alternative interpretation of attempts to demonstrate mimetic advantage in nature. *Evolution*, **29,** 650–658.

Waldbauer, G. P. and Sternburg, J. G. 1983. A pitfall in using painted insects in studies of protective coloration. *Evolutionary Biology*, **37,** 1085–1086.

Waldbauer, G. P. and Sternburg, J. G. 1987. Experimental field demonstration that two aposematic butterfly color patterns do not confer protection against birds in Northern Michigan. *American Midland Naturalist*, **118,** 145–152.

Waldbauer, G. P., Sternburg, J. G. and Maier, C. T. 1977. Phenological relationships of wasps, bumblebees, their mimics, and insectivorous birds in an Illinois sand area. *Ecology*, **58,** 583–591.

Wallace, A. R. 1865. On the phenomena of variation and geographical distribution as illustrated by the

Papilionidae of the Malayan region. *Transactions of the Linnean Society of London*, **25,** 1–71.

Wallace, A. R. 1867. *Proceedings of the Entomological Society of London*, **March 4th,** IXXX–IXXXi.

Wallace, A. R. 1871. *Contributions to the theory of natural selection. 2nd edn.* London: MacMillan & Co.

Wallace, A. R. 1889. *Darwinism—an exposition of the theory of natural selection with some of its applications.* London: MacMillan and Co.

Wartzok, D. and Ketten, D. R. 1999. Marine mammal sensory systems. In: *Biology of marine mammals* (Ed. by Reynolds, J. E. I. and Rommel, S. A.), pp. 117–175. Washington: Smithsonian Institution Press.

Weldon, P. J. and Rappole, J. H. 1997. A survey of birds odorous or unpalatable to humans: possible indications of chemical defense. *Journal of Chemical Ecology*, **23,** 2609–2633.

Weller, S. J., Jacobson, N. L. and Conner, W. E. 1999. The evolution of chemical defences and mating systems in tiger moths (Lepidoptera : Arctiidae). *Biological Journal of the Linnean Society*, **68,** 557–578.

West, D. A. 1994. Unimodal Batesian polymorphism in the Neotropical swallowtail butterfly *Eurytides lysithous* (Hbn). *Biological Journal of the Linnean Society*, **52,** 197–224.

Whitman, D. W., Blum, M. S. and Alsop, D. W. 1990. Allomones: Chemicals for defense. In: *Insect defenses: adaptive mechanisms and strategies of prey and predators* (Ed. by Evans, D. L. and Schmidt, J. O.), pp. 229–251. New York: State University of New York Press.

Whoriskey, F. G. 1991. Stickleback distraction displays: sexual or foraging deception against egg cannibalism. *Animal Behaviour*, **41,** 989–995.

Whoriskey, F. G. and FitzGerald, G. J. 1985. Sex, cannibalism and sticklebacks. *Behavioral Ecology and Sociobiology*, **18,** 15–18.

Wickler, W. 1968. *Mimicry in Plants and Animals.* London: Weidenfeld and Nicholson.

Widder, E. A. 1999. Bioluminescence. In: *Adaptive Mechanisms in the Ecology of Vision* (Ed. by Archer, S. N., Djamgoz, M. B. A., Lorw, E. R., Partridge, J. C. and Vallerga, S.), pp. 555–581. New York: Kluwer Academic Publishers.

Widder, E. A. 2002. Bioluminescence and the pelagic visual environment. *Marine and Freshwater Behaviour and Physiology*, **35,** 1–26.

Wiklund, C. and Järvi, T. 1982. Survival of distasteful insects after being attacked by naive birds—a reappraisal of the theory of aposematic coloration evolving through individual selection. *Evolution*, **36,** 998–1002.

Wiklund, C. and Sillén-Tullberg, B. 1985. Why distasteful butterflies have aposematic larvae and adults, but cryptic pupae—evidence from predation experiments on the monarch and the european swallowtail. *Evolution*, **39,** 1155–1158.

Wilkinson, D. M., Sherratt, T. N., Phillip, D. M., Wratten, S. D., Dixon, A. F. G. and Young, A. J. 2002. The adaptive significance of autumn leaf colours. *Oikos*, **99,** 402–407.

Williams, D. 2001. *Naval camouflage 1914–1945: a complete visual reference.* London: Catham Publishing.

Williamson, C. E. 1980. The predatory behaviour of *Mesocyclops edax*: predator preferences, prey defences and starvation-induced changes. *Limnology and Oceanography*, **25,** 903–909.

Williamson, G. B. and Black, E. M. 1981. Mimicry in hummingbird-pollinated plants? *Ecology*, **62,** 494–496.

Williamson, G. B. and Nelson, C. E. 1972. Fitness set analysis of mimetic adaptive strategies. *The American Naturalist*, **106,** 525–537.

Wilson, K., Cotter, S. C., Reeson, A. F. and Pell, J. K. 2001. Melanism and disease resistance in insects. *Ecology Letters*, **4,** 637–649.

Wilson, K., Thomas, M. B., Blanford, S., Doggett, M., Simpson, S. J. and Moore, S. L. 2002. Coping with crowds: Density-dependent disease resistance in desert locusts. *Proceedings of the National Academy of Sciences of the United States of America*, **99,** 5471–5475.

Winemiller, K. O. 1990. Caudal eyespots as deterrents against fin predation in the neotropical cichlid *Astronotos ocellatus*. *Copeia*, **1990,** 665–673.

Wise, K. K., Conover, M. R. and Knowlton, F. F. 1999. Responses of coyotes to avian distress calls: testing the startle-predator and predator-attraction hypotheses. *Behaviour*, **136,** 935–949.

Wong, B. B. M. and Schiestl, F. P. 2002. How an orchid harms its pollinator. *Proceedings of the Royal Society of London, Series B—Biological Sciences*, **269,** 1529–1532.

Woodland, D. J., Jaafar, Z. and Knight, M.-L. 1980. The 'pursuit deterrent' function of alarm calls. *American Naturalist*, **115,** 748–753.

Woolfson, A. and Rothschild, M. 1990. Speculating About Pyrazines. *Proceedings of the Royal Society of London Series B—Biological Sciences*, **242,** 113–119.

Wourms, M. K. and Wasserman, F. E. 1985. Butterfly wing markings are more advantageous during handling than during initial strike of an avian predator. *Evolution*, **39,** 845–851.

Wrazidlo, I. 1986. Untersuchungen zur Reaktion von Fasanen- und Wachtelküken auf Warnfbige Beute. Göttingen: University of Göttingen.

Wright, D. I. and O'Brien, W. J. 1982. Differential location of chaborus larvae and daphnia by fish: the importance of motion and visible size. *The American Midland Naturalist*, **108,** 68–73.

Yachi, S. and Higashi, M. 1998. The evolution of warning signals. *Nature*, **394,** 882–884.

Yamauchi, A. 1993. A population-dynamic model of Batesian mimicry. *Researches on Population Ecology*, **35,** 295–315.

Yeargan, K. V. 1994. Biology of Bolas spiders. *Annual Review of Entomology*, **39,** 81–99.

Yosef, R., Carrel, J. E. and Eisner, T. 1996. Contrasting reactions of loggerhead shrikes to two types of chemically defended insect prey. *Journal of Chemical Ecology*, **22,** 173–181.

Yotsu, M., Yamazaki, T., Meguro, Y., Endo, A., Murata, M., Naoki, H. and Yasumoto, T. 1987. Production of tetrodotoxin and Its derivatives by pseudomonas Sp. Isolated from the skin of a pufferfish. *Toxicon*, **25,** 225–228.

Young, A. M. 1971. Wing coloration and reflectance in *Morpho* butterflies as related to reproductive behaviour and escape from avain predators. *Oecologia*, **7,** 209–222.

Young, B. A., Solomon, J. and Abishahin, G. 1999. How many ways can a snake growl? The morphology of sound production in *Ptyas mucosus* and its potential mimicry of *Ophiophagus*. *Herpetological Journal*, **9,** 89–94.

Young, R. E. 1983. Ocean bioluminescence: an overview of general functions. *Bulletin of Marine Science*, **33,** 829–845.

Young, R. E. and Arnold, J. M. 1982. The functional morphology of a ventral photophore from the mesophelagic squid Abralia trigonura. *Malacology*, **23,** 135–163.

Young, R. E. and Roper, C. F. E. 1976. Intensity regulation of bioluminescence during countershading in living animals. *Fishery Bulletin*, **75,** 239–252.

Zahavi, A. 1977. Reliability of communication systems and the evolution of altruism. In: *Evolutionary Ecology* (Ed. by Stonehouse, B. and Perrins, C. M.), pp. 253–259. London: McMillan.

Zalucki, M. P., Brower, L. P. and Alonso, A. 2001. Detrimental effects of latex and cardiac glycosides on survival and growth of first-instar monarch butterfly larvae Danaus plexippus feeding on the sandhill milkweed Asclepias humistrata. *Ecological Entomology*, **26,** 212–224.

Zalucki, M. P., Malcolm, S. B., Paine, T. D., Hanlon, C. C., Brower, L. P. and Clarke, A. R. 2001. It's the first bites that count: survival of first-instar monarchs on milkweeds. *Austral Ecology*, **26,** 547–555.

Zaret, T. M. 1972. Predators, invisible prey, and the nature of polymorphism in the cladocera (class crustacea). *Limnology and Oceanography*, **17,** 171–184.

Zaret, T. M. and Kerfoot, W. C. 1975. Fishing predation on *Bosima longirostris*: body-size selection versus visibility selection. *Ecology*, **56,** 232–237.

Zrzavy, J. and Nedved, O. 1999. Evolution of mimicry in the New World *Dysdercus* (Hemiptera : Pyrrhocoridae). *Journal of Evolutionary Biology*, **12,** 956–969.

Zuberbuhler, K., Noe, R. and Seyfarth, R. M. 1997. Diana monkey long-distance calls: messages for conspecifics and predators. *Animal Behaviour*, **53,** 589–604.

# Author index

# Species index

# Subject index